Biology, Cl
and Appli
of Apocar(

Biology, Chemistry, and Applications of Apocarotenoids

Edited By

Siva Ramamoorthy
Renata Rivera-Madrid
C. George Priya Doss

CRC Press
Taylor & Francis Group
Boca Raton London New York

CRC Press is an imprint of the
Taylor & Francis Group, an **informa** business

First edition published 2021
by CRC Press
6000 Broken Sound Parkway NW, Suite 300, Boca Raton, FL 33487-2742

and by CRC Press
2 Park Square, Milton Park, Abingdon, Oxon, OX14 4RN

Library of Congress Cataloging-in-Publication Data
Names: Ramamoorthy, Siva, editor. | Madrid, Renata Rivera, editor. | Doss, C George Priya, editor.
Title: Biology, chemistry and applications of apocarotenoids / edited by Siva Ramamoorthy, Renata Rivera Madrid, C George Priya Doss
Description: First edition. | Boca Raton, FL : CRC Press, 2020. | Includes bibliographical references and index.
Identifiers: LCCN 2020033596 | ISBN 9780367361600 (hardback) | ISBN 9780429344206 (ebook)
Subjects: LCSH: Carotenoids. | Carotenoids--Biotechnology. | Carotenoids--Health aspects. | Carotenoids--Analysis.
Classification: LCC QK898.C34 B565 2020 | DDC 575.4/9--dc23
LC record available at https://lccn.loc.gov/2020033596

ISBN 13: 978-0-367-36160-0 (hbk)
ISBN 13: 978-0-429-34420-6 (ebk)

Typeset in Times
by SPi Global, India

Contents

Preface

Carotenoids are a large class of isoprenoid pigments produced by plants and certain microbes. More than 700 naturally occurring carotenoids have been identified. Apocarotenoids are tailored from carotenoids by oxidative enzymes. This reaction is initiated by carotenoid cleavage enzymes that cleave carotenoids into apocarotenoids. These carotenoid cleavage oxygenases (CCOs) can act via a single cleavage of C40 carotenoid substrates or via several sequential cleavage events, leading to multiple products with various functions in entire taxa, including acting as visual or volatile signals to attract pollinating and seed dispersal agents. They also play a key role in allelopathic interactions and plant defence. Some apocarotenoids are produced from plants in high quantities to be used as pigments, aromas, etc. Recent studies show that the loss of carotenoid cleavage enzymes induces the development of axillary branches, indicating that apocarotenoids convey signals that regulate plant architecture. Biologically and commercially important apocarotenoids include abscisic acid (plant hormone), bixin, apocarotenal, crocetin, food orange 7, ionones, strigolactones and peridinin.

This book has been written by leading experts in apocarotenoid research and gives a comprehensive overview on the diversity of apocarotenoid compounds and the latest discoveries on their nature and functions; the potential uses in agriculture, pharmacy, food industry, and apocarotenoid production at industrial level. *Biology, Chemistry, and Applications of Apocarotenoids* provides detailed accounts of the fundamental chemistry of apocarotenoids and the basic methods used in carotenoid research, and critical discussions of the biochemistry, functions, and applications of these important compounds. Topics covered in the proposed book include various aspects of the roles of apocarotenoids in colour and colouration, photosynthesis and other photofunctions, and protection. The formation and roles of carotenoid metabolites and breakdown products as perfume/aroma compounds are also outlined.

The book is intended for use by graduate and advanced undergraduate students, and researchers in plant physiology, molecular biology, biochemistry, biophysics and medicine, and will also be useful as a general reference on apocarotenoids. The ultimate aim of the book is to provide up-to-date and critical research surveys by experts in the field, incorporating some background material to make the chapters accessible to apocarotenoid researchers who are not specialist on the particular topic.

Editors

Dr. Ramamoorthy Siva is a professor and head of the Department of Biotechnology, Vellore Institute of Technology (VIT), India. He earned his undergraduate and postgraduate degrees in plant science from Madurai Kamaraj University, India. He earned his PhD in *Plant Genetic Diversity* at Bharathidasan University, India. He did his postdoctoral research at Ben Gurion University (Israel) and Gyeongsang National University (South Korea). His research interests are plant biotechnology and pigments biology. For more than 15 years he has carried out extensive research on plant dyes and pigments. He is the recipient of prestigious MASHAV fellowship from Ministry of Foreign Affairs, Israel at the Institute of Applied Research, Ben Gurion University of the Negev, Israel (2004). Dr. Siva has been recognised for his contributions at the international level in the field of Biotechnology and has been honoured by University of Colombo, Sri Lanka (2011) and recipient of "Young Scientist" project supported by DST, India. He is involved in popularizing science in India by organising conferences and conducting lecture workshops on various biological aspects with the support of Science Academies, India since, 2009. Dr. Siva has been successful to obtain various national and international research grants and several funded research projects. Dr. Siva is an IUCN species commission member for medicinal plants. He was nominated as Fellow of Linnean Society (FLS), London in 2013; Elected as a Fellow of Royal Society of Biology (UK) in 2015, and Fellow of Royal Society of Chemistry (UK) in 2020. He was selected as the "Young Scientist of the Year"(2017) by Tamil Nadu Academy of Sciences, Chennai and was awarded the International Visiting Fellow, University of Essex, UK (2017). He has published more than 100 manuscripts in reputed peer reviewed journals with high impact factor. Currently, he is the Associate Editor of *BMC Plant Biology*, *BMC Biotechnology* and Journal of Natural Products and Bioprospecting (Springer) and the Academic Editor of PLoS One and Peer J.

Dr. Renata Rivera-Madrid earned her PhD in plant physiology and molecular biology from the Université de Perpignan, France. Dr. Rivera is senior researcher in molecular biology of plant carotenoids at the Biochemistry and Molecular Biology Unit of the Scientific Research Centre of Yucatán (Centro de Investigación Científica de Yucatán, A.C, CICY). Her research interests are metabolic engineering and plant genetics. For the last two decades she has been working mainly with *Bixa orellana* (achiote), a plant which produces bixin, a worldwide valued natural pigment. Dr. Rivera's team is currently working on the regulatory mechanisms that control carotenoid production in higher plants, as well as on genetic variability, germplasm conservation, and genetic improvement using traditional methods and biotechnological tools. Dr. Rivera's main contributions to the knowledge of the molecular genetics of *Bixa orellana* and the molecular bases for its improvement are acknowledged at national and international level. Her publications, and her studies on the expression and regulation of the genes involved in the synthesis of carotenoids and bixin, are considered the basis for molecular genetic characterization. She has been in charge of more than ten *Bixa orellana* projects funded by national and international institutions and has authored more than 30 peer reviewed papers in specialized journals. As an active professor at the centre's postgraduate division, she trains MD, PhD, and postgraduate students, has directed 27 concluded pre-graduate theses, 17 master theses, four doctorate theses, and three post-doctorates. She has evaluated projects summited to the National Council for

Science and Technology (Consejo Nacional de Ciencia y Tecnología, CONACYT), as well as international projects. Dr. Rivera is a member of the Mexican Science Academy and a level II member of the National Researcher's System (Sistema Nacional de Investigadores, SNI). She is currently Guest Editor in *Frontiers in Plant Science* for the special issue on "Interactions between Biochemical Pathways Producing Plant Colors and Scents".

Dr. C. George Priya Doss earned his PhD from Vellore Institute of Technology (VIT), India and presently positioned as HOD, Department of Integrative Biology, School of Biosciences and Technology. His research interest spans both computational biology and bioinformatics, with prior in-depth knowledge and experience in molecular modelling, molecular simulation analysis, drug discovery, nano-informatics and development and construction of various biological databases. He has authored and co-authored more than 170 scientific publications in international peer reviewed journals and book chapters.

Contributors

Margarita Aguilar-Espinosa
Unidad de Bioquímica y Biología Molecular de Plantas
Centro de Investigación Científica de Yucatán A.C.
Mérida, Yucatán, México

Víctor Manuel Carballo-Uicab
Unidad de Bioquímica y Biología Molecular de Plantas
Centro de Investigación Científica de Yucatán A.C.
Mérida, Yucatán, México

Yair Cárdenas-Conejo
Laboratorio de Agrobiotecnología, CONACYT-Universidad de Colima
Colima, Mexico

David Chan-Rodriguez
Unidad de Bioquímica y Biología Molecular de Plantas
Centro de Investigación Científica de Yucatán A.C.
Mérida, Yucatán, México

Jonathan Dzib-Cauich
Unidad de Bioquímica y Biología Molecular de Plantas
Centro de Investigación Científica de Yucatán A.C.
Mérida, Yucatán, México

Leepica Kapoor
Department of Biotechnology, School of Bio Sciences and Technology
Vellore Institute of Technology, Vellore, India

Hebron Kirubai Raj
Department of Biotechnology, School of Bio Sciences and Technology
Vellore Institute of Technology, Vellore, India

Siva Ramamoorthy
Department of Biotechnology, School of Bio Sciences and Technology
Vellore Institute of Technology, Vellore, India

Renata Rivera-Madrid
Unidad de Bioquímica y Biología Molecular de Plantas
Centro de Investigación Científica de Yucatán A.C.
Mérida, Yucatán, México

Edith E. Uresti-Rivera
Centro de Investigación en Ciencias de la Salud y Biomedicina, Universidad Autónoma de San Luis Potosí, San Luis Potosí, México

Rosa Us-Camas
Unidad de Bioquímica y Biología Molecular de Plantas
Centro de Investigación Científica de Yucatán A.C.
Mérida, Yucatán, México

Michael H. Walter
Leibniz Institute of Plant Biochemistry, Department of Cell and Metabolic Biology
Halle (Saale), Germany

1 Biology and Chemistry of Apocarotenoids

Margarita Aguilar-Espinosa, Víctor Manuel Carballo-Uicab and Renata Rivera-Madrid

CONTENTS

INTRODUCTION

Apocarotenoids are terpenoid compounds with defined molecular structures. Carotenoid-specific cleavage oxygenases (CCOs) are responsible for the oxidative cleavage of carotenoids by acting on specific double bonds on this molecule (Tran et al. 2012; Priya and Siva 2014). This enzymatic reaction leads to the formation of apocarotenoids. In the plant science community, CCOs are also known as carotenoid cleavage dioxygenases (CCDs), while at the same time in the animal section the term carotenoid monooxygenases (CMOs) is still practiced (Walter and Strack 2011). Apocarotenoids such as vitamin A and abscisic acid (ABA) are essential for developmental processes of both animals and plants, respectively (Ohmiya 2009). Use of plant model systems (e.g., *Arabidopsis thaliana*; *Zea mays*; *Crocus sativus*; *Bixa orellana*) has contributed to the scientific knowledge of apocarotenoid biosynthesis through CCDs (Schwartz et al. 1997; Bouvier et al. 2003a; Tan et al. 2003; Gómez-Gómez et al. 2010). In Arabidopsis, CCDs are classified into two groups: 9-cis-epoxycarotenoid dioxygenases (NCEDs) and CCDs. Five members of the NCED (NCED2, NCED3, NCED5, NCED6, and NCED9) group participate exclusively in the ABA synthesis pathway while four members of CCD (CCD1, CCD4, CCD7, and CCD8) have a broader range of action (Ohmiya 2009). Apocarotenoids can also be formed through non-enzymatic oxidative cleavage of carotenoids by singlet oxygen ($^1O_2$)-mediated photo-oxidation or by lipoxygenase- and peroxidase-mediated co-oxidation (Leenhardt et al. 2006; González-Pérez et al. 2011; Hou et al. 2016).

BIOLOGY OF APOCAROTENOIDS

As mentioned above, apocarotenoids play important biological roles in the growth of vegetables (Sankari et al. 2018). In plants, the apocarotenoids abscisic acid (ABA) and strigolactone (SL) function as hormones dictating developmental processes. ABA contributes to grain dormancy or environmental abiotic stress responses, while SL is involved in growth inhibition and plant ramification (Walter and Strack 2011; Liu et al. 2013). Another role of apocarotenoids is they act as signalling molecules (ACSs) that play regulatory roles throughout plant development beyond those ascribed to ABA, SL (Hou et al. 2016), volatile apocarotenoids (including β-ionone, β-cyclocitral, theaspirone, β-damescenone, and α-damescenone) (Walter and Strack 2011), and chromophores (e.g., bixin and crocetin) (Bouvier et al. 2003b). Bixin – a red orange pigment – is synthesized in high quantities in the seeds of the tropical tree *B. orellana*. Bixin is highly valuable in the food and cosmetics industry. The proposed biosynthesis pathway of bixin from C40 carotenoid lycopene involves three enzymes: lycopene cleavage dioxygenase (*BoLCD*); bixin aldehyde dehydrogenase (*BoBADH*); and norbixin methyltransferase (*BoBMT*) (Bouvier et al. 2003a). The scientific authors suggest introducing these three genes into tomato fruit – containing high levels of the proposed precursor lycopene – as a strategy to produce bixin (Bouvier et al. 2003a).

The role apocarotenoids play in herbivore–plant communication is very important to agricultural chemical ecology (Cáceres et al. 2016). Apocarotenoids too bind in allelopathy intercommunication and plant protection system (Bouvier et al. 2003b). They also play a role in carotenogenesis, plant development, and environmental responses, in part via changes in nuclear gene expression and retrograde (plastid-to-nucleus) signalling (Pogson et al. 2008). Many apocarotenoids produced by CCDs provide fruits and flowers with distinctive colour, aroma, and flavour. Apocarotenoids have a lot of importance in health because they are not synthesized in animals (including humans) by *de novo* but instead these compounds are obtained from their diet for nutrient supplementation and health benefits. Apocarotenoid products, such as retinal derivatives, vitamin A and retinoic acid, are involved in nutrition and animal development (Giuliano et al. 2003).

CHEMISTRY OF APOCAROTENOIDS

Apocarotenoid is a term commonly used within carotenoid researchers to refer to any molecule produced during the cleavage of carotenoids. This definition differs from the IUPAC. The IUPAC defines apocarotenoids as 'carotenoids in which the carbon skeleton has been shortened by the removal of fragments from one end or both ends' (Britton et al. 2004). According to the IUPAC guidelines to name a compound, a carotenoid must contain the central methyl group intact at the C (20) or C (20′). Retinoids do not meet these requirements to be classified by the IUPAC as carotenoids (Figure 1.1). Officially, apocarotenoids should meet this requirement if they are to be considered within the group of molecules classified as carotenoids (Meléndez-Martínez 2019).

In diverse organisms, CCDs act in a region-specific manner on double bonds of carotenoids, leading to the formation of apocarotenoids with defined structures. Apocarotenoids have essential functions in the plant metabolism and play important ecological roles. The CCD mechanism action involves the incorporation of oxygen between adjacent carbon atoms along the carotenoid backbone. When CCDs act on bicyclic carotenoids, a single centric or eccentric cleavage produces two monocyclic structures; however, a double symmetrical cleavage yields diapocarotenoids (Walter and Strack 2011). When a sufficient number of conjugated double bonds are conserved in the products generated by the action of the CCDs, they can have pigment character like their parent carotenoids, while the chomophore is occasionally complemented by carboxyl chemical or keto groups Various monocyclic scission products with a reduced aliphatic chain are uncoloured and could act as growth regulators (e.g., ABA, C_{15}) or strong fragrance compounds (C_{13}, C_{10}) (Schlicht et al. 2008). All conjugated double bonds within the carotenoid backbone are potential sites for CCDs to act on (Walter and Strack 2011).

The enzymatic pathways giving rise to the apocarotenoid plant hormones and apocarotenoid signals in animals are thought to be similar. In animal cells, the carotenoid precursor is converted into a thermostable isomer to favour the enzymatic reaction. The isomerization step is accepted to be conserved in plants (Figure 1.2) (Hu et al. 2006).

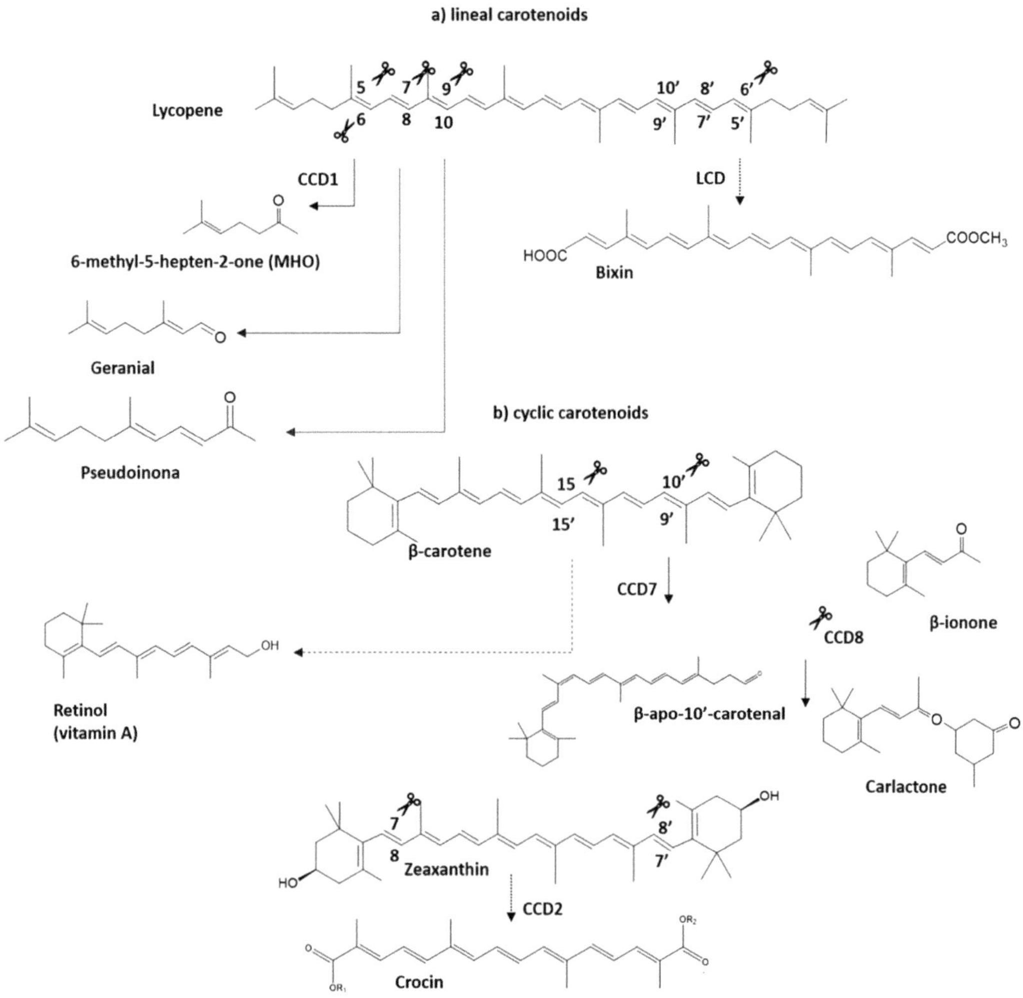

FIGURE 1.1 Diversity of carotenoid cleavage products of the symmetric and asymmetric cleavages of in (a) Lineal carotenoids: lycopene, (b) Cyclic carotenoids: β-carotene and zeaxanthin. Cleavage sites in substrates are indicated by scissors: carotenoid cleavage dioxygenase 1,2,7,8 (CCD), lycopene cleavage dioxygenase (LCD), β-carotene 15,15′-oxigenase (BCO).

GENERAL ASPECTS OF APOCAROTENOID PRECURSORS

Carotenoids are natural pigments of a subgroup of lipophilic isoprenoid compounds (Britton et al. 2004). At present, of 1,117 known natural carotenoids, 683 were identified with a known source organism (Yabuzaki 2017) to display a wide range of colours such as orange, red, and yellow (Yuan et al. 2015). In general, carotenoids are obtained from phytoene – 40-carbon backbone with conjugated double bonds (Hirschberg 2001; Maresca et al. 2008). However, C30 and C50 carotenoids could be produced by determined bacteria via various channels (Walter and Strack 2011). Carotenoids are synthesized by phototrophic (such as plant and algae) and some non-phototrophic (bacteria and fungi) organisms. Animals do not possess the ability to synthesize carotenoids (Fraser and Bramley 2004; Cazzonelli 2011).

The biological role of carotenoids as photoprotective and light-absorbing and antioxidant molecules is related to their polyene backbone. Moreover, carotenoids provide fruits, flowers, and

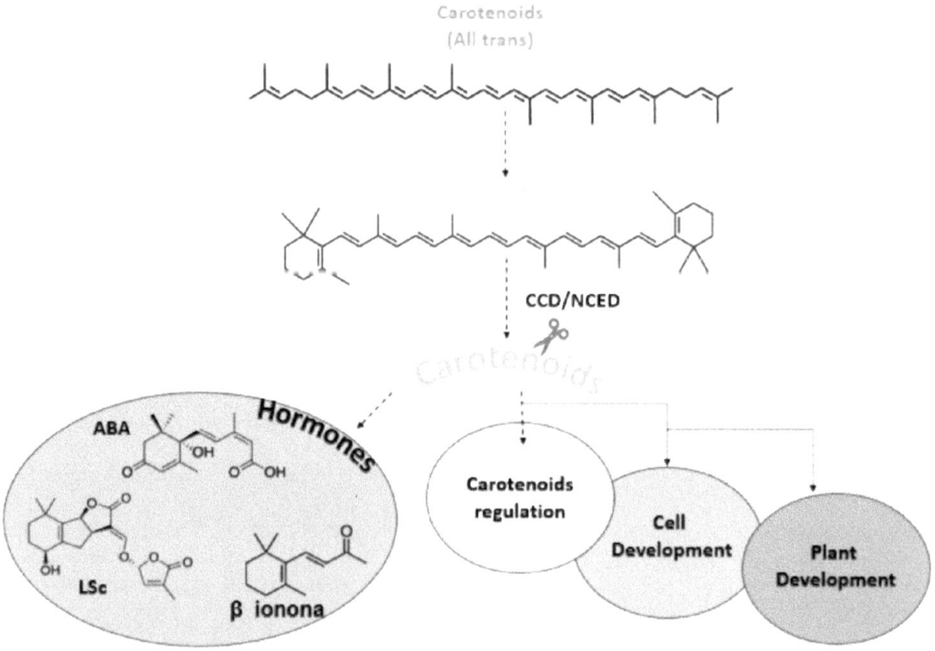

FIGURE 1.2 General schematic for apocarotenoids signal biosynthesis in plants.

vegetables with red, yellow, and orange colour. Indirectly, carotenoids play a role in plant reproduction by helping attract pollinators (Nisar et al. 2015). Carotenoids are synthesized in plastids of fruit, flowers, root, and seeds. In the green photosynthetic plastid chloroplast, carotenoids function as structural and accessory pigments as well as membrane stabilizers, thereby protecting from photo-oxidative stress (Nisar et al. 2015). Carotenoids are also synthesized and stored in chromoplasts (coloured plastids) in crystalline form as oil droplets (Bartley and Scolnik 1995), amyloplasts (starch-storing plastids), leucoplasts (colourless plastids), etioplasts (dark-grown precursors of the chloroplast), and elaioplasts (lipid-storing plastids) (Cazzonelli and Pogson 2010). Carotenoids also promote human health benefits associated with antioxidant activity, reducing age-linked macular degeneration of the eye, and are believed to be precursors for vitamin A (Cazzonelli 2011).

GENERAL CHEMICAL ASPECTS OF APOCAROTENOID PRECURSORS

The majority of carotenoids are derivatives of the C40 tetraterpenoid pigment phytoene, which are biologically synthesized by head-to-head bonding of two C20 geranylgeranyl diphosphate molecules. The carotenoids are formed from a phytoene molecule (40C) upon undergoing various enzymatic reactions. They are a type of isoprenoid structure molecule, which means they have a variable amount of chemical compounds with conjugated double bonds, and therefore have the property of absorbing visible light at different wavelengths (from 380 to 500 nm) which gives them colours that range from yellow to red and are widely used as dyes. A greater number of double bonds capture longer wavelengths (more towards red). Thus, with only three conjugated bonds, phytoene can only capture ultraviolet light (being colourless), whereas lycopene (tomato's red colour), with 11 conjugated double bonds, absorbs from ultraviolet to red. The carotenoids are classified into two groups based on functional groups: carotenes and xanthophylls. Carotenes have only hydrocarbon chains such as α-carotene, β-carotene, and lycopene, and the second group is xanthophylls, containing oxygen atoms as a functional group. This can be present in the form of an alcohol (—OH) (e.g., lutein), a ketone (=O) (e.g., canthaxanthin), a combination of alcohol and ketone (e.g., astaxanthin), or as esters of alcohol (e.g., fucoxanthin). Polar groups (epoxy, hydroxyl, and keto) are most commonly attached to the rings or the ends of the molecule, but infrequently to the centre that changes the polarity of carotenoids and modify the biological functions (Heath et al. 2013; Saini et al. 2015).

The pigment character of carotenoids is attributed to the uncoloured basal C40 structure of phytoene by adding double bonds. Phytoene desaturases introduce 4 desaturations in phytoene leading to the red lycopene (13 double bonds), that give colour to ripe tomatoes. Additional desaturation up to 15 fully conjugated double bonds has been carried out artificially by molecular breeding, resulting in pink 3,4-3′4′-tetradehydro-lycopene (Schmidt-Dannert et al. 2000).

In hydrogenous chain carotenoids, each conjugated double bonds are feasible scission sites by CCDs and needed dioxygen (O^2) producing considerable diversity of apocarotenoids join in numerous physiological or environmental processes (Walter and Strack 2011). The carotenoid biosynthesis pathway for the production of apocarotenoids has been well investigated and has led to many successful genetic engineering attempts (Giuliano et al. 2008; Goo et al. 2015).

HISTORY OF CARETONOID CLEAVAGE DIOXYGENASES

The conversion of β-carotene to vitamin A was first described by Moore in 1930 who fed β-carotene to vitamin A-deficient rats. The initial confirmation recommends the continuance of a characteristic enzyme and can be investigated in 1965 for both association of researchers; Goodman and Huang (1965) and Olson and Hayaishi (1965) individually description the enzymatic split up of β-carotene to form retinal using natural enzyme fraction separate of rat liver and intestine. However, more than 30 years later, the principal member of this group, named *Viviparous* 14 (*vp14*), was cloned and molecularly identified in a screen for viviparous maize seeds that showed a decreased level of abscisic acid (ABA) resulting from *vp*14 mutation. A screen for viviparous maize seeds led to the cloning of *VP14*, a gene encoding a protein that has sequence homology to lignostilbene dioxygenase (LZD) from *Pseudomonas paucimobilis* (Kamoda and Saburi 1993). Vivipary results from a lack of ABA which had been postulated to be derived from the splitting of a carotenoid catalysed by LSD. Enzymatic analysis of recombinant VP14 verified that it was 9-cis-epoxycarotenoids, the first committed step in ABA biosynthesis (Auldridge et al. 2006).

Homology-based analysis using VP14 that led to identification of numerous enzymes involved in the synthesis of apocarotenoids has been identified. Based on the recognized substrates and procedure of catalysis, these enzymes are referred to as carotenoid-cleaved dioxygenases (CCDs). The CCD family has members present in bacteria, plants, and animals. Representatives of the CCD progeny contribution have three characteristics: first, they require a Fe^{2+} for catalytic activity; second, they contain four conserved histidines that are thought to coordinate iron binding; and third, they contain a conserved peptide sequence at their carboxyl terminus that minimally constitutes a signature sequence for the family (Auldridge et al. 2006).

ROLES OF CARETONOID CLEAVAGE DIOXYGENASES

CCDs present specificity for the double bond that they split but are indiscriminate in their substance selection as the substrate (Auldridge et al. 2006). In plants, CCD enzymes are distributed into five classes: CCD1, CCD2, CCD4, CCD7, and CCD8 (Auldridge et al. 2006; Walter and Strack 2011; Frusciante et al. 2014). New specific CCDs from species with unknown functions have been identified by transcriptome and large-scale genome sequencing programmes in numerous species of plants. Actually, biochemical studies on CCDs for a limited number of plants and their functions in different organisms have been reported. The CCD4 group is present in all flowering plants, generally composed of several members (Gómez-Gómez et al. 2010; Rubio-Moraga et al. 2014). Its functions are related to the pigments and volatile aroma in flowers and fruit tissues (Rubio et al. 2008; Ohmiya 2009; Lashbrooke et al. 2013; Rodrigo et al. 2013). CCD4 activity in seeds, leaves, and roots, associated with carotenoid rotation with apocarotenoid production, has also been reported (Gonzalez-Jorge et al. 2013; Avendaño-Vázquez et al. 2014; Lätari et al. 2015).

Scientific researchers have found in plants that CCD1 and CCD7 split the 9-10 double bonds of their carotenoid substrate, although 9-cis-epoxycarotenoid dioxygenases (NCEDs) cause excision at the position 11,12 double bonds. Additionally, BoLCD from *B. orellana* splits at the lycopene 5,6 double bond (Bouvier et al. 2003a), and CsZCD from *Crocus sativus* (a monocot) was found to cleave at the 7,8 double bond of zeaxanthin (Bouvier et al. 2003b). BoLCD and CsZCD have been reported to be shorter than any CCD isolate to date, and although the CsZCD was immunohistochemically shown to reside in plastids, it had no identifiable transit peptide. In addition, CsZCD and BoLCD share high sequence identity at the nucleotide and amino-acid levels (97%), which is incredible for enzymes from different species that catalyse excision at different double bonds (Auldridge et al. 2006). Also in *B. orellana*, Rodríguez-Ávila et al. (2011) isolated a new member of the CCD family (BoCCD1). The putative BoCCD1 open reading frame (ORF) sequence differs from the CCD isolated in this plant by Bouvier et al. (2003a) and shares a high homology with other plant CCD1s involved in the formation of apocarotenoids. The BoCCD1 expression pattern was correlated with carotenoid accumulation in this plant (mainly bixin). *In situ* hybridization methodology was used for the first time in this plant to investigate the expression of BoCCD1 and pigment accumulation in different tissues of *B. orellana* (Rodríguez-Ávila et al. 2011).

Bixa orellana samples were collected from Abidjan (Ivory Coast Africa) by Bouvier et al. (2003a), and they found a CCD4-type *B. orellana* dioxygenase catalysing the simultaneous oxidation of the 5,6 and (5′,6′) double bond of lycopene (BoLCD). This enzyme is considered to participate in bixin biosynthesis, split the bixin aldehyde, which after being oxidized to norbixin is methylated to produce bixin. The corresponding bixin aldehyde dehydrogenase and norbixin methyltransferase involved in these reactions are identified (Bouvier et al. 2003a). Nevertheless, Cárdenas-Conejo et al. (2015) investigate a photosynthetic (leaf) and nonphotosynthetic (immature and mature seeds) transcriptome from a bixin-producing *B. orellana* Yucatecan accession, and none of the genes putatively involved in the process were found. Recently, two BoCCDs enzymes (BoCCD1 and BoCCD4-3) involved in bixin synthesis were identified, and enzymatic expression of both in *E. coli* cells revealed the capacity to generate bixin aldehyde from lycopene, confirming their 5,6-5′,6′ excision activity (Carballo-Uicab et al. 2019).

In Arabidopsis, there are nine CCDs categorized according to their enzymatic specificities, and they determine the CCD nomenclature in other species. Five CCD members are 9-*cis* epoxy carotenoid dioxygenase subfamily (NCED2, NCED3, NCED5, NCED6, and NCED9) and participate in ABA synthesis. CCD1, CCD4, and CCD7 have broad substrate specificity while CCD8 may be specific for SL synthesis. One additional CCD subfamily, dubbed 'CCD2' *in Crocus sativus*, produces saffron-flavour and fragrance apocarotenoids (Hou et al. 2016).

In the scientific literature, there is information of apocarothenoids in numerous plant species for application in the therapeutic industry, food industry, and cosmetics industry. The use of new colours, flavours, and spices obtained from flowers, fruits, leaves, and seeds have gained increasing interest, replacing synthetics as many of them have carcinogenic properties.

ACKNOWLEDGEMENTS

This work was supported by the Consejo Nacional de Ciencia y Tecnología (CONACYT) (Fronteras de la Ciencia No. 2016-01-1716).

REFERENCES

Auldridge, Michele E, Donald R McCarty, and Harry J Klee. 2006. "Plant Carotenoid Cleavage Oxygenases and Their Apocarotenoid Products." *Current Opinion in Plant Biology* 9 (3): 315–321. https://doi.org/10.1016/j.pbi.2006.03.005.

Avendaño-Vázquez, Aida-Odette, Elizabeth Cordoba, Ernesto Llamas, Carolina San Román, Nazia Nisar, Susana De la Torre, Maricela Ramos-Vega, María de la Luz Gutiérrez-Nava, Christopher Ian Cazzonelli, and Barry James Pogson. 2014. "An Uncharacterized Apocarotenoid-Derived Signal Generated in ζ-Carotene Desaturase Mutants Regulates Leaf Development and the Expression of Chloroplast and Nuclear Genes in Arabidopsis." *The Plant Cell* 26 (6): 2524–2537.

Bartley, Glenn E, and Pablo A Scolnik. 1995. "Plant Carotenoids: Pigments for Photoprotection, Visual Attraction, and Human Health." *The Plant Cell* 7 (7): 1027.

Bouvier, Florence, Odette Dogbo, and Bilal Camara. 2003a. "Biosynthesis of the Food and Cosmetic Plant Pigment Bixin (Annatto)." *Science* 300 (5628): 2089–2091.

Bouvier, Florence, Claude Suire, Jérôme Mutterer, and Bilal Camara. 2003b. "Oxidative Remodeling of Chromoplast Carotenoids: Identification of the Carotenoid Dioxygenase CsCCD and CsZCD Genes Involved in Crocus Secondary Metabolite Biogenesis." *The Plant Cell* 15 (1): 47–62.

Britton, G, S Liaaen-Jensen, and H Pfander. 2004. "Carotenoids with a C40 Skeleton (a) Hydrocarbons." *Carotenoids Handbook, Main List* 31: 1–50.

Cáceres, LA, S Lakshminarayan, KK-C Yeung, BD McGarvey, A Hannoufa, MW Sumarah, X Benitez, and IM Scott. 2016. "Repellent and Attractive Effects of α-, β-, and Dihydro-β-Ionone to Generalist and Specialist Herbivores." *Journal of Chemical Ecology* 42 (2): 107–117.

Carballo-Uicab, Víctor Manuel, Yair Cárdenas-Conejo, Alba Adriana Vallejo-Cardona, Margarita Aguilar-Espinosa, Jacobo Rodríguez-Campos, Hugo Serrano-Posada, José Alberto Narváez-Zapata, Felipe Vázquez-Flota, and Renata Rivera-Madrid. 2019. "Isolation and Functional Characterization of Two Dioxygenasese Putatively Involved in Bixin Biosynthesis in Annatto (*Bixa orellana* L.)." *PeerJ* 7 (June): e7064. https://doi.org/10.7717/peerj.7064.

Cárdenas-Conejo, Yair, Víctor Carballo-Uicab, Meric Lieberman, Margarita Aguilar-Espinosa, Luca Comai, and Renata Rivera-Madrid. 2015. "De Novo Transcriptome Sequencing in Bixa Orellana to Identify Genes Involved in Methylerythritol Phosphate, Carotenoid and Bixin Biosynthesis." *BMC Genomics* 16 (1): 877. https://doi.org/10.1186/s12864-015-2065-4.

Cazzonelli, Christopher I. 2011. "Carotenoids in Nature: Insights from Plants and Beyond." *Functional Plant Biology* 38 (11): 833–847.

Cazzonelli, Christopher I, and Barry J Pogson. 2010. "Source to Sink: Regulation of Carotenoid Biosynthesis in Plants." *Trends in Plant Science* 15 (5): 266–274.

Fraser, Paul D, and Peter M Bramley. 2004. "The Biosynthesis and Nutritional Uses of Carotenoids." *Progress in Lipid Research* 43 (3): 228–265. https://doi.org/10.1016/j.plipres.2003.10.002.

Frusciante, Sarah, Gianfranco Diretto, Mark Bruno, Paola Ferrante, Marco Pietrella, Alfonso Prado-Cabrero, Angela Rubio-Moraga, Peter Beyer, Lourdes Gómez-Gómez, and Salim Al-Babili. 2014. "Novel Carotenoid Cleavage Dioxygenase Catalyzes the First Dedicated Step in Saffron Crocin Biosynthesis." *Proceedings of the National Academy of Sciences of the United States of America* 111 (33): 12246–12251.

Giuliano, Giovanni, Salim Al-Babili, and Johannes Von Lintig. 2003. "Carotenoid Oxygenases: Cleave It or Leave It." *Trends in Plant Science* 8 (4): 145–149.

Giuliano, Giovanni, Raffaela Tavazza, Gianfranco Diretto, Peter Beyer, and Mark A Taylor. 2008. "Metabolic Engineering of Carotenoid Biosynthesis in Plants." *Trends in Biotechnology* 26 (3): 139–145.

Gómez-Gómez, Lourdes, Angela Rubio-Moraga, and Oussama Ahrazem. 2010. "Understanding Carotenoid Metabolism in Saffron Stigmas: Unravelling Aroma and Colour Formation." *Functional Plant Science and Biotechnology* 4: 56–63.

Gonzalez-Jorge, Sabrina, Sun-Hwa Ha, María Magallanes-Lundback, Laura Ullrich Gilliland, Ailing Zhou, Alexander E Lipka, Yen-Nhu Nguyen, Ruthie Angelovici, Haining Lin, and Jason Cepela. 2013. "Carotenoid Cleavage Dioxygenase 4 Is a Negative Regulator of β-Carotene Content in Arabidopsis Seeds." *The Plant Cell* 25 (12): 4812–4826.

González-Pérez, Sergio, Jorge Gutiérrez, Francisco García-García, Daniel Osuna, Joaquín Dopazo, Óscar Lorenzo, José L Revuelta, and Juan B Arellano. 2011. "Early Transcriptional Defense Responses in Arabidopsis Cell Suspension Culture under High-Light Conditions." *Plant Physiology* 156 (3): 1439–1456.

Goo, Young-Min, Eun-Heui Han, Jae Cheol Jeong, Sang-Soo Kwak, Jaeju Yu, Yun-Hee Kim, Mi-Jeong Ahn, and Shin-Woo Lee. 2015. "Overexpression of the Sweet Potato IbOr Gene Results in the Increased Accumulation of Carotenoid and Confers Tolerance to Environmental Stresses in Transgenic Potato." *Comptes Rendus Biologies* 338 (1): 12–20.

Goodman, DeWitt S, and Helen S Huang. 1965. "Biosynthesis of Vitamin A with Rat Intestinal Enzymes." *Science* 149 (3686): 879–880.

Grant, MW. 1949. *Vitamin A and Carotene*. Nature Publishing Group.

Heath, Jeremy J, Don F Cipollini, and John O Stireman. 2013. "The Role of Carotenoids and Their Derivatives in Mediating Interactions between Insects and Their Environment." *Arthropod-Plant Interactions* 7 (1). https://doi.org/10.1007/s11829-012-9239-7.

Hirschberg, Joseph. 2001. "Carotenoid Biosynthesis in Flowering Plants." *Current Opinion in Plant Biology* 4 (3): 210–218.

Hou, Xin, John Rivers, Patricia León, Ryan P McQuinn, and Barry J Pogson. 2016. "Synthesis and Function of Apocarotenoid Signals in Plants." *Trends in Plant Science* 21 (9): 792–803.

Hu, Kang-Quan, Chun Liu, Hansgeorg Ernst, Norman I Krinsky, Robert M Russell, and Xiang-Dong Wang. 2006. "The Biochemical Characterization of Ferret Carotene-9′, 10′-Monooxygenase Catalyzing Cleavage of Carotenoids In Vitro and In Vivo." *Journal of Biological Chemistry* 281 (28): 19327–19338.

Kamoda, Shigehiro, and Yoshimasa Saburi. 1993. "Cloning, Expression, and Sequence Analysis of a Lignostilbene-α, β-Dioxygenase Gene from Pseudomonas Paucimobilis TMY1009." *Bioscience, Biotechnology, and Biochemistry* 57 (6): 926–930.

Lashbrooke, Justin G, Philip R Young, Samantha J Dockrall, Krishnan Vasanth, and Melané A Vivier. 2013. "Functional Characterisation of Three Members of the Vitis Vinifera L. Carotenoid Cleavage Dioxygenase Gene Family." *BMC Plant Biology* 13 (1): 156.

Lätari, Kira, Florian Wüst, Michaela Hübner, Patrick Schaub, Kim Gabriele Beisel, Shizue Matsubara, Peter Beyer, and Ralf Welsch. 2015. "Tissue-Specific Apocarotenoid Glycosylation Contributes to Carotenoid Homeostasis in Arabidopsis Leaves." *Plant Physiology* 168 (4): 1550–1562.

Leenhardt, F, B Lyan, E Rock, A Boussard, J Potus, E Chanliaud, and C Remesy. 2006. "Genetic Variability of Carotenoid Concentration, and Lipoxygenase and Peroxidase Activities among Cultivated Wheat Species and Bread Wheat Varieties." *European Journal of Agronomy* 25 (2): 170–176.

Liu, Junwei, Mara Novero, Tatsiana Charnikhova, Alessandra Ferrandino, Andrea Schubert, Carolien Ruyter-Spira, Paola Bonfante, Claudio Lovisolo, Harro J Bouwmeester, and Francesca Cardinale. 2013. "Carotenoid Cleavage Dioxygenase 7 Modulates Plant Growth, Reproduction, Senescence, and Determinate Nodulation in the Model Legume Lotus Japonicus." *Journal of Experimental Botany* 64 (7): 1967–1981.

Maresca, Julia A, Joel E Graham, and Donald A Bryant. 2008. "The Biochemical Basis for Structural Diversity in the Carotenoids of Chlorophototrophic Bacteria." *Photosynthesis Research* 97 (2): 121–140.

Meléndez-Martínez, Antonio J. 2019. "An Overview of Carotenoids, Apocarotenoids, and Vitamin A in Agro-Food, Nutrition, Health, and Disease." *Molecular Nutrition & Food Research* 63 (15): 1801045.

Moore, Thomas. 1930. "Vitamin A and carotene." *Biochemical Journal* 24 (3): 692–702 doi:10.1042/bj0240692.

Nisar, Nazia, Li Li, Shan Lu, Nay Chi Khin, and Barry J Pogson. 2015. "Carotenoid Metabolism in Plants." *Molecular Plant* 8 (1): 68–82.

Ohmiya, Akemi. 2009. "Carotenoid Cleavage Dioxygenases and Their Apocarotenoid Products in Plants." *Plant Biotechnology* 26 (4): 351–358.

Olson, J A, and O Hayaishi. 1965. "The Enzymatic Cleavage of Beta-Carotene into Vitamin A by Soluble Enzymes of Rat Liver and Intestine." *Proceedings of the National Academy of Sciences of the United States of America* 54 (5): 1364–1370. https://doi.org/10.1073/pnas.54.5.1364.

Pogson, Barry J, Nick S Woo, Britta Förster, and Ian D Small. 2008. "Plastid Signalling to the Nucleus and Beyond." *Trends in Plant Science* 13 (11): 602–609.

Priya, R, and R Siva. 2014. "Phylogenetic Analysis and Evolutionary Studies of Plant Carotenoid Cleavage Dioxygenase Gene." *Gene* 548 (2): 223–233.

Rodrigo, María J, Berta Alquézar, Enriqueta Alós, Víctor Medina, Lourdes Carmona, Mark Bruno, Salim Al-Babili, and Lorenzo Zacarías. 2013. "A Novel Carotenoid Cleavage Activity Involved in the Biosynthesis of Citrus Fruit-Specific Apocarotenoid Pigments." *Journal of Experimental Botany* 64 (14): 4461–4478.

Rodríguez-Ávila, NL, JA Narváez-Zapata, JE Ramírez-Benítez, ML Aguilar-Espinosa, and R Rivera-Madrid. 2011. "Identification and Expression Pattern of a New Carotenoid Cleavage Dioxygenase Gene Member from Bixa Orellana." *Journal of Experimental Botany* 62 (15): 5385–5395.

Rubio-Moraga, Angela, José Luís Rambla, Asun Fernández-de-Carmen, Almudena Trapero-Mozos, Oussama Ahrazem, Diego Orzáez, Antonio Granell, and Lourdes Gómez-Gómez. 2014. "New Target Carotenoids for CCD4 Enzymes Are Revealed with the Characterization of a Novel Stress-Induced Carotenoid Cleavage Dioxygenase Gene from Crocus Sativus." *Plant Molecular Biology* 86 (4–5): 555–569.

Rubio, Angela, José Luís Rambla, Marcella Santaella, M Dolores Gómez, Diego Orzáez, Antonio Granell, and Lourdes Gómez-Gómez. 2008. "Cytosolic and Plastoglobule-Targeted Carotenoid Dioxygenases from Crocus Sativus Are Both Involved in β-Ionone Release." *Journal of Biological Chemistry* 283 (36): 24816–24825.

Saini, Ramesh Kumar, Shivraj Hariram Nile, and Se Won Park. 2015. "Carotenoids from Fruits and Vegetables: Chemistry, Analysis, Occurrence, Bioavailability and Biological Activities." *Food Research International* 76 (August): 735–750. https://doi.org/10.1016/j.foodres.2015.07.047.

Sankari, Mohan, Priya Rajendra Rao, Hridya Hemachandran, Phani Kumar Pullela, Iftikhar Aslam Tayubi, Babu Subramanian, KM Gothandam, Pooja Singh, and Siva Ramamoorthy. 2018. "Prospects and Progress in the Production of Valuable Carotenoids: Insights from Metabolic Engineering, Synthetic Biology, and Computational Approaches." *Journal of Biotechnology* 266: 89–101.

Schlicht, Markus, Olga Šamajová, Doreen Schachtschabel, Stefano Mancuso, Diedrik Menzel, Wilhelm Boland, and František Baluška. 2008. "D'orenone Blocks Polarized Tip Growth of Root Hairs by Interfering with the PIN2-mediated Auxin Transport Network in the Root Apex." *The Plant Journal* 55 (4): 709–717.

Schmidt-Dannert, Claudia, Daisuke Umeno, and Frances H Arnold. 2000. "Molecular Breeding of Carotenoid Biosynthetic Pathways." *Nature Biotechnology* 18 (7): 750–753.

Schwartz, Steven H, Bao Cai Tan, Douglas A Gage, Jan AD Zeevaart, and Donald R McCarty. 1997. "Specific Oxidative Cleavage of Carotenoids by VP14 of Maize." *Science* 276 (5320): 1872–1874.

Tan, Bao-Cai, Leina M Joseph, Wen-Tao Deng, Lijuan Liu, Qin-Bao Li, Kenneth Cline, and Donald R McCarty. 2003. "Molecular Characterization of the Arabidopsis 9-cis Epoxycarotenoid Dioxygenase Gene Family." *The Plant Journal* 35 (1): 44–56.

Tran, Phoi T, Marina N Sharifi, Subhajit Poddar, Rachel M Dent, and Krishna K Niyogi. 2012. "Intragenic Enhancers and Suppressors of Phytoene Desaturase Mutations in Chlamydomonas Reinhardtii." *PLoS One* 7 (8).

Walter, Michael H, and Dieter Strack. 2011. "Carotenoids and Their Cleavage Products: Biosynthesis and Functions." *Natural Product Reports* 28 (4): 663–692.

Yabuzaki, Junko. 2017. "Carotenoids Database: Structures, Chemical Fingerprints and Distribution among Organisms." *Database* [Online]. https://doi.org/10.1093/database/bax004.

Yuan, Hui, Junxiang Zhang, Divyashree Nageswaran, and Li Li. 2015. "Carotenoid Metabolism and Regulation in Horticultural Crops." *Horticulture Research* 2 (1): 1–11.

2 Natural Sources of Apocarotenoids and Their Applications

Jonathan Dzib-Cauich, Rosa Us-Camas and Renata Rivera-Madrid

CONTENTS

INTRODUCTION

Historically, carotenoids were known as natural pigments. Subsequently, the study of the nature of these molecules revealed their importance for life on earth. Carotenoids are tetraterpenes (40 carbons), with conjugated double bonds in their polienic chain. They are obtained naturally from plants, bacteria, algae, and non-photosynthetic fungi. Animals cannot synthesize carotenoids and they get them through their diet. An exception is a species of aphids, which can synthesize torulene, a red carotenoid (Moran and Jarvik, 2010). Carotenoids fulfil important functions for the survival of organisms: in plants they capture light in ranges where chlorophyll is not functional, defend the photosynthetic apparatus from photooxidative deterioration, and provide colours to flowers, fruits, and seeds ensuring reproductive success. Additionally, carotenoids are precursors of apocarotenoids such as abscisic acid (ABA) and strigolactones (SL) (Fiorilli et al., 2019), important regulators of plant growth and development. Apocarotenoids (**apo** means **from**) are the product of the oxidative cleavage of carotenoids. The structure of the carotenoids is shortened by removing fragments of one or both ends (Harrison and Bugg, 2014). In plants, the biosynthesis of apocarotenoids can be enzymatic or non-enzymatic (Hou et al., 2016). Carotenoid cleavage dioxygenases (CCDs) cleave the double bonds in carotenoids, allowing the formation of specific apocarotenoids (Schwartz et al., 1997; Auldridge et al., 2006; Vogel et al., 2008), while non-enzymatic apocarotenoid biosynthesis involves the participation of reactive oxygen species (ROS), such as the oxidation of β-carotene via singlet oxygen (O_2) (Havaux, 2014). CCDs are in all taxa as an evolutionarily conserved family of enzymes not dependent on Fe^{2+} heme (Harrison and Bugg, 2014). So far, seven families of the CCDs have been reported: NCED, CCD1, CCD2, CCD4, CCD7, CCD8, and ZAS, each of these families is classified by the types of bonds and carotenoids that cleave (Frusciante et al., 2014; Wang et al., 2019). Thus, NCED enzymes participate in the cleavage of the 11,12 / 11′,12′ bonds of the 9-Z-epoxycarotenoids to produce xanthoxin, a precursor of ABA, an important regulator of plant growth (Tan et al., 2003). Apocarotenoids are important for different biological processes; for example in plants, they function as molecules that regulate growth and development, they function as communication signals between surrounding organisms and are important for the response to biotic and abiotic stress (Tan et al., 2003; Wang et al., 2019). Additionally, apocarotenoids provide colours to flowers, fruits, and seeds for successful reproduction and protect plants from pathogen infections. In animals, retinal is one of the most important apocarotenoids because it is necessary for the detection of light in the retina and vision. In humans, apocarotenoids, such as vitamin A, in addition to its important role in vision, have antioxidant properties that prevent the formation of free radicals and prevent the formation of diseases such as cancer (Harrison and Quadro, 2018). In fungi, apocarotenoids, such as trisporic acid, function as the initial signal system that mediates the recognition between mating partners, induce morphogenesis, and cause early sexual development, and they also function as a signal for carotenogenesis between fungi. In bacteria, the apocarotenoid staphyloxanthin confers virulence to the bacterium *Staphylococcus aureus*, a well-known pathogenic agent.

Due to the importance of the functions of apocarotenoids in animals, bacteria, fungi, and plants, coupled with the fact that new apocarotenoid molecules and their functions are discovered every day, the pharmaceutical, food, and cosmetic industries have become interested in apocarotenoids for their potential applications in these sectors. Thus, in this chapter, we will focus on the review of the main apocarotenoids produced by plants, animals, fungi, and bacteria, their biosynthesis, and its biological function in these organisms, as well as their current and future applications.

PLANT APOCAROTENOIDS

Carotenoid Cleavage Dioxygenases (CCDs)

Apocarotenoids are molecules with less than 40C that are derived from the oxidative cleveage of carotenoids. The cleavage can be non-enzymatic or nonspecific when it involves the participation of by reactive oxygen species and can be enzymatic or specific when CCDs are involved (Hou et al., 2016). CCDs are in all taxa as an evolutionarily conserved family of enzymes not dependent on Fe^{2+} heme (Harrison and Bugg, 2014) and are responsible for the synthesis of the vast majority of existing apocaroteonoids. Recently in the tomato genome a new type of lipoxygenase enzyme (TomLoxC) involved in the accumulation of apocarotenoids has been discovered (Gao et al., 2019).

In plants, seven families of CCD enzymes have been reported: NCED, CCD1, CCD2, CCD4, CCD7, CCD8, and ZAS, each of which is classified based on the types of bonds and the carotenoids they cut (Frusciante et al., 2014; Wang et al., 2019). NCED enzymes are characterized by excision in the 11,12 / 11′,12′ bonds of the 9-Z-epoxycarotenoids giving rise to the xanthoxin precursor of ABA. The enzymes of the CCD1 family are located in the cytosol of plant cells and cut cyclic and linear carotenoids at the 5,6 / 5′,6′, 7,8 / 7′,8′, and 9,10 / 9′,10′ bonds. Similarly, the enzymes of the CCD2 family have been reported in the plastids of *Crocus sativus* cells and are responsible for the production of crocetin. The CCD2 family cleaves carotenoids and apocarotenoids at positions 7,8 / 7′,8′ near the β-ionone ring (Frusciante et al., 2014; Ahrazem et al., 2016a). The CCD4 family is divided into two types of enzymes: (I) those that cut into non hydroxylated bicyclic carotenoids at positions 9,10 or 9′,10′ such as trans-β-carotene to give rise to apo-10′-carotenoids (27C) and cyclohexenone (13C), corresponding to β-ionone, and (II) the enzymes that cut hydroxylated bicyclic carotenoids at positions 7,8 or 7′,8′, like those that produce citraurine (3-OH-β-apo-8′-carotenoid) (Huang et al., 2009; Rodrigo et al., 2013). Additionally, in this family of CCDs, a type of CCD4 has been reported that seems to cut linear carotenoids at positions 5,6/5′,6′ (Carballo-Uicab et al., 2019). On the other hand, the enzymes of the CCD7 and CCD8 family participate in the synthesis of strigolactones in plastids, where they cut trans-carotenoids to give rise to 27C apocarotenoids, which are converted into 18C and 9C apocarotenoids (Alder et al., 2012). Finally, the ZAS constitutes a new family of CCDs that participate in the production of zaxinone, an apocarotenoid that releases and promotes root growth, decreases the production of strigolactones (SL), and inhibits the germination of *Striga* (Wang et al., 2019).

β-Apocarotenoids

In plants, β-carotene is cleaved at different positions in its chain carbon by enzymes of the CCDs family or by ROS to produce apocarotenoids generally referred to as β-apocarotenoids, including ionone, β-cyclocitral and strigolactones (SL), among others (Figure 2.1.) (Han et al., 2019).

Ionones

The ionones are a set of 13-carbon apocarotenoids (13C) that can derive from the degradation of β-carotene and α-carotene (β-ionones and α-ionones). This occurs mainly in flowers, where they have an important role in the attraction of pollinating insects. Some of the main sources of β-ionones

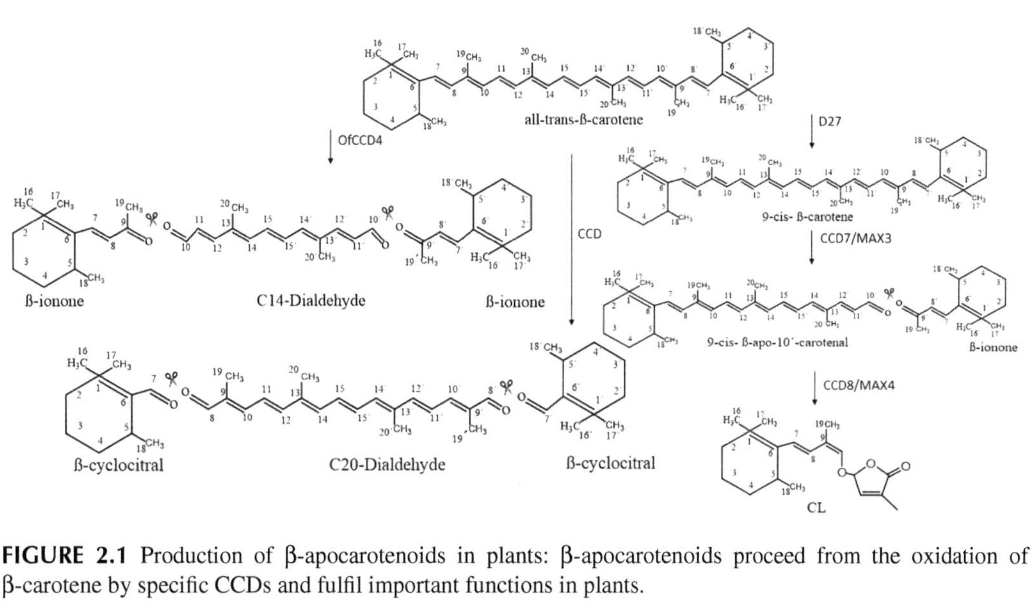

FIGURE 2.1 Production of β-apocarotenoids in plants: β-apocarotenoids proceed from the oxidation of β-carotene by specific CCDs and fulfil important functions in plants.

are the flowers of *Osmanthus fragrans* cultivar Yingui, a species of shrub of the Oleaceae family native to Asia, found from the Himalayas to China (Han et al., 2019). In *O. fragrans*, β-ionone is a key aromatic component, derived mainly from the symmetrical cut on the 9,10 / 9′,10′ bonds of the β-carotene made by the OfCCD4 enzyme that is transcriptionally regulated by ERF, a transcription factor which belongs to a large family of transcription factors in plants (Han et al., 2019). Likewise, in the flowers of other species such as *Chrysantremum* × *morifolium*, *C. sativus*, and *Malus domestica*, the family of CCD4 enzymes are responsible for the cleavage of β-carotene in the same sites as OfCCD4 to produce β-ionone (Huang et al., 2009).

Due to their extremely low odour-detection threshold, they are used in the fragrance industry, and their production in simple microorganisms has been considered as an innovative topic (Keasling, 2012; López et al., 2015; Zhang et al., 2018).

β-Cyclocitral

β-cyclocitral is another important β-apocarotenoid. One of the first functions attributed to this apocarotenoid was as signal molecule with the direction of chloroplast to the nucleus in response to high light stress conditions. Under excess light conditions, ROS such as singlet oxygen (O^2) can cause damage to proteins and lipids resulting in cellular alterations, decreased photosynthetic efficiency, and impaired growth (Gonzales-Pérez et al., 2011). In *Arabidopsis thaliana*, it was reported that β-cyclocitral acts as a second messenger by altering the expression of genes that respond to the presence of O^2. Likewise, experiments with mutants showed that β-cyclocitral interacts with MBS (METHYLENE BLUE SENSITIVITY) proteins during O^2 signalling, leading to photoacclimation (Shumbe et al., 2017). Recently, β-cyclocitral was found to increase cell divisions in root meristems, stimulating lateral branching and the growth of monocot and dicot species. Also, exogenous treatment with β-cyclocitral improved vigour and promoted the growth of roots and shoots of rice plants exposed to soils contaminated with salt (Dickinson et al., 2019). It is known that natural compounds capable of increasing root growth and branching are desirable tools to improve stress tolerance in crops, and therefore, β-cyclocytral can be a good ally to increase the vigour of plants with agronomic importance in soils with high salinity and low water availability. Until now, research has focused solely on understanding the signalling mechanisms of this apocarotenoid, but in the future, the large-scale production of β-cyclocitral could allow its commercialization as a product to improve the tolerance to different types of stress of crops of agronomic importance (Dickinson et al., 2019; Felemban et al., 2019).

Strigolactones

Strigolactones (SL) is one of the most studied β-apocarotenoids. So far, 25 different molecules have been identified (Felemban et al., 2019). The biosynthesis of the SL begins in the plastids, through a conversion reaction of trans-β-carotene to 9-cis-β-carotene catalyzed by the enzyme D27, in turn, the 9-cis-β-carotene is cleaved by the enzyme CCD7/MAX3 to give rise to the formation of 9-cis-β-apo-10′-carotenal and β-inone. Then the enzyme CCD8/MAX4 converts 9-cis-β-apo-10′-carotene into carlactone (CL) and W-OH-(4-CH3) heptanal. The carlactone from which all known SLs are derived so far is transported from the plastids to the cytosol (Alder et al., 2012). In the cytosol, the carlactone can be converted to carlactonic acid (CLA) by the enzyme MAX1/carlactone oxidase. Likewise, CLA can be methylated to give rise to methyl carlactonoate (MeCLA) through a reaction catalyzed by methyltransferase enzymes. All these reactions give rise to the diversity of SL that are known (Felemban et al., 2019).

All strigolactones come from the enzymatic cleavage of β-carotene. However, in a recent study, it is presumed that CCD7/CCD8 enzymes cut hydroxylated carotenoids to produce 3-OH-CL a possible SL precursor (Baz et al., 2018). The strigolactones can be classified into canonical SL and non-canonical SL. The diversity of the SL is a consequence of the structural variability of the non-canonical SL and the modifications of the α and β rings in the canonical ones, which include methylation, hydroxylation, ketolation, and epoxidation (Felemban et al., 2019).

Strigolactones are phytohormones that perform important functions in model plants as well as in plants of agronomic importance. They participate in development processes, in branching outbreaks, in root architecture, in improving the senescence of leaves in the presence of ethylene, and in providing response to biotic and abiotic stress conditions (Kapulnik and Koltai, 2014; Ueda and Kusaba, 2015; Van Rongen et al., 2019). Besides, SL allows the assimilation of phosphorus in plants by directly participating in symbiosis with arbuscular mycorrhizal fungi (AM). However, in addition to their beneficial functions, SL can also cause losses in cultivars by causing the germination of seeds of parasitic plants such as *Striga* (Kapulnik and Koltai, 2014).

Abscisic Acid (ABA)

The abscisic acid is another amply studied apocarotenoid in plants. ABA is a phytohormone that coordinates different developmental processes and growth and responds to biotic and abiotic stress in plants. The ABA biosynthesis pathway begins with zeaxanthin generated by the hydroxylation of β-carotene by the β-carotene hydroxylase family enzymes (BCH1/2). Zeaxanthin is epoxidated by zeaxanthin epoxidase (ZEP) to give rise to 9-cis-anteraxanthin and 9-cis-violaxanthin (Cazzonelli and Pogson, 2010). In turn, 9-cis-violaxanthin is converted to 9-cis-neoxanthin by neoxanthin synthase (NYS). Both 9-cis-violaxanthin and 9-cis-neoxanthin are converted to xanthoxin by enzymes of the family 9-cis-epoxycarotenoid dioxygenases (NCEDs) (Sah et al., 2016). The step catalyzed by NCED has been reported as limiting the production of ABA in several plant species such as *Zea mays*, *Sorgo bicolor*, and *Oryza sativa* (Vallabhaneni et al., 2010). In *A. thaliana*, five NCED genes that participate in the synthesis of ABA has been reported (Tan et al., 2003). It is presumed that the AtNCED3 paralogue is directly involved in the biosynthesis of ABA under abiotic stress conditions (Sussmilch et al., 2017). Subsequently, xanthoxin (15C) is transported from the plastid to the cytosol of the cell where it is converted into ABA aldehyde by the enzyme xanthoxin dehydrogenase. In turn, the ABA aldehyde is converted into ABA by ABA aldehyde oxidase (AAO) (Asad et al., 2019).

ABA signalling pathways involve three main components: the ABA family of PYR/PYL/RCAR, or PYLs receptors (pyrobactin resistance); the negative regulator type 2C protein phosphatases (PP2Cs); and the positive regulator class III sucrose non-fermenting-1-related protein kinase 2 (SnRK2) (Ma et al., 2009; Park et al., 2009; Shinozawa et al., 2019). After exposure to abiotic stress (drought, salinity, or temperature), ABA binds to the PYR/PY2/RCAR receptors, and then to PP2C, forming the ABA-PYLs-PP2C ternary complex. This inhibits the phosphatase activity of PP2Cs and prevents dephosphorylation of SnRK2, maintaing it in an active state (Ma et al., 2009; Park et al., 2009; Shinozawa et al., 2019). Activated SnRK2 phosphorylates and activates the transcription factors of ABA-responsive elements (AREB/ABF) that in turn bind to promoter regions of ABA-responsive genes that trigger the stress response in plants (Gómez-Porras et al., 2007; Ma et al., 2009).

In plants, ABA controls the maturation of seeds and is responsible for their dormancy, which plays an important role in homogeneous germination. It has been observed that ABA is in the early and middle stages of seed development, and its absence causes germination. Therefore, seed dormancy is controlled by the biosynthesis and degradation of ABA, which is regulated by the enzyme NCED3 and CYP707A, respectively. Additionally, seed dormancy involves the balance between ABA and gibberellic acid (GA), which act as antagonists in the signalling and biosynthesis pathways (Sah et al., 2016). The seed is a crucial organ in higher plants and the transition from dormancy to germination is an important ecological and economic feature. Because ABA has a central role in the induction and maintenance of seed dormancy, ABA could have applications for the maintenance of seeds in germplasm banks. On the other hand, the phytohormone ABA is a central mediator to stress (Yamaguchi-Shinozaki and Shinozaki, 2006) by drought, salinity, cold (Xue-Xuan et al., 2010; Sah et al., 2016), and UV radiation (Berli et al., 2011). ABA constituting a homeostasis system during stress which includes interactions with the ubiquitin-proteasome system to alter the degradation of transcriptional regulators, stomatal closure which restricts cell growth and the induction of genes involved with senescence (Sah et al., 2016; Mirer et al., 2018; Arif et al., 2019). The moss *Physcomitrella patens* is highly tolerant to drought, UV-B radiation, salt, and osmotic stress (Frank et al., 2005; Wolf et al., 2010). This tolerance is due in part to the increase in ABA levels under these conditions of abiotic stress. Similarly, the core components of ABA signalling and the ABA response are conserved between *P. patens* and angioesperms (Machuka et al., 1999; Kamisugi and Cuming, 2005; Marella et al., 2006; Wang et al., 2010; Bhyan et al., 2012; Amagai et al., 2018; Zhao et al., 2018). A transcriptomic analysis after the exogenous application of ABA (100 μM) to *P. patens* at different times (30, 60, and 180 min) identified 1,030 differentially expressed genes, almost all members of ABA-dependent signalling and genes involved in cell wall modifications due to morphological and physiological changes during diaspora formation. The exogenous application of ABA also triggered the formation of vegetative diasporas (rounded cells without central vacuole, due to a greater storage of starch and lipids that prepares these cells for growth in unfavourable environmental conditions), which allow the survival of plants in unfavourable environmental conditions (Arif et al., 2019).

Due to its central role in the response of abiotic stress, ABA is an apocarotenoid with the potential to improve crops of agronomic importance under conditions of abiotic stress such as salinity and drought. In recent decades, the world population of human beings has increased exponentially and is expected to increase to more than 1 billion by 2030, and more than 2.4 billion by 2050; therefore, the production of food to supply the world population must grow at a similar rate, increasing by 70% by 2050 (Sah et al., 2016). However, the main limitations are land with high salinity and the lack of water availability in some regions, which compromise the production of inputs for human consumption. Researchers are looking for alternatives to counteract these limitations through the genetic engineering of phytohormones, with ABA being a good candidate. Therefore, a greater understanding of the biosynthesis of ABA and its signalling pathways in different plant species will allow obtaining crops that are more tolerant or resistant to abiotic stress conditions.

β-Citraurin

β-Citraurin (3-hydroxy-β-apo-8′-carotenal) is an apocarotenoid that is directly involved in the citrus peel colour, where it represents up to 30% of the total carotenoids present in the peel. The external colour of citrus fruits is a crucial characteristic of quality, and the uniform and attractive bright orange colour is probably the main factor that determines consumer acceptance (Lado et al., 2018). The β-citraurine, was discovered for the first time in 1936 by Zechmeister and Tucson and until now very little is known about its role and its biosynthesis, although it is presumed that it is produced by the asymmetric cutting of the enzyme CCD4b1 in the 7′,8′ bonds of zeaxanthin and β-cryptoxanthin (Ma et al., 2013; Rodrigo et al., 2013). In recent work, it was found that light exerts a positive effect on the colour of citrus peel by increasing the expression of the carotenoid pathway genes that give rise to β-citraurine precursors (Lado et al., 2019). The β-citraurin has potential for application as natural dye. Due to its structure of polienic chain with conjugated bonds, it could also have antioxidant properties beneficial for health. So far there are few studies on β-citraurine, but future research will likely demonstrate new functions of these apocarotenoids in plants and also in mammals.

Zaxinone

Zaxinone is a recently discovered apocarotenoid. Zaxinone is part of the ubiquitous family of hormones and signalling molecules derived from the oxidative cleavage of carotenoids (Wang et al., 2019). The expression of the *Loc.Os09g15240a* cDNA fused to thioredoxin in *Escherichia coli* bacteria and the incubation of the soluble fraction of these cells with different carotenoids and apocarotenoids *in vitro* allowed the identification of zaxinone synthase (ZAS). It was found that ZAS cleaves apo-8′-(30C), apo-10′ (27C) and apo-12′-zeaxanthinal (25C), at the C13, C14 double bond, producing apo-13-zeaxanthinone or zaxinone (3-OH-β-apo-13-carotene (18C)). Additional experiments identified that ZAS had a preference for the apo-10′-zeaxanthinal (27C) substrate to produce zaxinone (C_{18}) (Wang et al., 2019). The *in vitro* identification of zaxinone allowed to confirm its presence in rice, tobacco and *A. thaliana*. Additionally, a phylogenetic analysis using 748 CCDs protein sequences from 69 species of organisms, including mosses, ferns, monocotyledons, and dicotyledons, showed that the ZAS correspond to a new subfamily of CCDs present mainly in plant species that interact with arbuscular mycorrhizal fungi (Wang et al., 2019). In rice, three ZAS paralogues were identified (ZAS-L1, ZAS-L2, and ZAS-L3) (Wang et al., 2019).

The mutant *zas* seedlings found in rice indicated a reduction of the main crown root length, shoot length, as well as the root and shoot biomass. However, the exogenous application of zaxinone (2.5 μM) to the *zas* mutant allowed plant roots to recover, suggesting that zaxinone plays an important role in the growth and development and that is required at certain concentrations for normal rice development (Wang et al., 2019). In the same work, it was also demonstrated that zaxinone is a negative regulator of biosynthesis of SL at the transcriptional level under limiting conditions of Pi (phosphorus). In *zas* mutants and wild-type seedlings exposed to Pi starvation, the application of zaxinone (2.5–5 μM) decreased the transcription levels of the SL biosynthetic enzymes DWARF27 (D27), CCD7 (D17), CCD8 (D10), and the 4-deoxyorobanchol-forming carlactone oxidase (CO). It is known that SL favour the germination of the seeds of parasitic plants such as *Striga*, which represents a problem in world food. The germination analysis of the *Striga* in the *zas* mutant showed a 20–30% increase in germination compared to the wild-type plant. However, when 5 μM of zaxinone was added to the plants, a reduction in SL levels and a 40% reduction in *Striga* germination was observed compared to wild-type plants, suggesting that zaxinone not only regulates negatively the biosynthesis of SL but also participates in the inhibition of the germination of the parasite plant *Striga*. Finally, it was found that zaxinone is not essential to establish symbiosis of arbuscular mycorrhizal fungi (AM), but is required for optimal colonization levels (Wang et al., 2019).

The newly discovered apocarotenoid zaxinone has future applications for both the strengthening of crops of agronomic importance, as well as an excellent molecule for the regulation of symbiosis between plants and arbuscular mycorrhizal fungi (Felemban et al., 2019; Fiorilli et al., 2019). Additionally, parasitic plants represent a serious problem for world food, as it is the main cause of the sequestration of nutrients important for the good development and growth of plants. The application of low concentrations (10 μM) of zaxinone to *Striga* susceptible cv. rice IAC-165, grown in pots with *Striga* infested soil, clearly reduced the number of emerging parasitic plants (by about 80%) compared to untreated control, showing the potential of zaxinone for the elimination of parasitic plants that grow alongside the crops of interest (Wang et al., 2019).

Anchorene

Recently, several types of diapocarotenoids have been found in the *A. thaliana* model plant and their analysis has allowed the identification of a new apocarotenoid called anchorene that derives from all the carotenoids downstream of Z-carotene with cleavage in the 11,12 / 11′,12′ bonds (Jia et al., 2019a). In *A. thaliana*, the anchorene is a diapocarotenoid that promotes the growth of anchor roots (ANR). At high concentrations of anchorene (10 μM) a weak inhibition in the length of the primary roots was observed, but surprisingly, the amount of ANR increased (Jia et al., 2019a). The architecture of the roots is of importance in crops since they depend on the anchoring and absorption of nutrients (Jia et al., 2019), therefore, the anchorene has promising applications for the root architecture of species of agronomic importance. However, further research on its biosynthesis pathway and its biological functions in plants is still needed.

Before anchorene, monocarboxylate apocarotenoids appeared to be the only ones with signalling functions in plants. The instability, the low abundance and its low ionization efficiency in mass spectrometry, as well as the absence of efficient protocols to detect the presence of apocarotenoids dialdehydes (diapocarotenoids) had contributed to their passing unnoticed (Mi et al., 2019). Until now, the only molecules with these characteristics had been bixina from *Bixa orellana* (Vilar et al., 2014) and crocetin from *C. sativus* (Moratalla-López et al., 2019).

Bixin

B. orellana is the main source of bixin, a linear apocarotenoid of 25C with nine double bonds, with an empirical molecular formula $C_{25}H_{30}O_4$ and a molecular weight of 394.5 g/mol (Vilar et al., 2014). The bixin biosynthesis pathway was first proposed by Jako et al. (2002) through a library of expressed sequence tags (EST). Later Bouvier et al. (2003) based on homology with the genes of the crocetin biosynthesis pathway in *Crocus sativus*, isolated from *B. orellana* the genes of lycopene excision dioxygenase (*BoLCD*), bixin aldehyde dehydrogenase (*BoBADH*) and norbixin methyltransferase (*BonBMT*), which encode the enzymes involved in bixin biosynthesis. It is presumed that the heterologous expression of these genes in *E. coli* cells that produce lycopene generate bixin. Subsequently, with the development of the first *de novo* transcriptome made from young leaf, mature seed and immature seed of *B. orellana*, several genes of the family of carotenoid cleavage dioxygenases (BoCCD1-1, BoCCD1-2, BoCCD1-3, BoCCD1-4, BoCCD4-1, BoCCD4-2, BoCCD4-3, and BoCCD4-4), aldehyde dehydrogenases (BoADH2B4, BoADH2B7-1, BoADH2B7-2, BoALDH2C4, BoADH3H1-1, BoADH3H1-2, BoADH3I1, BoADH3F1, BoADH3F2, BoADH5F1, BoADH6B2-1, BoADH6B2-2, BoADH6B3, BoADH7B4, BoADH10A8, BoADH11A3, BoADH12A1, BoADH18B-1, BoADH18B-2, and BoADH22A1) and methyltransferases (BoSABATH1, BoSABATH2, BoSABATH3, BoSABATH4, BoSABATH5, BoSABATH6, BoSABATH7, BoSABATH8, BoSABATH9, BoSABATH10, BoSABATH11, and BoSABATH12) were identified (Cárdenas-Conejo et al., 2015). Based on the phylogenetic analysis and the quantification of gene expression, a new bixin biosynthetic pathway located in the plastids was proposed. The proposed route begins with the condensation of pyruvate and glyceraldehyde 3-phosphate to give isopentenyl pyrophosphate (IPP) through the methylerythritol 4-phosphate (MEP) pathway. In turn, four molecules of IPP are condensed to form geranylgeranyl pyrophosphate (GGPP) (20C). The condensation of two GGPP molecules forms phytoene (40C), a colourless carotenoid, which is converted to trans-lycopene after several desaturations and isomerization reactions. In turn, lycopene is cleaved by BoCCD4-3 at 5,6 / 5′,6′ bonds to form bixin aldehyde, additionally, the substitution reactions of aldehyde groups by carboxylic groups catalyzed by ALDH3H1-1 gives rise to norbixin. Finally, norbixin is methylated with the addition of a methyl group through a reaction catalyzed by BoSABATH4 to give rise to bixin production (Cárdenas-Conejo et al., 2015).

After saffron crocetin, bixin is the second pigment of world importance (Giridhar and Parimalan, 2010) due to its application in the food, pharmaceutical, cosmetic, and textile industries (Vilar et al., 2014). Bixin is also used as a substitute for the dangerous, toxic and carcinogenic synthetic pigment tartrazine (E102) (Salinas et al., 2005). Additionally, bixin is accepted by The Food and Agriculture Organization of the United Nations (FAO) and the World Health Organization (WHO) as a pigment for human consumption. Studies in laboratory animals indicate that its consumption does not generate a toxicological effect and does not promote cancer (Haveland-Smith, 1981; Lauro and Francis, 2000).

The main importers of bixin are the United States, Canada, United Kingdom, Denmark, and Japan, who buy bixin mainly from Kenya, India, Peru, Brazil, and Mexico (Vedavathy, 2003). One of the forms of commercialization of bixin is through a ground seed extract of *B. orellana*. A kilogram of dry seed costs \$5–10, depending on the percentage of bixin in the seed (Raddatz-Mota et al., 2017). Farmers report that each hectare of crop (approximately 1,200 plants of *B. orellana*) produces about 3,000 kg of dry seed, which represents about \$30,000 per hectare. Bixin is used by the food industry in the preparation of beverages, cheeses, yogurts, butter, margarine, mayonnaise, ice cream, sweets and bread, and as a condiment for traditional dishes from several Latin American countries. Also, in the textile industry, it is used as a natural non-polluting pigment, and in the cosmetic industry for the elaboration of lipsticks, blushes, and makeup. The pharmaceutical industry began the use of bixin for drug pigmentation and later as a food supplement for its antioxidant, anticancer, hypoglycaemic, antibiotic and anti-inflammatory properties. The Joint FAO/WHO Expert Committee on Food Additives (JECFA) estimated that the ADI (acceptable daily intake) for bixin is 0–2.5 mg/kg of body weight per day, expressed as pure pigment. Research groups recommend the consumption of achiote seeds and pigment in the diet for their medicinal, nutraceutical and nutritional properties, and therefore promote their cultivation and production. The direct applications of bixin as a natural pigment for human consumption has encouraged the study of its biosynthesis route and its beneficial properties for health. Even research groups have proposed bixin heterologous biosynthesis as an alternative to increase their production (Rivera-Madrid et al., 2016; Teixeira da Silva et al., 2018).

Crocetin

Crocetin is an important natural pigment for the industry. Crocetin is generated by *C. sativus* L. or saffron, a triploid species, sterile and whose red stigmas, where crocetin accumulates, making it one of the most expensive species on earth. The colour, flavour, and aroma of the species are due to the accumulation of saffron-specific apocarotenoids such as crocetin, picocrocitin, and safranal (Moratalla-López et al., 2019). Crocetin is a diterpenic compound. Its elemental composition is $C_{20}H_{24}O_4$ and its molecular weight is 328.4 g/mol. It is slightly soluble in aqueous solution, and soluble in organic bases (Alonso et al., 1996). It is known that the carotenoid precursors of saffron apocarotenoids is zeaxanthin. The biosynthesis pathway of crocetin was reported by Bouvier et al. (2003b), where it was suggested that the enzyme zeaxanthin cleavage dioxygenase (ZCD) is responsible for the symmetric cutting of zeaxanthin at positions 7,8 / 7′,8′. However, in another study, it was shown that the ZCD enzyme reported by Bouvier et al. (2003b) is a truncated form of the CCD4 enzyme located in the plastoglobules, without cleavage activity, and that the whole enzyme cuts the β-carotene at the 9,10 / 9′,10′ bonds, producing β-ionone and not zeaxanthin as previously thought (Rubio et al., 2008).

Subsequently, in another work, seven different transcripts of the CCDs were identified using the deep transcriptome sequencing of six-stage of development of saffron stigma. It was found that CCD2 is responsible for the cleavage of zeaxanthin at positions 7,8 / 7′,8′. The enzymatic activity of CCD2 on zeaxanthin was demonstrated using heterologous systems such as *E. coli* and *Zea mays*, and in both cases zeaxanthin and 3-OH-β-apo-8′-carotenal are the preferred substrates, for the first and second cleavage reaction, respectively, to produce crocetin dialdehyde (20C) and 3-OH-β-cyclocitral (10C) (Frusciante et al., 2014; Ahrazem et al., 2016a). Crocetin aldehyde, in turn, is converted to crocetin by aldehyde dehydrogenase (ALDH), and later into crocins by at least two UDPG-glucosyltransferases. The remaining product, 3-OH-β-cyclocitral, is converted into picrocrocin by a UDPG-glucosyltransferase, and then into safranal by an enzyme still unknown (Frusciante et al., 2014).

Like bixin, crocetin is considered a harmless natural pigment for its consumers and the environment. Also, for its antitumor, antioxidant, antihypertensive, anti-atherosclerotic and antidepressant properties, it is considered as a molecule with potential use in the medical and food field (Gutheil et al., 2012). Studies with cell suspensions showed that crocetin has significant effects against neurodegenerative diseases, in the decrease of blood pressure and the maintenance of renal function. Besides, saffron extract has properties to combat various types of cancers. It has been observed that crocetin decreases cell proliferation and increases apoptosis in breast cancer cells (MCF-7 and MDA-MB-231 cells) (Chryssanthi et al., 2007; Mousavi et al., 2009). Also, it has positive effects on the inhibition of other types of cancer, such as lung cancer, cervical cancer, colon-rectal cancer, leukaemia, liver cancer, pancreatic cancer and skin cancer (Gutheil et al., 2012). On the other hand, the main use of saffron is in the kitchen, due to its ability to impart colour, flavour, and aroma to food and beverages. But recently due to its medicinal properties, it has been considered a functional or nutraceutical food and although there are reports that the consumption of saffron at high concentrations by pregnant women can cause abortions or effects on the kidneys of their children, consumption at a concentration of 1.5 g/day does not present any toxic effects for humans. Animal tests have shown that the non-toxic consumption of crocetin is 0.1–5 g/kg body weight (Abdullaev, 2002), but more studies are needed to stipulate the amount of non-harmful consumption of saffron in humans.

The demand for crocetin and other apocarotenoids of saffron grows exponentially and its main limitations are the recovery of apocarotenoids from the stigma of the flower and problems associated with obtaining large numbers of flowers (150,0000 flowers are required to obtain a kg of stigma). For this reason, the synthesis of crocetin in microorganisms has been proposed as one of the alternatives for obtaining crocetin. Crocetin production in *E. coli* was reported by Bouvier et al. (2003). Additionally, a patent on the production of crocetin in *Chlorella vulgaris* has been generated (Lou et al., 2016). The transformation of *S. cerevisiae* strains producing β-carotene with a combination of CrtZ and CCD genes generated crocetin at a concentration of 223 μg/L. Besides, when ALD gene of *Synechocystis* sp. was added to the CrtZ and CCD2 cassette, and the yeasts were grown in 5 L bioreactors, the amount of crocetin produced increased, being the highest reported (6,278 μg/L). The synthesis of crocetin by microorganisms is one of the most viable alternatives to counteract the problems of its production and meeting demand. Crocetin offers great opportunities for commercialization as a natural food pigment, and as a functional and nutraceutical food, possessing benefits associated with the prevention and reduction of cancer and depression, among other diseases (Chai et al., 2017).

ANIMAL APOCAROTENOIDS

In animals such as mammals, carotenoids are obtained through their diet and serve as an antioxidant and signalling molecules and as retinol precursors (Badyaev et al., 2015; Harrison and Quadro, 2018). Carotenoid production has only been reported in a species of aphids, which have inherited the carotenogenic genes of a species of fungus by horizontal transfer. These insects synthesize the red torulene carotenoid for energy and as camouflage (Moran and Jarvik, 2010). In animals, carotenoids are oxidized by the enzymes β-carotene-15,15′-oxygenase (CMOI) and β-carotene-9′,10′-oxygenase (CMOII) (Kiefer et al., 2001; Mengual et al., 2013; Harrison and Quadro, 2018) to give rise to apocarotenoids. Only a part of the apocarotenoids found in mammal tissues are absorbed intact in the intestine (Mercadante et al., 1997; Gajic et al., 2006; Hu et al., 2006; Kopec et al., 2010; Amengual et al., 2013; Arscott, 2013).

Vitamin A is one of the most studied apocarotenoids in mammals due to its health benefits (Kiefer et al., 2001). Vitamin A deficiency in developing countries represents a global health problem affecting mainly pregnant women, breastfeeding mothers, and preschoolers (Grune et al., 2010; Tanumihardjo et al., 2016; Wirth et al., 2017; Bastos Maia et al., 2019). Low levels of vitamin A cause vision problems, increase the risks of colon cancer, increase the mortality rate in children with measles, and affect the epithelial surfaces of the human body (Harrison and Quadro, 2018). New apocarotenoids such as β-apocarotenoids and apolycopenoids have been discovered in mammalian tissues. A large part of these apocarotenoids is found in foods rich in β-carotene and lycopene (Al-Hasani, 1968; Al-Hasani and Parrish, 1972; He et al., 1997; Johnson et al., 2002; Riebeling et al., 2002; Tatman and Mo, 2002; Liu et al., 2004; Lian et al., 2007; Ziouzenkova et al., 2007; Lian and Wang, 2008; Spiegler et al., 2018; Ip et al., 2013; Sun et al., 2014; Costabile et al., 2016). β-apocarotenoids and apolycopenoids in animals are produced through the oxidation of β-carotene and lycopene by the family of CMOI and CMOII enzymes (Kiefer et al., 2001; Kopec et al., 2010). β-apocarotenoids and apolycopenoids improve embryonic development and can reduce the incidence of cancer. The presence of apocarotenoids and the elucidation of their functions in humans barely come to light, so future research on this topic will help us better understand the role of these molecules and their future applications against health problems.

Vitamin A

Vitamin A is one of the most studied and important apocarotenoids in mammals and humans. Vitamin A is a 20C molecule that contains a β-ionone ring at one end (Karrer et al., 1931; Holmes and Corbet, 1937). Mammals are unable to synthesize vitamin precursors. In their first month, mammals get vitamin A as preformed vitamin A through breast milk. In the adult stage, vitamin A is obtained through the diet as a preformed vitamin A when consuming products of animal origin such as meat, milk, fish, and eggs, or as provitamin A when consuming vegetables that contain β-carotene, α-carotene, and β-cryptoxanthin (Arscott, 2013). The pathway of vitamin A biosynthesis in mammals begins with the enzyme β-carotene-15,15′-oxygenase (CMOI) that primarily cleaves β-carotene symmetrically in its bonds 15,15′ to produce two retinaldehyde molecules, a molecule used primarily as a rhodopsin chromophore, which is of utmost importance in the vision (Olson and Hayaishi, 1965; O'Byrne et al., 2005). Retinaldehyde by the action of the retinaldehyde dehydrogenase enzymes (RALDH or ALDH1 family) can be oxidized to form polar retinoic acid (AR) (Chambon, 1996). In turn, AR can increase its polarity by being converted to 4-hydroxy acid or 4-oxo retinoic acid by the CYP26 enzyme. On the other hand, retinaldehyde can be reduced to form retinol, the alcohol form of vitamin A, and then reversibly stored in the liver as a retinol ester in a reaction catalyzed by the enzyme retinol acyltransferase (LRAT) (D'Ambrosio et al., 2011). Similarly, the cleavage of β-carotene in an enzymatic and non-enzymatic way causes the formation of β-apocarotenoids such as β-apo-8′-carotenal, β-apo-10′-carotenal, β-apo-12′-carotenal, and β-apo-14′-carotenal which can be converted into retinaldehyde by the action of the enzyme CMOI. In addition to β-carotene, the CMOI enzyme can cut other provitamin A carotenoids such as α-carotene and β-cryptoxanthin to form retinaldehyde. On the other hand, CMOII can produce retinaldehyde molecules from provitamin A carotenoids although it is less efficient than CMOI. CMOII also cleaves non-provitamin A carotenoids such as zeaxanthin and lutein and generates apo-10′-carotenoids (Kiefer et al., 2001; Mengual et al., 2013; Harrison and Quadro, 2018). The digestion and absorption of preformed vitamin A and its precursors are associated with lipids in the small intestine, so a low diet in these can cause deficiency in vitamin A (DVA). The expression of the CMOI enzyme in the human intestine is an indicator that part of the β-carotene is converted into retinoids before entering the bloodstream. Almost 17–45% of β-carotene passes directly into the bloodstream and is transported by lipoproteins such as chylomicrons, VLDL (very low-density lipoproteins), IDL (intermediate density lipoproteins), and LDL (low-density lipoproteins) along with other lipids such as cholesteryl esters and retinyl esters, travelling from the intestinal barrier to all body tissues (Shete and Quadro, 2013). Vitamin A and its retinoid derivatives are relevant to human health, even the WHO (World Health Organization) considered it a priority for public health. WHO recommends levels in serum above 0.70 μmol/L of vitamin A to avoid health problems caused by its deficiency (Tanumihardjo et al., 2016). DVA together with protein deficiency is one of the biggest health problems in the world, mainly in developing countries where it is estimated that more than 250 million preschoolers have DVA and that another 5 million are clinically affected due to this deficiency (Tanumihardjo et al., 2016). In addition to children, pregnant and nursing women and newborns, are affected by DVA. In pregnant women, the lack of vitamin A has systemic effects on several fetal organs and the fetal skeleton, so its deficiency can be harmful (Bastos Maia et al., 2019). Adequate vitamin A intake is recommended for early lung development, alveolar formation, maintenance and tissue regeneration since vitamin deficiency appears to contribute to the risk of bronchopulmonary dysplasia (BPD) in the newborn (Timoneda et al., 2018). On the contrary, excess of vitamin A during pregnancy can have teratogenic effects in the first 60 days after conception (Bastos Maia et al., 2019). An increase in vitamin A intake may help reduce the risk of graft versus host disease (GVHD), through better intestinal permeability, a reduction in mucosal injury and a reduction in lymphocyte localization in the intestine, suggesting an important role of vitamin A in the immune development of the intestine (Lounder et al., 2017). Vitamin A deficiency in developing countries presents health problems because natural sources of vitamin A such as fruits and vegetables have high costs (Grune et al., 2010). On the contrary, in developed countries, DVA is very low in the population mainly due to easy access to sources of vitamin A. However, more than 20% of the population in these countries has less vitamin A, and the main cause is poor diet, the low consumption of healthy fats, and strict vegetarian diets. Vitamin A plays an important role in embryonic development, cell differentiation and proliferation, vision, immunity, is involved in insulin resistance, lipid metabolism, energy balance, redox signalling, and has antioxidant properties (Harrison and Quadro, 2018). Because of its importance in human health, vitamin A deficiency is fought with multivitamins or supplements with β-carotene mainly in the most marginalized regions of the world where access to natural sources of vitamin A is difficult. However, despite reducing the mortality and morbidity rate in children with DVA, an excess of this vitamin can cause bone problems. In patients with cystic fibrosis (CF) with vitamin A deficiency, multivitamins and β-carotene supplements do not seem to benefit patients' health (De Vries et al., 2018). One of the best ways to combat DVA in developing countries is with a diet rich in β-carotene, which is why plant biofortification programmes through genetic engineering focus on increasing this carotenoid (Yao et al., 2018).

β-Apocarotenoids

The precursor of β-apocarotenoids found in animals, including mammals such as humans, is β-carotene, which is obtained mainly from the consumption of vegetables (Amengual et al., 2013). Its non-detection in the bloodstream suggests that β-apocarotenoids are rapidly absorbed or metabolized by the intestine. The biosynthesis of β-apocarotenoids specifically in animal tissues is attributed to the enzyme CMOII that cuts the β-carotene from the diet at positions 9′,10′ generating β-apo-10′-carotenal and β-ionone (Kiefer et al., 2001; Amengual et al., 2013). β-apocarotenoids of different lengths seem to come from a non-enzymatic route (Harrison and Quadro, 2018). The first reports of the function of these compounds were as growth stabilizers in rats and other animals (Al-Hasani, 1968; Al-Hasani and Parrish, 1972). However, it was not clarified whether this function is due to its action *per se* or its role as precursors of retinoids since β-apocarotenoids of different lengths act as substrates of the CMOI enzyme, which in turn converts them into vitamin A and its derivatives (Spiegler et al., 2018). In a study on embryonic development in pregnant rats with vitamin A deficiency, it was observed that β-apo-10′-carotenal acts as a source of vitamin A, maintaining embryonic development through the activity of the enzyme CMOII. It was proposed that supplementation with β-apo-10′-carotene in pregnant women deprived of vitamin A would prevent intoxication with β-carotene and could help embryonic development (Spiegler et al., 2018). Similarly, Sun et al. (2014) demonstrated that β-apo-13′-carotenonal antagonizes 9-*cis*-retinoic acid (9cRA) and inhibits the transcriptional activity of RXRα (X retinoid receptors). Similarly, β-apo-14′-carotenal effectively inhibited the activation of RXRα (Ziouzenkova et al., 2007). Therefore, the antagonistic role of these derivatives would help to counteract the harmful effects of an excess of β-carotene.

β-ionone is one of the best-known β-apocarotenoids. Although its mechanism is not clear, in mammals it appears to be involved in the inhibition of growth and cell proliferation, and the induction of apoptosis. A study in gastric cancer cells (SGC-7901) found a decrease in cell proliferation relative to the dose of β-ionone. It was observed that the seven-day treatment with 200 mmol/L of β-ionone reduced cell proliferation to 94.15% and induced apoptosis (Liu et al., 2004). Similar effects were observed in melanoma and meningioma cells and breast cancer (He et al., 1997; Johnson et al., 2002; Riebeling et al., 2002; Tatman and Mo, 2002). Additionally, β-ionone stimulates melanogenesis and dendritogenesis (Gelis et al., 2016). The role of β-apocarotenoids in mammals is largely unknown (Harrison and Quadro, 2018). However, there is increasing interest in β-apocarotenoids as their importance in the development and health of humans is discovered.

APOLICOPENOIDS

Apolicopenoids are apocarotenoids derived from lycopene cleavage. Apolicopenoids can be generated by autooxidation, by oxidation mediated by free radicals and singlet oxygen and by enzymes. Apolicopenoids have been found in different plant tissues. In *B. orellana* seeds, apolycopenoids such as methyl (7Z, 9Z, 9′Z)-apo-6′-lycopenoate, methyl (9Z)-apo-8′-lycopenoate and methyl (all-E)-apo-8′-lycopenoate have been reported (Mercadante et al., 1997). In tomato, red grapefruit, watermelon, and other fruits and vegetables that contain lycopene, apo-6′- ,8′- ,10′- ,12′- and 14′-lycopenal have been found (Kopec et al., 2010). On the other hand, mammals can obtain apolycopenoids enzymatically, using lycopene as a substrate or preformed apolycopenoids through their diet. Lycopene produces apolycopenoids through reactions catalyzed by CMOI enzymes that generate cycloretinal (Harrison and Quadro, 2018) and CMOII that generate apo-10′-lycopenal, apo-10′-lycopenol and apo-10′-lycopenoic (Kopec et al., 2010). In the lung tissue of ferrets fed for 9 months with a diet rich in lycopene, apo-10′-lycopenol was found (Hu et al. 2006). Something similar was observed in rats fed for 30 days with high doses of lycopene, in the liver tissue were found apo-8′-lycopenal and apo-12′-lycopenal (Gajic et al. 2006). In the serum of pre-ruminating bovines with 2 weeks of lycopene supplementation, 5,6-dihydrolycopene and 5,6-dihydro-5-cis-lycopene were detected (Sicilia et al. 2005). The first apolycopenoids reported in human plasma was 5,6-dihydroxy-5,6-dihydrolycopene (Khachik et al., 1992). Subsequently, epimeric 2,6-cyclolycopene-1,5-diols were identified in human milk and serum (Khachik et al., 1997). In independent work, seven people who consumed tomato juice daily for 8 weeks presented levels apo-6′- ,8′-,10′-,12′- and 14′-lycopenal (Kopec et al., 2010). However, recently Cichon et al. (2018) stated that the administration to eight people with olive oil supplemented with 10.2 mg of lycopene labelled with 13C did not present the previously reported aldehyde cleavage products; instead, it has identified a new candidate oxidative metabolite: 1,2-lycopene epoxide.

Lycopene has a beneficial impact on several chronic diseases including cancer. However, the metabolites derived from lycopene cleavage are likely truly responsible for some of the functions attributed to lycopene (Mein et al., 2008). Thus apo-10′-lycopenoic acid (APO10LA), an apolycopenoids generated from the cleavage of lycopene at its double bond 9′,10′ by CMOII, can effectively inhibit liver tumorigenesis by stimulating SIRT1 signalling. SIRT1 encodes a NAD + dependent protein deacetylase that catalyzes the deacetylation of many non-histone proteins and has been implicated as a key regulator of lipid metabolism (Chung et al., 2012). Similarly, male mice with CMOII deficient fed for 12 weeks on a diet high in saturated fat (HSFD) and supplemented with 10 mg/kg of APO10LA, showed a reduction of 72% in the incidence of hepatic steatosis (Ip et al., 2015), probably due to the positive regulation of the expression and activity of SIRT1 (Chung et al., 2012). APO10LA also inhibited the growth of normal human bronchial epithelial cells (NHBE), immortalized normal bronchial epithelial cells (BEAS-2B) and non-small cell lung cancer A549 cells, demonstrating that this apolycopenoid is a potential chemopreventive agent against pulmonary tumorigenesis. Inhibition was associated with a decrease in cyclin E, inhibition of cell cycle progression from phase G (1) to phase S, and increased protein levels of cell cycle regulators p21 and p27 (Lian et al., 2007). Like lycopene, apo-10′-lycopenoic acid has antioxidant properties. Treatment with apo-10′-lycopenoic acid, dependent on time and dose, increases the nuclear levels of transcription factor Nrf2 (factor 2 related to factor E2) in BEAS-2B cells. Activated Nrf2 binds to the antioxidant response element (ARE) of the promoter regions of the genes encoding detoxifying/antioxidant enzymes that result in the inactivation of ROS (Lian and Wang, 2008). Therefore, it is strongly suggested that the anticancer and antioxidant functions of lycopene are mediated by apo-10′-lycopeneides through the activation of Nrf2 and the induction of the expression of genes that encode for detoxifying/antioxidant enzymes.

Because of the important role that apolycopenoids have for the benefit of health, their identification and the discovery of their mechanism of action are an innovative topic. Many of these apolycopenoids have not been able to be studied *in vivo*, and so their function on health remains uncertain, but as the detection methods improve and deeper studies are carried out, new functions of these amazing molecules could come to light.

BACTERIAL APOCAROTENOIDS

Very little is known of the function of apocarotenoids in bacteria. However, in cyanobacteria, it is presumed that they act as photoprotective pigments and accessories in the thylakoid membrane (Ahrazem et al., 2016b). Likewise, they are responsible for the production of aromas, participate as signalling molecules, are important virulence factors, provide integrity to the cell membrane, act as stabilizers of ROS, and participate in in the proton pump (Clauditz, 2006; Liu and Nizet, 2009; Mishra et al., 2011). *In vitro* studies have shown that bacterial apocarotenoids are biologically active molecules that have potent anticancer, antimicrobial and antioxidant activities, which promises their use as potential therapeutic agents (Barretto and Vootla, 2018). However, so far, research has focused on the elucidation of the biosynthetic pathway of the natural apocarotenoids of bacteria, mainly in the CCDs involved. Thus, enzymes of the family type CCD1, CCD7 and β-apocarotenoid-15,15′-oxygenase (APCO) have been identified in cyanobacteria genomes, suggesting that these CCDs were present before the separation between bacteria and plants (Cui et al., 2012).

Bacterial Carotene Oxygenases

One of the first carotene oxygenases reported was the β-carotene oxygenase of *Cyanobacterium microcystis*, which cleaves β-carotene and zeaxanthin at positions 7,8 / 7′,8′, producing the β-cyclocitral, crocetindial and hydroxy-β-cyclocitral apocarotenoids (Jüttner and Höflacher, 1985). Subsequently, the apocarotenoid cleavage oxygenase (ACO) from the *Cyanobacterium synechocystis* was the first CCD to be crystallized (Kloer et al., 2005).

Bacterial CCDs retain the universal structure of a propeller with seven blades, of which five are formed by four strands, and two blades consist of five strands. The active centre contains an Fe^{2+} ion as a cofactor, essential for excision activity. Its main role is to activate the oxygen involved in the cleavage reaction. The Fe^{2+} ion is coordinated by four histidine residues, which are conserved in all CCDs of all taxa. A second coordination group is formed by three glutamate residues, which are involved in the interaction through hydrogen bonds with histidine residues (Ahrazem et al., 2016).

On the other hand, compared to CCDs of plants that are specific, some CCDs of bacteria cleave carotenoids and apocarotenoids in different positions. For example, DioX1 of *Synechocystis* sp. PCC 6803 is an ACO that cleaves at the 15,15′ double bonds of apo-β-carotenals to produce retinal. In mammals, DioX1 used trans-apo-β-carotenals of different lengths and functional groups to produce *trans*-retinal (Ruch et al., 2005). In addition to the carotenoid cleavage function, bacterial CCDs can perform isomerizations and even cleave other molecules such as stilbenes (Marasco and Schmidt-Dannert, 2008).

Recently, the analysis of 37 cyanobacterial genomes showed a total of 61 putative CCD sequences, which were abundant in *Acaryochloris marina*, in filamentous nitrogen-fixing cyanobacteria and the unicellular cyanobacterial *Cyanothece* (Cui et al., 2012). The phylogenetic analysis indicated that the subfamily CCD1, CCD7, and APCO may have originated before the division of cyanobacteria and plants. In turn, the CCD4 and NCED families seem to originate recently in plants (Cui et al., 2012).

Staphyloxanthin

Staphyloxanthin is a type of membrane-bound triterpenoid apocarotenoid pigment. It consists of a 30C polienic chain with conjugated double bonds typical of carotenoid pigments. It is produced by strains of *Staphylococcus* ssp., a genus of staphylococcal bacteria of the cocci class. Staphylococcal bacteria comprises microorganisms that are present in the mucosa and the skin of humans, other mammals, and birds. This type of bacteria includes *Staphylococcus aureus*, one of the most virulent and well-known members of this genus. Staphyloxanthin is a virulence factor in *S. aureus* and is responsible for its golden colour. The removal of *CrtM* encoding the staphyloxanthin biosynthesis enzyme causes colorlessness and susceptibility to death by human and mouse neutrophils (Liu and Nizet, 2009). The loss of the staphyloxanthin pigment also results in a loss of the virulence of *S. aureus* in murine skin abscess (Liu et al., 2005) and greater susceptibility to hydrogen peroxide, superoxide radical, hydroxyl radical, hypochloride, and singlet oxygen (Clauditz et al., 2006). Finally, it was demonstrated that the transfer of a fragment of the CrtM and CrtN genes involved in the biosynthesis of staphyloxanthin to *Streptococcus*, gave the transformants pigmentation and the ability to produce lesions in a mouse skin model, demonstrating that this apocarotenoid confers pathogenicity (Wieland et al., 1994; Liu et al., 2005; Liu and Nizet, 2009).

On the other hand, Mishra et al. (2011), reports that staphyloxanthin provides resistance to the *S. aureus* cell membrane against non-oxidative host defences, mainly cationic microbial peptides. The study used a set of isogenic clinical strains of *S. aureus*, and the strains supplemented with *CrtM* showed a higher production of carotenoids, a greater rigidity of the cell membrane and reduced susceptibility to death by cationic peptides. It was suggested that branched-chain fatty acids (BCFA) and staphyloxanthin regulate membrane fluidity and staphylococcal virulence. Braungardt and Singh (2019), reported that the mutation in the lpdA gene that encodes the functional enzyme branched-chain α-ketoacid dehydrogenase (BCKDH), generated a decrease in BCFA, causing greater membrane fluidity, slow growth, and a low *in vivo* survival of *S. aureus*. These effects were exacerbated when the CrtM gene was also mutated. The results found in this work showed that both BCFA and staphyloxanthin represent key points to control *S. aureus* infection in humans. *S. aureus* bacteria are a human health problem that cannot be easily fought due to antibiotic resistance. Abbas (2019), used glyceryl trinitrate (GTN), which surprisingly inhibited the production of staphyloxanthin, biofilms and the oxidative stress tolerance of bacteria, suggesting that this molecule is a powerful antivirulence candidate for the attenuation of the pathogenicity of *S. aureus*.

On the other hand, staphyloxanthin has important biological properties for human health. The astafiloxanthin obtained from the *Staphylococcus gallinarum* bacteria isolated from the intestinal microflora of the *Bombyx mori* insect, showed *in vitro* antimicrobial activity against the pathogens *S. aureus*, *E. coli*, and *Candida albicans*. Besides, it presented antioxidant activity, protective activity against DNA damage and anticancer activity against four different types of cancer (Barretto and Vootla, 2018). Based on the above, staphyloxanthin, in addition to being used as a target against *S. aureus* infections in humans, can also be used as a potential therapeutic agent, especially due to its anticancer property.

FUNGAL APOCAROTENOID

Carotenoids are also produced by fungi. In fungi carotenoids are mainly produced from the mevalonate pathway and, until phytoene, the pathway is very similar to that reported in photosynthetic organisms (Ávalos and Carmen Limón, 2015; Ahrazem et al., 2016). However, unlike photosynthetic organisms, the formation of phytoene to lycopene involves four desaturations performed by a single desaturase (CrtI) (Ávalos and Carmen Limón, 2015; Ahrazem et al., 2016; Liang et al., 2018). In fungi, lycopene is the precursor of cyclic carotenoids such as γ-carotene and β-carotene, which give rise to the main apocarotenoids (Liang et al., 2018). Another difference is that the formation of β-carotene from lycopene involved two β-cyclizations performed by a bifunctional enzyme with lycopene cyclase/phytoene synthase (CrtYB) activities (Ávalos and Carmen Limón, 2015; Ahrazem et al., 2016; Liang et al., 2018). CrtYB bifunctional activity has been reported in the fungi *Xanthophyllomyces dendrorhous*, *Phycomyces blakesleeanus*, *Mucor circinelloides* and *Rhodosporidium diobovatum* (Ávalos et al., 2014).

The carotenes mainly produced by fungi are β-carotene, astaxanthin and neurosporaxanthin, the last one is an apocarotenoid (Ávalos and Carmen Limón, 2015). The production of β-carotene occurs mainly in zygomycetes (order of the Mucorales), basidiomycetes and ascomicetes (Ávalos and Carmen Limón, 2015). In fungi, β-carotene also works as a precursor of other carotenoids such as astaxanthin, a red/orange xanthophyll widely distributed in nature and in aquatic environments (Higuera-Ciapara et al., 2006). Astaxanthin is the main pigment that provides the orange-reddish colour of fish and crustaceans such as lobsters and salmon (Higuera-Ciapara et al., 2006). It also has antioxidant activity with potential benefits in nutrition and human health. In humans, the antioxidant activity of astaxanthin counteracts diseases such as cancer, diabetes, ocular problems, and ageing, presents anti-inflammatory properties, and improves the immune system (Hussein et al., 2006; Davinelli et al., 2018). The basidiomycete *X. dendrorhous* (also know as *Phaffia rhodozyma*) produces astaxanthin in large quantities and is used for industrial production (Johnson, 2003; Rodríguez-Sáiz et al., 2010). In recent years, the industrial production of astaxanthin has taken importance not only for its applications in human health but also in aquaculture because pigmentation is an attraction for the consumer, as well as being an essential nutrient for proper growth and reproduction of certain aquatic animals (Johnson, 2003; Rodríguez-Sáiz et al., 2010; Davinelli et al., 2018). It has been suggested that in fungi carotenoids do not seem to be so indispensable because their absence only alters pigmentation but does not generate apparent phenotypic alterations (Ávalos and Carmen Limón, 2015).

Apocarotenoids are also produced by fungi. There are three main apocarotenoids produced by fungi: retinal, trisporic acids, and neurosporaxanthin (Ahrazem et al., 2016) and all of them are enzymatically synthesized by specific carotenoid cleavage oxygenases (CCOs) (Ahrazem et al., 2016; Liang et al., 2018). Retinal and trisporic acids are derived from β-carotene, while neurosporaxanthin is derived from torulene (Ávalos and Carmen Limón, 2015; Ahrazem et al., 2016; Liang et al., 2018).

Retinal

In fungi, retinal is a 20C compound that is derived from the symmetrical cleavage of β-carotene (Prado-Cabrero et al., 2007a). The retinal is a chromophore that exerts its function by binding with rhodopsins (Prado-Cabrero et al., 2007a). Rhodopsins are a class of photoreceptor protein that contain seven transmembrane helices with conserved lysine residues, through which the retinal covalently binds (Prado-Cabrero et al., 2007a). The light generates the isomerization of the apocarotenoid attached to the rhodopsins, and this triggers the reactions that drive various processes ranging from ion pumps to protein–protein interactions to protein signalling (Brown, 2004; Spudich, 2006; Prado-Cabrero et al., 2007a). Rhodopsins are widely distributed in the domain of the Archaea, Bacteria, and Eukarya (Spudich, 2006). In fungi, these proteins showed a high analogy with the transmembrane regions of bacteriorhodopsin and have been classified as fungal rhodopsins (RDs) and opsin-related proteins (ORPs) (Brown, 2004). NOP-1 was the first rhodopsin identified in fungi, it was identified in *N. crassa*. The analysis of the amino acid sequence of NOP-1 showed that it had the characteristic structure of rhodopsins and that shared an 81.8% homology with archaeological rhodopsins (Bieszke et al., 1999). Ops is another rhopodsins identified in the ascomycete *Leptosphaeria maculans*. Ops present a high similarity (70.3%) and identity (53.3%) with NOP-1 of *N. crassa* (Idnurm and Howlett, 2001). Another ortholog of NOP-1 is OpsA from *F. fujikuroi* (Estrada and Ávalos, 2009). CarO from *F. fujikuroi* is another well-characterized rhodopsin (Prado et al., 2004). Rhodopsin functionality has only been demonstrated for Ops (Waschuk et al., 2005), NOP-1 (Bieszke et al., 1999), and CarO (García-Martínez et al., 2015).

The following steps for retinal production after β-carotene include the partition of carotenoid cleavage oxygenases. In the ascomycete *F. fujikuroi* the retinal production is catalyzed by CarX, a carotenoid-cleaving oxygenase (Thewes et al., 2005; Prado-Cabrero et al., 2007a). CarX was the first retinal synthesizing enzyme reported in fungi. Heterologous expression of CarX in *E. coli* cells that accumulate β-carotene led to retinal formation. *In vitro* assays also confirmed its ability to produce retinal from substrates other than β-carotene, such as γ-carotene, torulene and β-apo-8′-carotene. CarX cleaves at the 15,15′ double bond and requires the presence of at least one ionone ring in the substrate to performs its cleavage reaction (Prado-Cabrero et al., 2007a). In *N. crassa* no CarX orthologs were found but NcCAO1, a putative carotenoid cleavage oxygenase was a candidate involved in retinal production (Díaz-Sánchez et al., 2013). However, CAO-1 did not cleave β-carotene and other tested apocarotenoids, but efficiently cleaved resveratrol and its derivative piceatannol (Díaz-Sánchez et al., 2013). So, more research is needed to find the retinal provider for NOP-1 rhodopsin in *N. crassa*. On the other hand, in the basidiomycete *Ustilago maydis*, Cco1 is responsible for retinal formation, which catalyzes the symmetric cleavage of β-carotene to produce two retinal molecules (Estrada et al., 2009).

In animals, retinal is one of the most important apocarotenoids because it is necessary for the detection of light in the retina and vision (Blomhoff and Blomhoff, 2006). In humans, the β-carotene-15,15′-dioxygenase (BCDO) catalyzes the oxidative cleavage of β-carotene into two retinal molecules (Yan et al., 2001). In humans, the retinal covalently binds to a G protein-coupled receptor (GPCR) called opsin. A single photon of light energy activates 11-*cis*-retinal that undergoes photoisomerization to give rise to all-*trans*-retinal. This leads to a conformational change of GPCR opsin, generating a signal transduction cascade that triggers nerve impulses that our brain interprets as the vision (Tsin et al., 2018). It has been reported that genetic and structural changes in the proteins involved in vision generate several ocular diseases (Tsin et al., 2018). Humans are unable to synthesize β-carotene, so they need to acquire it through diet. Vitamin A deficiency and its derived retinoids represent a global health problem (Tanumihardjo et al., 2016).

Vitamin A is a term used to designate any compound that has retinol biological activity, which include retinol, retinal, and retinoic acid (Blomhoff and Blomhoff, 2006; Tanumihardjo et al., 2016). In developed countries, the deficiency of vitamin A and retinoids is mainly due to a diet low in carotenoids. The problem is even more serious in undeveloped countries, where β-carotene sources such as fruits and vegetables have high costs (Grune et al., 2010; Tanumihardjo et al., 2016). Therefore, direct administration of retinoids such as retinal represents a good option to help counteract β-carotene deficiency not only in developed but also in undeveloped countries. Thus, the high production of retina in fungi makes them an alternative for the industrial production of this apocarotenoid, and a possible solution to the health problems caused by vitamin A deficiency.

Trisporic Acid

Trisporoids are apocarotenoids generated by the double oxidative cleavage of β-carotene. They are compounds of 18C or 19C, if they contain a methyl ester group (Barrero et al., 2017). The trisporoid group is classified in an A–E series, and each serial can have five groups according to the oxidation state of R1 (trisporins when R1 is a CH_3, trisporols when R1 is a CH_2OH, trisporic acids when R1 is a COOH and methyl trisporates and 4-dihydro-methyl trisporates when R1 is COOMe) (Barrero et al., 2017). Trisporic acid was isolated for the first time from submerged cultures of *Choanephora trispora* (also known as *Blakeslea trispora*) (Caglioti et al., 1966) and it is considered one of the most important apocarotenoid of the order of the Mucorales (*Phycomyces blakesleeanus*, *B. trispora*, and *Mucor mucedo*) (Barrero et al., 2017). Trisporic acid constitutes the initial signal system that mediates the recognition between mating partners, can induce morphogenesis and early sexual development, as well as carotenogenesis and also can act as a signal for carotenogenesis between fungi (Rao and Modi, 1977; Schimek and Wöstemeyer, 2009; Ávalos and Carmen Limón, 2015; Ahrazem et al., 2016; Barrero et al., 2017). Trisporic acid induces carotenogenesis in minus (–) strains but not in the plus (+) strain of *B. trispora* by acting as a derepressor of enzymes involved in the carotene biosynthesis pathway (Rao and Modi, 1977). It also induces the expression of genes involved in carotenogenesis (Schmidt et al., 2005; Kuzina et al., 2008) and sexual development (Schimek and Wöstemeyer, 2009).

The initial step for trisporic acid biosynthesis involves the oxidative cleavage of β-carotene by mono- or di-oxygenases at specific double bonds. TSP3 and TSP4 were the first putative carotene oxygenases identified in the synthesis of trisporic acid. They were identified in the zygomycete *Rhizopus oryzae* (Burmester et al., 2007). TSP3 was also identified in *B. trispora* (Burmester et al., 2007). In *R. oryzae*, trisporic acid induced expression of the *tsp3* but not *tsp4* gene. The expression of the *tsp3* was also induced in *B. trispora*. In both species, the expression of *tsp3* correlated with sexual development. TSP3 cleavage activity of *B. trispora* on β-carotene was verified by the decoloration of *E. coli* strains that produce β-carotene ans express *tsp3* (Burmester et al., 2007). Subsequently in the mucoral *Phycomyces blakesleeanus* two molecules of 7C ((2E, 4E) 6-hydroxy-5-methylhexa-2,4-dienoic acid and 2-methyl isomer) were identified, possibly derived from the cutting at the 13,14 and 11′,12′ double bonds of β-carotene. The split of β-carotene generates 7C, 18C, and 15C molecules that form the skeleton of the three main families of apocarotenoids in Mucorales, methylhexanoids, trisporoids, and cyclopharynesoids, respectively (Polaino et al., 2010; Barrero et al., 2017). Recently two carotene oxygenases CarS (ortholog of TSP3) and acaA were identified as responsible for the production of triposric acid by the sequential cleavage of β-carotene. According to the results, the cleavage activity of CarS at the 11′,12′ double bond of β-carotene results in the formation of β-apo-12′-carotenal, a compound of 25C and 15C apocarotenals used for the synthesis of cyclofarnesoids. AcaA cleaves at the 13, 14 double bond of β-apo-12′-carotenal, resulting in β-apo-13-carotenone, an 18C compound that gives rise to triporoids and trisporic acid. The expression level of CarS was abundant in the course of the mating of *P. blakesleeanus* (Medina et al., 2011).

Trisporic acid induces carotenogenesis in the minus (−) strains but not in the plus (+) strain of *B. trispora*, acting as a depressant of enzymes involved in the carotene biosynthesis pathway (Rao and Modi, 1977). In minus (−) strains of *B. trispora*, the synthesis of carotenes is stimulated and trisporic acid is produced in very low quantities (Sutter et al., 1973; Rao and Modi, 1977). Trisporic acid also improves the synthesis of β-carotene in the mating of different species, as between plus (+) strain of *B. trispora* and minus (−) strain of *Mucor mucedo* (Van Den Ende, 1967). However, during mating, the minus (−) strain of *M. mucedo* produce carotene but does not produce trisporic acid (Nieuwenhuis and Van Den Ende, 1975), something similar happens with the minus (−) strain of *P. blakesleeanus* (Schimek and Wöstemeyer, 2009). Similarly, Govind and Cerdá-Olmedo (1986) found that synthetic racemic (9Z)-methyl trisporate B, a trisporate precursor, made by a plus mating (+) strain, can stimulate carotenogenesis in minus (−) mating strain of *P.blakesleeanus*.

Currently, *B. trispora*, is used as a natural source for industrial and commercial production of β-carotene (Nelis and De Leenheer, 1991) and lycopene (Martínez-Cámara et al., 2018). It has been reported that in *B. trispora*, trisporic acid induces the expression of genes involved in carotenogenesis such as carA and carB (Schmidt et al., 2005; Kuzina et al., 2008). Other results indicate that trisporic acid is a global regulator in *B. trispora*, generating changes at the metabolic level with changes in the content of various fatty acids, carbohydrates and amino acids (Sun et al., 2012). In fungi, trisporic acid constitutes the initial signal system for carotenogenesis between couples in mating and individually; therefore, this characteristic can be exploited and thus increase production levels of β-carotene in *B. trispora* or in other zygomycetes from the order Mucorales. Besides, information on the regulatory mechanisms of trisporic acid in *B. trispora* at the metabolic level, have the potential to improve the production of β-carotene naturally, or in genetically modified microbes.

Neurosporaxanthin

Neurosporaxanthin is a 35C apocaroenoid chemically defined as β-apo-40′-carotenoic acid that derives from the torulene (40C) (Aasen and Jensen, 1965; Ávalos et al., 2017; Liang et al., 2018). It was first identified from an acidic mixture of pigments of *Neurospora crassa* (Zalokar, 1957) and then in *Fusarium aquaeductuum* (Aasen and Jensen, 1965). Now it is known that neurosporaxanthin is the major apocarotenoid produced by the genus *Neurospora* and *Fusarium* (Ávalos et al., 2017). Neurosporaxanthin has also been found in the genus *Podospora* (Strobel et al., 2009) and *Verticillium* (Guy Valadon et al., 1982). The biosynthesis pathway of the neurosporaxanthin has been well studied in *N. crassa* and *Fusarium* genus (Ávalos et al., 2017). In *N. crassa*, its biosynthesis is performed in a similar way from the GGPP to β-carotene, however instead of four desaturations there are five to produce 3,4-didehydrolycopene, and instead of two cyclization only one cyclization is introduced to the 40C skeleton to generate the torulene (Díaz-Sánchez et al., 2011b; Ávalos and Carmen Limón, 2015; Ávalos et al., 2017). In *N. crassa* AL-1 performs all five unsaturations, giving rise to 3,4-didehydrolycopene (Schmidhauser et al., 1990; Hausmann and Sandmann, 2000). The enzyme AL-2 performs the cyclization step of 3,4-didehydrolycopene to produce torulene (Hausmann and Sandmann, 2000; Díaz-Sánchez et al., 2011b). The cyclase domain of AL-2 is highly conserved and has a high homology with CrtYB of *X. dendrorhous* and CarRA of *F. fujikuroi*, suggesting that they perform the same function (Ávalos et al., 2017). In the *Fusarium* genus, CarB performs the five desaturations of the neurosporaxanthin biosynthesis pathway and cyclization can be achieved either in neurosporene or lycopene, depending on the species. Cyclization gives rise to γ-carotene, the precursor to torulene (Ávalos et al., 2017). The last steps of the biosynthesis of neurosporaxanthin involve the participation of CCDs. CAO2 in *N. crassa* (Saelices et al., 2007) and CarT in *Fusarium* (Prado-Cabrero et al., 2007b; Jin et al., 2010) are responsible for the asymmetric cleavage of 5C (at the 4′,5′ double bond from toruleno) for the production of β-apo-4′-carotenal (35C). The reaction is very specific, γ-carotene which differs from the torulene in that it only has four insaturations, could not be cleaved by CAO2 (Saelices et al., 2007). Finally, in *N. crassa* the aldehyde dehydrogenase YLO-1 oxidizes the terminal aldehyde of β-apo-4′-carotenal to convert it into a carboxyl group, producing the neurosporaxanthin (Estrada et al., 2008). In *F. fujikuroi* the ortholog of YLO-1 is CarD (Díaz-Sánchez et al., 2011a).

The orange colour of the fungi of the genus of *Neurospora* and *Fusarium* is mainly due to the accumulation of neurosporaxanthin. Both fungi present a similar pathway of neurosporaxanthin biosynthesis. Now that the enzymes involved in the biosynthesis of neurosporaxanthin are known, the possible biotechnological applications for the industrial production of this apocarotenoid in fungi are being investigated. Ávalos et al. (2012) found that the production of large amounts of neurosporaxanthin over other carotenoids depended on the culture conditions and/or genetic background of *Neurospora* and *Fusarium*. For the high production of neurosporaxanthin, they recommended cultivating the *CarS* strain of *F. fujikuro* in the dark and low nitrogen broth, while for *N. crassa* the wild strain was sufficient to generate an overproduction through lighting at a low temperature (Ávalos et al., 2012). However, in other studies, it has been found that the production of neurosporaxanthin is highly induced by light not only in *N. crassa* but also in wild strains of *F. fujikuroi* (Ávalos and Estrada, 2010). On the other hand, studies in *N. crassa* have focused on their photobiology (Chen and Loros, 2009; Chen et al., 2010), while studies in *F. fujikuro* have focused in gibberellins and secondary metabolism (Ávalos et al., 2007), so studies on the application of neurosporaxanthin are still scarce. Recently, it was found that torulene, the neurosporaxanthin precursor, has strong antioxidant, antimicrobial and anticancer properties in rats and mice (Kot et al., 2018). Also, its use was suggested as an additive for food, feedstock, and cosmetics (Kot et al., 2018). Therefore, obtaining and growing mutant strains of *CAO2* in *N. crassa* and *CarT* in *F. fujikuroi* could be a good option for the production of torulene in great quantities.

ACKNOWLEDGEMENTS

The authors thank Consejo Nacional de Ciencia y Tecnología (CONACYT) for their support with the project: Fronteras de la Ciencia No. 2016-01-1716. JDC and RUC co-authors, PhD grant (No.815226) and postdoctoral position (FC No. 2016-01-1716), respectively, were financially supported by CONACYT.

BIBLIOGRAPHY

Aasen, A. J., & Jensen, S. L. (1965). Fungal carotenoids. II: The structure of the carotenoid acid neurosporaxanthin. *Acta Chemica Scandinavica*, 19(8), 1843–1853.

Abbas, H. A., Elsherbini, A. M., & Shaldam, M. A. (2019). Glyceryl trinitrate blocks staphyloxanthin and biofilm formation in *Staphylococcus aureus*. *African Health Sciences*, 19(1), 1376–1384. https://doi.org/10.4314/ahs.v19i1.10

Abdullaev, F. (2002). Cancer chemopreventive and tumoricidal properties of saffron (*Crocus sativus* L.). *Experimental Biology and Medicine*, 227(1), 20–25. doi:10.1177/153537020222700104

Ahrazem, O., Rubio-Moraga, A., Berman, J., Capell, T., Christou, P., Zhu, C., & Gómez-Gómez, L. (2016a). The carotenoid cleavage dioxygenase CCD2 catalysing the synthesis of crocetin in spring crocuses and saffron is a plastidial enzyme. *The New phytologist*, 209(2), 650–663. https://doi.org/10.1111/nph.13609

Ahrazem, O., Gómez-Gómez, L., Rodrigo, M. J., Avalos, J., & Limón, M. C. (2016b). Carotenoid cleavage oxygenases from microbes and photosynthetic organisms: Features and functions. *International Journal of Molecular Sciences*, 17(11), 1781. doi:10.3390/ijms17111781

Alder, A., Jamil, M., Marzorati, M., Bruno, M., Vermathen, M., Bigler, P., … Al-Babili, S. (2012). The path from β-carotene to carlactone, a strigolactone-like plant hormone. *Science*, 335(6074), 1348–1351.

Al-Hasani, S. M. (1968). Vitamin A activity of beta-apo-carotenals in *Coturnix coturnix* japonica. *Journal of Nutrition*, 94(3), 402–406.

Al-Hasani, S. M., & Parrish, D. B. (1972). Forms of vitamin A and of carotenoids in tissues, blood serum and yolk of eggs from *Cotournix coturnix* japonica fed-apo-carotenals. *Journal of Nutrition*, 102(11), 1437–1440.

Alonso, G. L., Salinas, M. R., Esteban-Infantes, F. J., & Sanchez-Fernandez, M. A. (1996). Determination of safranal from saffron (*Crocus sativus* L.) by thermal desorption-gas chromatography. *Journal of Agricultural and Food Chemistry*, 44, 185–188.

Amagai, A., Honda, Y., Ishikawa, S., Hara, Y., Kuwamura, M., Shinozawa, A., … Umezawa, T. (2018). Phosphoproteomic profiling reveals ABA-responsive phosphosignaling pathways in Physcomitrella patens. *The Plant Journal*, 94(4), 699–708. doi:10.1111/tpj.13891

Amengual, J., Widjaja-Adhi, M. A., Rodriguez-Santiago, S., Hessel, S., Golczak, M., Palczewski, K., & von Lintig, J. (2013). Two carotenoid oxygenases contribute to mammalian provitamin A metabolism. *The Journal of Biological Chemistry*, 288(47), 34081–34096. doi:10.1074/jbc.M113.501049

Arif, M. A., Hiss, M., Tomek, M., Busch, H., Meyberg, R., Tintelnot, S., … Frank, W. (2019). ABA-induced vegetative diaspore formation in Physcomitrella patens. *Frontiers in Plant Science*, 10, 315. doi:10.3389/fpls.2019.00315

Arscott, S. A. (2013). Carotenoids and human health. In S. Tanumihardjo (Ed.), *Food sources of carotenoids* (pp. 3–19). New York, NY: Springer Science + Business Media.

Asad, M., Zakari, S. A., Zhao, Q., Zhou, L., Ye, Y., & Cheng, F. (2019). Abiotic stresses intervene with ABA signaling to induce destructive metabolic pathways leading to death: Premature leaf senescence in plants. *International Journal of Molecular Sciences*, 20(2), 256. doi:10.3390/ijms20020256

Auldridge, M. E., McCarty, D. R., & Klee, H. J. (2006). Plant carotenoid cleavage oxygenases and their apocarotenoid products. *Current Opinion in Plant Biology*, 9(3), 315–321.

Ávalos, J., & Carmen Limón, M. (2015). Biological roles of fungal carotenoids. *Current Genetics*, 61(3), 309 324.

Ávalos, J., Cerda-Olmedo, E., Reyes, F., & Barrero, A. F. (2007). Gibberellins and other metabolites of Fusarium fujikuroi and related fungi. *Current Organic Chemistry*, 11(8), 721–737.

Ávalos, J., Díaz-Sánchez, V., García-Martínez, J., Castrillo, M., Ruger-Herreros, M., & Limón, M. C. (2014). Carotenoids. In: J. F. Martín, C. García-Estrada, & S. Zeilinger (Eds)., *Biosynthesis and molecular genetics of fungal secondary metabolites* (pp. 149–185). New York, NY: Springer Science + Business Media.

Ávalos, J., & Estrada, A. F. (2010). Regulation by light in Fusarium. *Fungal Genetics and Biology*, 47(11), 930–938.

Ávalos, J., Prado-Cabrero, A., & Estrada, A. F. (2012). Neurosporaxanthin production by Neurospora and Fusarium. In J. L. Barredo (Ed.), *Microbial carotenoids from fungi: Methods and protocols* (pp. 263–274). Totowa, NJ: Humana Press.

Ávalos, J., Pardo-Medina, J., Parra-Rivero, O., Ruger-Herreros, M., Rodríguez-Ortiz, R., Hornero-Méndez, D., & Limón, M. (2017). Carotenoid biosynthesis in Fusarium. *Journal of Fungi*, 3(3), 39.

Badyaev, A. V., Morrison, E. S., Belloni, V., & Sanderson, M. J. (2015). Tradeoff between robustness and elaboration in carotenoid networks produces cycles of avian color diversification. *Biology Direct*, 10, 45. doi:10.1186/s13062-015-0073-6

Barrero, A. F., Del Pino, M. M. H., Arteaga, J. F., & González-Delgado, J. A. (2017). Occurrence and chemical synthesis of apocarotenoids from mucorales: A review. *Natural Product Communications*, 12(5), 733–741.

Barretto, D. A., & Vootla, S. K. (2018). In Vitro Anticancer Activity of Staphyloxanthin Pigment Extracted from Staphylococcus gallinarum KX912244, a Gut Microbe of Bombyx mori. *Indian Journal of Microbiology*, 58(2), 146–158. https://doi.org/10.1007/s12088-018-0718-0

Bastos Maia, S., Rolland Souza, A. S., Costa Caminha, M. F., Lins da Silva, S., Callou Cruz, R., Carvalho Dos Santos, C., & Batista Filho, M. (2019). Vitamin A and pregnancy: A narrative review. *Nutrients*, 11(3), 681. doi:10.3390/nu11030681

Baz, L., Mori, N., Mi, J., Jamil, M., Kountche, B. A., Guo, X., … Al-Babili, S. (2018). 3-hydroxycarlactone, a novel product of the strigolactone biosynthesis core pathway. *Molecular Plant*, 11(10), 1312–1314. doi:10.1016/j.molp.2018.06.008

Berli, F. J., Fanzone, M., Piccoli, P., & Bottini, R. (2011). Solar UV-B and ABA are involved in phenol metabolism of *Vitis vinifera* L. increasing biosynthesis of berry skin polyphenols. *Journal of Agricultural and Food Chemistry*, 59(9), 4874–4884.

Bhyan, S. B., Minami, A., Kaneko, Y., Suzuki, S., Arakawa, K., Sakata, Y., & Takezawa, D. (2012). Cold acclimation in the moss Physcomitrella patens involves abscisic acid-dependent signaling. *Journal of Plant Physiology*, 169(2), 137–145. doi:10.1016/j.jplph.2011.08.004

Bieszke, J. A., Braun, E. L., Bean, L. E., Kang, S., Natvig, D. O., & Borkovich, K. A. (1999). The nop-1 gene of *Neurospora crassa* encodes a seven transmembrane helix retinal-binding protein homologous to archaeal rhodopsins. *Proceedings of the National Academy of Sciences of the United States of America*, 96(14), 8034–8039. doi:10.1073/pnas.96.14.8034

Blomhoff, R., & Blomhoff, H. K. (2006). Overview of retinoid metabolism and function. *Journal of Neurobiology*, 66(7), 606–630.

Bouvier, F., Dogbo, O., & Camara, B. (2003a). Biosynthesis of the food and cosmetic plant pigment bixin (annatto). *Science*, 300(5628), 2089–2091.

Bouvier, F., Suire, C., Mutterer, J., & Camara, B. (2003b). Oxidative remodeling of chromoplast carotenoids: Identification of the carotenoid dioxygenase CsCCD and CsZCD genes involved in Crocus secondary metabolite biogenesis. *The Plant Cell*, 15(1), 47–62. doi:10.1105/tpc.006536

Braungardt, H., & Singh, V. K. (2019). Impact of Deficiencies in Branched-Chain Fatty Acids and Staphyloxanthin in *Staphylococcus aureus*. *BioMed Research International*, 2019, 2603435. https://doi.org/10.1155/2019/2603435

Brown, L. S. (2004). Fungal rhodopsins and opsin-related proteins: Eukaryotic homologs of bacteriorhodopsin with unknown functions. *Photochemical Photobiological Sciences*, 3(6), 555–565.

Burmester, A., Richter, M., Schultze, K., Voelz, K., Schachtschabel, D., Boland, W., … Schimek, C. (2007). Cleavage of β-carotene as the first step in sexual hormone synthesis in zygomycetes is mediated by a trisporic acid regulated β-carotene oxygenase. *Fungal Genetics and Biology*, 44(11), 1096–1108.

Cárdenas-Conejo, Y., Carballo-Uicab, V., Lieberman, M., Aguilar-Espinosa, M., Comai, L., & Rivera-Madrid, R. (2015). De novo transcriptome sequencing in Bixa orellana to identify genes involved in methylerythritol phosphate, carotenoid and bixin biosynthesis. *BMC Genomics*, 16, 877. https://doi.org/10.1186/s12864-015-2065-4

Caglioti, L., Cainelli, G., Camerino, B., Mondelli, R., Prieto, A., Quilico, A., … Selva, A. (1966). The structure of trisporic-C acid. *Tetrahedron*, 22(7), 175–187.

Carballo-Uicab, V. M., Cárdenas-Conejo, Y., Vallejo-Cardona, A. A., Aguilar-Espinosa, M., Rodríguez-Campos, J., Serrano-Posada, H., … Rivera-Madrid, R. (2019). Isolation and functional characterization of two dioxygenases putatively involved in bixin biosynthesis in annatto (*Bixa orellana* L.). *PeerJ*, 7, e7064. doi:10.7717/peerj.7064

Cazzonelli, C. I., & Pogson, B. J. (2010). Source to sink: Regulation of carotenoid biosynthesis in plants. *Trends in Plant Science*, 15(5), 266–274. doi:10.1016/j.tplants.2010.02.003

Chai, F., Wang, Y., Mei, X., Yao, M., Chen, Y., Liu, H., … Yuan, Y. (2017). Heterologous biosynthesis and manipulation of crocetin in *Saccharomyces cerevisiae*. *Microbial Cell Factories*, 16(1), 54. doi:10.1186/s12934-017-0665-1

Chambon, P. (1996). A decade of molecular biology of retinoic acid receptors. *The Faseb Journal*, 10(9), 940–954.

Chen, C. H., Dunlap, J. C., & Loros, J. J. (2010). Neurospora illuminates fungal photoreception. *Fungal Genetics and Biology*, 47(11), 922–929. doi:10.1016/j.fgb.2010.07.005

Chen, C. H., & Loros, J. J. (2009). Neurospora sees the light: Light signaling components in a model system. *Communicative & Integrative Biology*, 2(5), 448–451. doi:10.4161/cib.2.5.8835

Chryssanthi, D. G., Lamari, F. N., Iatrou, G., Pylara, A., Karamanos, N. K., & Cordopatis, P. (2007). Inhibition of breast cancer cell proliferation by style constituents of different Crocus species. *Anticancer Research*, 27, 357–362.

Chung, J., Koo, K., Lian, F., Hu, K. Q., Ernst, H., & Wang, X. D. (2012). Apo-10'-lycopenoic acid, a lycopene metabolite, increases sirtuin 1 mRNA and protein levels and decreases hepatic fat accumulation in ob/ob mice. *The Journal of Nutrition*, 142(3), 405–410. doi:10.3945/jn.111.150052

Cichon, M. J., Moran, N. E., Riedl, K. M., Schwartz, S. J., & Clinton, S. K. (2018). Identification of an epoxide metabolite of lycopene in human plasma using 13C-labeling and QTOF-MS. *Metabolites*, 8(1), 24. doi:10.3390/metabo8010024

Clauditz, A., Resch, A., Wieland, K. P., Peschel, A., & Götz, F. (2006). Staphyloxanthin plays a role in the fitness of *Staphylococcus aureus* and its ability to cope with oxidative stress. *Infection and Immunity*, 74(8), 4950–4953. doi:10.1128/IAI.00204-06

Costabile, B. K., Kim, Y. K., Iqbal, J., Zuccaro, M. V., Wassef, L., Narayanasamy, S., ... Quadro, L. (2016). β-apo-10'-carotenoids modulate placental microsomal triglyceride transfer protein expression and function to optimize transport of intact β-carotene to the embryo. *The Journal of Biological Chemistry*, 291(35), 18525–18535. doi:10.1074/jbc.M116.738336

Cui, H., Wang, Y., & Qin, S. (2012). Genomewide analysis of carotenoid cleavage dioxygenases in unicellular and filamentous cyanobacteria. *Comparative and Functional Genomics*, 2012, 164690. doi:10.1155/2012/164690

D'Ambrosio, D. N., Clugston, R. D., & Blaner, W. S. (2011). Vitamin A metabolism: An update. *Nutrients*, 3(1), 63–103. doi:10.3390/nu3010063

Davinelli, S., Nielsen, M. E., & Scapagnini, G. (2018). Astaxanthin in skin health, repair, and disease: A comprehensive review. *Nutrients*, 10(4), 522. doi:10.3390/nu10040522

De Vries, J. J., Chang, A. B., Bonifant, C. M., Shevill, E., & Marchant, J. M. (2018). Vitamin A and beta (β)-carotene supplementation for cystic fibrosis. *The Cochrane Database of Systematic Reviews*, 8(8), CD006751. doi:10.1002/14651858.CD006751.pub5

Díaz-Sánchez, V., Estrada, A. F., Limón, M. C., Al-Babili, S., & Avalos, J. (2013). The oxygenase CAO-1 of Neurospora crassa is a resveratrol cleavage enzyme. *Eukaryotic Cell*, 12(9), 1305–1314. doi:10.1128/EC.00084-13

Díaz-Sánchez, V., Estrada, A., Trautmann, D., Al-Babili, S., & Avalos, J. (2011a). The gene carD encodes the aldehyde dehydrogenase responsible for neurosporaxanthin biosynthesis in *Fusarium fujikuroi. Febs J*, 278(17), 3164–3176.

Díaz-Sánchez, V., Estrada, A. F., Trautmann, D., Limón, M. C., Al-Babili, S., & Avalos, J. (2011b). Analysis of al-2 mutations in neurospora. *PLos One*, 6(7), e21948.

Dickinson, A. J., Lehner, K., Mi, J., Jia, K. P., Mijar, M., Dinneny, J., ... Benfey, P. N. (2019). β-cyclocitral is a conserved root growth regulator. *Proceedings of the National Academy of Sciences of the United States of America*, 116(21), 10563–10567. doi:10.1073/pnas.1821445116

Estrada, A. F., & Ávalos, J. (2009). Regulation and targeted mutation of opsA, coding for the NOP-1 opsin orthologue in *Fusarium fujikuroi. Journal of Molecular Biology*, 387(1), 59–73.

Estrada, A. F., Brefort, T., Mengel, C., Díaz-Sánchez, V., Alder, A., Al-Babili, S., & Avalos, J. (2009). Ustilago maydis accumulate β-carotene at levels determined by a retinal-forming carotenoid oxygenase. *Fungal Genetics and Biology*, 46(10), 803–813.

Estrada, A. F., Youssar, L., Scherzinger, D., Al-Babili, S., & Avalos, J. (2008). The ylo-1 gene encodes an aldehyde dehydrogenase responsible for the last reaction in the Neurospora carotenoid pathway. *Molecular Microbiology*, 69(5), 1207–1220.

Felemban, A., Braguy, J., Zurbriggen, M. D., & Al-Babili, S. (2019). Apocarotenoids involved in plant development and stress response. *Frontiers in Plant Science*, 10, 1168. doi:10.3389/fpls.2019.01168

Fiorilli, V., Wang, J. Y., Bonfante, P., Lanfranco, L., & Al-Babili, S. (2019). Apocarotenoids: Old and new mediators of the arbuscular mycorrhizal symbiosis. *Frontiers in Plant Science*, 10, 1186. doi:10.3389/fpls.2019.01186

Fleshman, M. K., Lester, G. E., Riedl, K. M., Kopec, R. E., Narayanasamy, S., Curley Jr., R. W., ... Harrison, E. H. (2011). Carotene and novel apocarotenoid concentrations in orange-fleshed Cucumis melo melons: Determinations of β-carotene bioaccessibility and bioavailability. *Journal of Agricultural and Food Chemistry*, 59(9), 4448–4454. doi:10.1021/jf200416a

Frank, W., Ratnadewi, D., & Reski, R. (2005). Physcomitrella patens is highly tolerant against drought, salt and osmotic stress. *Planta*, 220(3), 384–394.

Frusciante, S., Diretto, G., Bruno, M., Ferrante, P., Pietrella, M., Prado-Cabrero, A., & Giuliano, G. (2014). Novel carotenoid cleavage dioxygenase catalyzes the first dedicated step in saffron crocin biosynthesis.

Proceedings of the National Academy of Sciences of the United States of America, 111(33), 12246–12251. doi:10.1073/pnas.1404629111
Fuentes, P., Pizarro, L., Moreno, J. C., Handford, M., Rodriguez-Concepcion, M., & Stange, C. (2012). Light-dependent changes in plastid differentiation influence carotenoid gene expression and accumulation in carrot roots. *Plant Molecular Biology*, 79(1–2), 47–59. doi:10.1007/s11103-012-9893-2
Gajic, M., Zaripheh, S., Sun, F., & Erdman, J. W., Jr. (2006). Apo-8'-lycopenal and apo-12'-lycopenal are metabolic products of lycopene in rat liver. *Journal of Nutrition*, 136(6), 1552–1557.
Gao, L., Gonda, I., Sun, H., Ma, Q., Bao, K., Tieman, D. M., … Fei, Z. (2019). The tomato pan-genome uncovers new genes and a rare allele regulating fruit flavor. *Nature Genetics*, 51(6), 1044–1051. doi:10.1038/s41588-019-0410-2
García-Martínez, J., Brunk, M., Avalos, J., & Terpitz, U. (2015). The CarO rhodopsin of the fungus Fusarium fujikuroi is a light-driven proton pump that retards spore germination. *Scientific Reports*, 5, 7798. doi:10.1038/srep07798
Gelis, L., Jovancevic, N., Veitinger, S., Mandal, B., Arndt, H. D., Neuhaus, E. M., & Hatt, H. (2016). Functional characterization of the odorant receptor 51E2 in human melanocytes. *Journal of Biological Chemistry*, 291(34), 17772–17786. doi:10.1074/jbc.M116.734517
Giridhar, P., & Parimalan, R. (2010). A biotechnological perspective towards improvement of annatto color production for value addition—the influence of biotic elicitors. *Asia-Pacific J Mol Biol Biotechnol*, 18(1), 77–79.
Gómez-Porras, J. L., Riaño-Pachón, D. M., Dreyer, I., Mayer, J. E., & Mueller-Roeber, B. (2007). Genome-wide analysis of ABA-responsive elements ABRE and CE3 reveals divergent patterns in Arabidopsis and rice. *BMC Genomics*, 8, 260. doi:10.1186/1471-2164-8-260
González-Pérez, S., Gutiérrez, J., García-García, F., Osuna, D., Dopazo, J., Lorenzo, Ó., … Arellano, J. B. (2011). Early transcriptional defense responses in Arabidopsis cell suspension culture under high-light conditions. *Plant Physiology*, 156(3), 1439–1456. doi:10.1104/pp.111.177766
Goodman, D. S., & Huang, H. S. (1965). Biosynthesis of vitamin A with rat intestinal enzymes. *Science*, 149(3686), 879–880.
Govind, N. S., & Cerdá-Olmedo, E. (1986). Sexual activation of carotenogenesis in *Phycomyces blakesleeanus*. *Microbiology*, 132(10), 2775–2780.
Grune, T., Lietz, G., Palou, A., Ross, A. C., Stahl, W., Tang, G., … Biesalski, H. K. (2010). Beta-carotene is an important vitamin A source for humans. *The Journal of Nutrition*, 140(12), 2268S–2285S. doi:10.3945/jn.109.119024
Guirola, C. (2010). Tintes naturales: Su uso en Mesoamérica desde la época prehispánica. *Asociación FLAAR Mesoamérica*, 14.
Gutheil, W. G., Reed, G., Ray, A., Anant, S., & Dhar, A. (2012). Crocetin: An agent derived from saffron for prevention and therapy for cancer. *Current Pharmaceutical Biotechnology*, 13(1), 173–179. doi:10.2174/138920112798868566
Guy Valadon, L. R., Osman, M., Mummery, R. S., Jerebzoff-Quintin, S., & Jerebzoff, S. (1982). The effect of monochromatic radiation in the range 350 to 750 nm on the carotenogenesis in Verticillium agaricinum. *Physiologia Plantarum*, 56(2), 199–203.
Han, Y., Wang, H., Wang, X., Li, K., Dong, M., Li, Y., … Shang, F. (2019). Mechanism of floral scent production in *Osmanthus fFragrans* and the production and regulation of its key floral constituents, β-ionone and linalool. *Horticulture Research*, 6(106). doi:10.1038/s41438-019-0189-4
Handelman, G. J., Van Kuijk, F. J., Chatterjee, A., & Krinsky, N. I. (1991). Characterization of products formed during the autoxidation of beta-carotene. *Free Radic Biol Med*, 10(6), 427–437.
Harrison, P. J., & Bugg, T. D. (2014). Enzymology of the carotenoid cleavage dioxygenases: Reaction mechanisms, inhibition and biochemical roles. *Archives of Biochemistry and Biophysics*, 544, 105–111.
Harrison, E. H., & Quadro, L. (2018). Apocarotenoids: Emerging roles in mammals. *Annual Review of Nutrition*, 38, 153–172. doi:10.1146/annurev-nutr-082117-051841
Hausenblas, H. A., Saha, D., Dubyak, P. J., & Anton, S. D. (2013). Saffron (*Crocus sativus* L.) and major depressive disorder: A meta-analysis of randomized clinical trials. *Journal of Integrative Medicine*, 11(6), 377–383. doi:10.3736/jintegrmed2013056
Hausmann, A., & Sandmann, G. (2000). A single five-step desaturase is involved in the carotenoid biosynthesis pathway to β-carotene and torulene in *Neurospora crassa*. *Fungal Genetics and Biology*, 30(2), 147–153.
Havaux, M. (2014). Carotenoid oxidation products as stress signals in plants. *The Plant Journal*, 79, 597–606.
Haveland-Smith, R. B. (1981). Evaluation of the genotoxicity of some natural food colours using bacterial assays. *Mutat Res*, 91, 285–290. doi:10.1016/0165-7992(81)90002-6

He, L., Mo, H., Hadisusilo, S., Qureshi, A. A., & Elson, C. E. (1997). Isoprenoids suppress the growth of murine B16 melanomas *in vitro* and *in vivo*. *The Journal of Nutrition*, 127(5), 668–674.
Higuera-Ciapara, I., Félix-Valenzuela, L., & Goycoolea, F. M. (2006). Astaxanthin: A review of its chemistry and applications. *Critical Reviews in Food Science and Nutrition*, 46(2), 185–196.
Holmes, H. N., & Corbet, R. E. (1937). The isolation of crystalline vitamin A. *Journal of the American Chemical Society*, 59(10), 2042–2047.
Hou, X., Rivers, J., León, P., McQuinn, R. P., & Pogson, B. J. (2016). Synthesis and function of apocarotenoid signals in plants. *Trends in Plant Science*, 21(9), 792–803.
Hu, K. Q., Liu, C., Ernst, H., Krinsky, N. I., Russell, R. M., & Wang, X. D. (2006). The biochemical characterization of ferret carotene-9',10'-monooxygenase catalyzing cleavage of carotenoids *in vitro* and *in vivo*. *The Journal of Biological Chemistry*, 281(28), 19327–19338. doi:10.1074/jbc.M512095200
Huang, F. C., Molnár, P., & Schwab, W. (2009). Cloning and functional characterization of carotenoid cleavage dioxygenase 4 genes. *Journal of Experimental Botany*, 60(11), 3011–3022. doi:10.1093/jxb/erp137
Hussein, G., Sankawa, U., Goto, H., Matsumoto, K., & Watanabe, H. (2006). Astaxanthin, a carotenoid with potential in human health and nutrition. *Journal of Natural Products*, 69(3), 443–449.
Idnurm, A., & Howlett, B. J. (2001). Characterization of an opsin gene from the ascomycete *Leptosphaeria maculans*. *Genome*, 44(2), 167–171.
Ip, B. C., Hu, K. Q., Liu, C., Smith, D. E., Obin, M. S., Ausman, L. M., & Wang, X. D. (2013). Lycopene metabolite, apo-10'-lycopenoic acid, inhibits diethylnitrosamine-initiated, high fat diet-promoted hepatic inflammation and tumorigenesis in mice. *Cancer Prevention Research*, 6(12), 1304–1316. doi:10.1158/1940-6207.CAPR-13-0178
Ip, B. C., Liu, C., Lichtenstein, A. H., von Lintig, J., & Wang, X. D. (2015). Lycopene and apo-10'-lycopenoic acid have differential mechanisms of protection against hepatic steatosis in β-carotene-9',10'-oxygenase knockout male mice. *The Journal of Nutrition*, 145(2), 268–276. doi:10.3945/jn.114.200238
Jako, C., Coutu, C., Roewer, I., Reed, D. W., Pelcher, L. E., & Covello, P. S. (2002). Probing carotenoid biosynthesis in developing seed coats of *Bixa orellana* (Bixaceae) through expressed sequence tag analysis. *Plant Science*, 163(1), 141–145. doi:10.1016/S0168-*9452*(02)00083-3
Jia, K. P., Dickinson, A. J., Mi, J., Cui, G., Xiao, T. T., Kharbatia, N. M., … Al-Babili, S. (2019a). Anchorene is a carotenoid-derived regulatory metabolite required for anchor root formation in Arabidopsis. *Science Advances*, 5(11), eaaw6787. doi:10.1126/sciadv.aaw6787
Jia, K. P., Li, C., Bouwmeester, H. J., & Al-Babili, S. (2019b). Strigolactone biosynthesis and signal transduction. *Strigolactones—Biology and Applications*, 1–45. doi:10.1007/978-3-030-12153-2_1
Jin, J. M., Lee, J., Lee, Y. W. (2010). Characterization of carotenoid biosynthetic genes in the ascomycete Gibberella zeae. *Fems Microbiology Letters*, 302(2), 197–202.
Johnson, E. A. (2003). *Phaffia rhodozyma*: Colorful odyssey. *International Microbiology*, 6(3), 169–174.
Johnson, M. D., Woodard, A., Okediji, E. J., Toms, S. A., & Allen, G. S. (2002). Lovastatin is a potent inhibitor of meningioma cell proliferation: Evidence for inhibition of a mitogen associated protein kinase. *The Journal of Neuro-Oncology*, 56(2), 133–142.
Jüttner, F., Höflacher, B. (1985). Evidence of β-carotene 7,8 (7′,8′) oxygenase (β-cyclocitral, crocetindial generating) in microcystis. *Archives of Microbiology*, 141(4), 337–343. doi:10.1007/BF00428846
Kamisugi, Y., & Cuming, A. C. (2005). The evolution of the abscisic acid-response in land plants: Comparative analysis of group 1 LEA gene expression in moss and cereals. *Plant Molecular Biology*, 59(5), 723–737.
Kane, M. A. (2012). Analysis, occurrence, and function of 9-cis-retinoic acid. *Biochimica Et Biophysica Acta*, 1821(1), 10–20.
Kapulnik, Y., & Koltai, H. (2014). Strigolactone involvement in root development, response to abiotic stress, and interactions with the biotic soil environment. *Plant Physiology*, 166(2), 560–569. doi:10.1104/pp.114.244939
Karrer, P., Morf, R., & Schoepp, K. (1931). Zur kenntnis des vitamins-A aus fischtranen [For knowledge of vitamin A from fish oil]. *Helv Chim Acta*, 14(5),1036–1040.
Keasling, J. D. (2012). Synthetic biology and the development of tools for metabolic engineering. *Metabolic Engineering*, 14(3), 189–195. doi:10.1016/j.ymben.2012.01.004
Khachik, F., Beecher, G. R., Goli, M. B., Lusby, W. R., & Daitch, C. E. (1992). Separation and quantification of carotenoids in human plasma. *Methods Enzymol*, 213, 205–219.
Khachik, F., Spangler, C. J., Smith Jr., J. C., Canfield, L.M., Steck, A., & Pfander, H. (1997). Identification, quantification, and relative concentrations of carotenoids and their metabolites in human milk and serum. *Analytical Chemistry*, 69(10), 1873–1881.
Kiefer, C., Hessel, S., Lampert, J. M., Vogt, K., Lederer, M. O., Breithaupt, D. E., & Von Lintig, J. (2001). Identification and characterization of a mammalian enzyme catalyzing the asymmetric oxidative cleavage of provitamin A. *Journal of Biological Chemistry*, 276(17), 14110–14116.

Kloer, D. P., Ruch, S., Al-Babili, S., Beyer, P., & Schulz, G. E. (2005). The structure of a retinal-forming carotenoid oxygenase. *Science*, 308(5719), 267–269.
Kloer, D. P., & Schulz, G. E. (2006). Structural and biological aspects of carotenoid cleavage. *Cellular and Molecular Life Sciences CMLS*, 63(19), 2291–2303.
Kopec, R. E., Riedl, K. M., Harrison, E. H., Curley Jr., R. W., Hruszkewycz, D. P., Clinton, S. K., & Schwartz, S. J. (2010). Identification and quantification of apo-lycopenals in fruits, vegetables, and human plasma. *Journal of Agricultural and Food Chemistry*, 58(6), 3290–3296. doi:10.1021/jf100415z
Kot, A. M., Błażejak, S., Gientka, I., Kieliszek, M., & Bryś, J. (2018). Torulene and torularhodin: 'New' fungal carotenoids for industry? *Microbial Cell Factories*, 17(1), 49.
Kushiro, T., Okamoto, M., Nakabayashi, K., Yamagishi, K., Kitamura, S., Asami, T., ... Nambara, E. (2004). The Arabidopsis cytochrome P450 CYP707A encodes ABA 8'-hydroxylases: Key enzymes in ABA catabolism. *The Embo Journal*, 23(7), 1647–1656. doi:10.1038/sj.emboj.7600121
Kuzina, V., Ramírez-Medina, H., Visser, H., Van Ooyen, A. J. J., Cerdá-Olmedo, E., & Van den Berg, J. A. (2008). Genes involved in carotene synthesis and mating in Blakeslea trispora. *Current Genetics*, 54(3), 143–152.
Lado, J., Alós, E., Manzi, M., Cronje, P., Gómez-Cadenas, A., Rodrigo, M. J., & Zacarías, L. (2019). Light regulation of carotenoid biosynthesis in the peel of mandarin and sweet orange fruits. *Frontiers in Plant Science*, 10, 1288. doi:10.3389/fpls.2019.01288
Lado, J., Gambetta, G., & Zacarias, L. (2018). Key determinants of citrus fruit quality: Metabolites and main changes during maturation. *Scientia Horticulturae*, 233, 238–248. doi:10.1016/j.scienta.2018.01.055
Lauro, G., & Francis, J. (2000). *Natural food colorants: Science and technology*. New York, NY: CRC Press.
Lian, F., Smith, D. E., Ernst, H., Russell, R. M., & Wang, X. D. (2007). Apo-10′-lycopenoic acid inhibits lung cancer cell growth *in vitro*, and suppresses lung tumorigenesis in the A/J mouse model *in vivo*. *Carcinogenesis*, 28(7), 1567–1574. doi:10.1093/carcin/bgm076
Lian, F., & Wang, X. D. (2008). Enzymatic metabolites of lycopene induce Nrf2-mediated expression of phase II detoxifying/antioxidant enzymes in human bronchial epithelial cells. *International Journal of Cancer*, 123(6), 1262–1268. doi:10.1002/ijc.23696
Liang, M. H., Zhu, J., & Jiang, J. G. (2018). Carotenoids biosynthesis and cleavage related genes from bacteria to plants. *Critical Reviews in Food Science and Nutrition*, 58(14), 2314–2333.
Liu, J. R., Chen, B. Q., Yang, B. F., Dong, H. W., Sun, C. H., Wang, Q., ... Song, Y. Q. (2004). Apoptosis of human gastric adenocarcinoma cells induced by beta-ionone. *World Journal of Gastroenterology*, 10(3), 348–351. doi:10.3748/wjg.v10.i3.348
Liu, G. Y., Essex, A., Buchanan, J. T., Datta, V., Hoffman, H. M., Bastian, J. F., ... Nizet, V. (2005). Staphylococcus aureus golden pigment impairs neutrophil killing and promotes virulence through its antioxidant activity. *The Journal of Experimental Medicine*, 202(2), 209–215. doi:10.1084/jem.20050846
Liu, G. Y., & Nizet, V. (2009). Color me bad: Microbial pigments as virulence factors. *Trends in Microbiology*, 17(9), 406–413. doi:10.1016/j.tim.2009.06.006
López, J., Essus, K., Kim, I. K., Pereira, R., Herzog, J., Siewers, V., ... Agosin, E. (2015). Production of β-ionone by combined expression of carotenogenic and plant CCD1 genes in *Saccharomyces Cerevisiae*. *Microbial Cell Factories*, 14, 84. doi:10.1186/s12934-015-0273-x
Lou, S., Wang, L., He, L., Wang, Z., Wang, G., & Lin, X. (2016). Production of crocetin in transgenic *Chlorella vulgaris* expressing genes crtRB and ZCD1. *Journal of Applied Phycology*, 28(3), 1657–1665. doi:10.1007/s10811-015-0730-2
Lounder, D. T., Khandelwal, P., Dandoy, C. E., Jodele, S., Grimley, M. S., Wallace, G., ... Davies, S. M. (2017). Lower levels of vitamin A are associated with increased gastrointestinal graft-versus-host disease in children. *Blood*, 129(20), 2801–2807. doi:10.1182/blood-2017-02-765826
Ma, Y., Szostkiewicz, I., Korte, A., Moes, D., Yang, Y., Christmann, A., & Grill, E. (2009). Regulators of PP2C phosphatase activity function as abscisic acid sensors. *Science*, 324, 1064–1068. doi:10.1126/science.1172408
Ma, G., Zhang, L., Matsuta, A., Matsutani, K., Yamawaki, K., Yahata, M., ... Kato, M. (2013). Enzymatic formation of β-citraurin from β-cryptoxanthin and Zeaxanthin by carotenoid cleavage dioxygenase4 in the flavedo of citrus fruit. *Plant Physiology*, 163(2), 682–695. doi:10.1104/pp.113.223297
Machuka, J., Bashiardes, S., Ruben, E., Spooner, K., Cuming, A., Knight, C., & Cove, D. (1999). Sequence analysis of expressed sequence tags from an ABA-treated cDNA library identifies stress response genes in the moss *Physcomitrella patens*. *Plant Cell Physiol*, 40(4), 378–387.
Marasco, E. K., & Schmidt-Dannert, C. (2008). Identification of bacterial carotenoid cleavage dioxygenase homologues that cleave the interphenyl alpha,beta double bond of stilbene derivatives via a monooxygenase reaction. *ChemBioChem: A European Journal of Chemical Biology*, 9(9), 1450–1461. doi:10.1002/cbic.200700724

Marella, H. H., Sakata, Y., & Quatrano, R. S. (2006). Characterization and functional analysis of ABSCISIC ACID INSENSITIVE3-like genes from *Physcomitrella patens*. *Plant Journal*, 46(6), 1032–1044.

Martínez-Cámara, S., Rubio, S., Del Río, H., Rodríguez-Sáiz, M., & Barredo, J. L. (2018). Lycopene production by mated fermentation of *Blakeslea trispora*. In: C. Barreiro & J. L. Barredo (Eds)., *Microbial carotenoids: Methods and protocols* (pp. 257–268). New York, NY: Springer Science + Business Media.

Medina, H. R., Cerdá-Olmedo, E., & Al-Babili, S. (2011). Cleavage oxygenases for the biosynthesis of trisporoids and other apocarotenoids in Phycomyces. *Molecular Microbiology*, 82(1), 199–208.

Mein, J. R., Lian, F., & Wang, X. D. (2008). Biological activity of lycopene metabolites: Implications for cancer prevention. *Nutrition Reviews*, 66(12), 667–683. doi:10.1111/j.1753-4887.2008.00120.x

Mercadante, A. Z., Steck, A., & Pfander, H. (1997). Isolation and identification of new apocarotenoids from annatto (*Bixa orellana*) seeds. *Journal of Agricultural and Food Chemistry*, 45, 1050–1054.

Mi, J., Jia, K. P., Balakrishna, A., Wang, J. Y., & Al-Babili, S. (2019). An LC-MS profiling method reveals a route for apocarotene glycosylation and shows its induction by high light stress in Arabidopsis. *Analyst*, 144(4), 1197–1204. doi:10.1039/C8AN02143K

Miret, J. A., Munné-Bosch, S., & Dijkwel, P. P. (2018). ABA signalling manipulation suppresses senescence of a leafy vegetable stored at room temperature. *Plant Biotechnology Journal*, 16(2), 530–544. doi:10.1111/pbi.12793

Mishra, N. N., Liu, G. Y., Yeaman, M. R., Nast, C. C., Proctor, R. A., McKinnell, J., & Bayer, A. S. (2011). Carotenoid-related alteration of cell membrane fluidity impacts Staphylococcus aureus susceptibility to host defense peptides. *Antimicrobial Agents and Chemotherapy*, 55(2), 526–531. doi:10.1128/AAC.00680-10

Moran, N. A., & Jarvik, T. (2010). Lateral transfer of genes from fungi underlies carotenoid production in aphids. *Science*, 328(5978), 624–627. doi:10.1126/science.1187113

Moratalla-López, N., Bagur, M. J., Lorenzo, C., Salinas, M., & Alonso, G. L. (2019). Bioactivity and bioavailability of the major metabolites of *Crocus sativus* L. Flower. *Molecules*, 24(15), 2827. doi:10.3390/molecules24152827

Mousavi, S. H., Tavakkol-Afshari, J., Brook, A., & Jafari-Anarkooli, I. (2009). Role of capases and Bax protein in saffron-induced apoptosis in MCF-7 cells. *Food and Chemical Toxicology*, 47, 1909–1913.

Narayanasamy, S., Sun, J., Pavlovicz, R. E., Eroglu, A., Rush, C. E., Sunkel, B. D., … Curley Jr., R. W. (2017). Synthesis of apo-13- and apo-15-lycopenoids, cleavage products of lycopene that are retinoic acid antagonists. *Journal of Lipid Research*, 58(5), 1021–1029. doi:10.1194/jlr.D073148

Nelis, H. J., & De Leenheer, A. P. (1991). Microbial sources of carotenoid pigments used in foods and feeds. *Journal of Applied Bacteriology*, 70(3), 181–191.

Nieuwenhuis, M., & Van Den Ende, H. (1975). Sex specificity of hormone synthesis in Mucor mucedo. *Archives of Microbiology*, 102(1), 167–169.

O'Byrne, S. M., Wongsiriroj, N., Libien, J., Vogel, S., Goldberg, I. J., Baehr, W., … Blaner, W. S. (2005). Retinoid absorption and storage is impaired in mice lacking lecithin:retinol acyltransferase (LRAT). *The Journal of Biological Chemistry*, 280(42), 35647–35657. doi:10.1074/jbc.M507924200

Olson, J. A., & Hayaishi, O. (1965). The enzymatic cleavage of beta-carotene into vitamin A by soluble enzymes of rat liver and intestine. *Proceedings of the National Academy of Sciences of the United States of America*, 54(5), 1364–1370. doi:10.1073/pnas.54.5.1364

Park, S. Y., Fung, P., Nishimura, N., Jensen, D. R., Fujii, H., Zhao, Y., … Cutler, S. R. (2009). Abscisic acid inhibits type 2C protein phosphatases via the PYR/PYL family of START proteins. *Science*, 324(5930), 1068–1071. doi:10.1126/science.1173041

Polaino, S., Herrador, M. M., Cerdá-Olmedo, E., & Barrero, A. F. (2010). Splitting of β-carotene in the sexual interaction of Phycomyces. *Organic & Biomolecular Chemistry*, 8(19), 4229–4231.

Prado, M. M., Prado-Cabrero, A., Fernández-Martín, R., & Avalos, J. (2004). A gene of the opsin family in the carotenoid gene cluster of Fusarium fujikuroi. *Current Genetics*, 46(1), 47–58.

Prado-Cabrero, A., Estrada, A. F., Al-Babili, S., & Avalos, J. (2007a). Identification and biochemical characterization of a novel carotenoid oxygenase: Elucidation of the cleavage step in the Fusarium carotenoid pathway. *Molecular Microbiology*, 64(2), 448–460.

Prado-Cabrero, A., Scherzinger, D., Avalos, J., & Al-Babili, S. (2007b). Retinal biosynthesis in fungi: Characterization of the carotenoid oxygenase CarX from Fusarium fujikuroi. *Eukaryotic Cell*, 6(4), 650.

Raddatz-Mota, D., Pérez-Flores, L. J., Carrari, F., Mendoza-Espinoza, J. A., de León-Sánchez, F. D., Pinzón-López, L. L., … Rivera-Cabrera, F. (2017). Achiote (*Bixa orellana* L.): A natural source of pigment and vitamin E. *Journal of Food Science and Technology*, 54(6), 1729–1741. doi:10.1007/s13197-017-2579-7

Raghavan, S., Hansen, J., Sonkar, S., Kumar, S., Kumar, K. K., Panchapagesa, M., … Hansen, K. R. (2013). *Methods and materials for recombinant production of saffron compounds*. WO/2013/021261 [A3]. Geneva: World Intellectual Property Organization.

Rao, S., & Modi, V. V. (1977). Carotenogenesis: Possible mechanism of action of trisporic acid in Blakeslea trispora. *Experientia*, 33(1), 31–33.
Riebeling, C., Forsea, A. M., Raisova, M., Orfanos, C. E., & Geilen, C. C. (2002). The bisphosphonate pamidronate induces apoptosis in human melanoma cells *in vitro*. *British Journal of Cancer*, 87(3), 366–371. doi:10.1038/sj.bjc.6600476
Rivera-Madrid, R., Aguilar-Espinosa, M., Cárdenas-Conejo, Y., & Garza-Caligaris, L. E. (2016). Carotenoid derivates in achiote (*Bixa orellana* L.) seeds: Synthesis and health promoting properties. *Frontiers in Plant Science*, 7, 1406. doi:10.3389/fpls.2016.01406
Rodrigo, M. J., Alquézar, B., Alós, E., Medina, V., Carmona, L., Bruno, M., ... Zacarías, L. (2013). A novel carotenoid cleavage activity involved in the biosynthesis of Citrus fruit-specific apocarotenoid pigments. *Journal of Experimental Botany*, 64(14), 4461–4478. doi:10.1093/jxb/ert260
Rodríguez-Ávila, N. L., Narváez-Zapata, J. A., Ramírez-Benítez, J. E., Aguilar-Espinosa, M. L., & Rivera-Madrid, R. (2011). Identification and expression pattern of a new carotenoid cleavage dioxygenase gene member from *Bixa orellana* L. . *Journal of Experimental Botany*, 62(15), 5385–5395. doi:10.1093/jxb/err201
Rodríguez-Sáiz, M., De la Fuente, J. L., & Barredo, J. L. (2010). Xanthophyllomyces dendrorhous for the industrial production of astaxanthin. *Applied Microbiology and Biotechnology*, 88(3), 645–658.
Rubio, A., Rambla, J. L., Santaella, M., Gómez, M. D., Orzaez, D., Granell, A., & Gómez-Gómez, L. (2008). Cytosolic and plastoglobule-targeted carotenoid dioxygenases from Crocus sativus are both involved in beta-ionone release. *The Journal of Biological Chemistry*, 283(36), 24816–24825. doi:10.1074/jbc.M804000200
Ruch, S., Beyer, P., Ernst, H., & Al-Babili, S. (2005). Retinal biosynthesis in Eubacteria: *In vitro* characterization of a novel carotenoid oxygenase from *Synechocystis* sp. PCC 6803. *Molecular Microbiology*, 55(4), 1015–1024.
Saelices, L., Youssar, L., Holdermann, I., Al-Babili, S., & Avalos, J. (2007). Identification of the gene responsible for torulene cleavage in the Neurospora carotenoid pathway. *Molecular Genetics and Genomics*, 278(5), 527–537.
Sah, S. K., Reddy, K. R., & Li, J. (2016). Abscisic acid and abiotic stress tolerance in crop plants. *Frontiers in Plant Science*, 7, 571. doi:10.3389/fpls.2016.00571
Saito, S., Hirai, N., Matsumoto, C., Ohigashi, H., Ohta, D., Sakata, K., & Mizutani, M. (2004). Arabidopsis CYP707As encode (+)– abscisic acid 8′-hydroxylase, a key enzyme in the oxidative catabolism of abscisic acid. *Plant Physiol*, 134, 1439–1449. doi:10.1104/pp.103.037614
Salinas, M. Y., Rubio, H. D., & Díaz, V. A. (2005). Extracción y uso de pigmentos del grano de maíz (*Zea mays*) como colorantes en yogur. *Archivos Latinoamericanos De Nutricion*, 55, 293–298.
Sathasivam, R., & Ki, J. S. (2018). A review of the biological activities of microalgal carotenoids and their potential use in healthcare and cosmetic industries. *Marine Drugs*, 16(1), 26. doi:10.3390/md16010026
Schimek, C., & Wöstemeyer, J. (2009). Carotene derivatives in sexual communication of zygomycete fungi. *Phytochemistry*, 70(15), 1867–1875.
Schmidhauser, T. J., Lauter, F. R., Russo, V. E., & Yanofsky, C. (1990). Cloning, sequence, and photoregulation of al-1, a carotenoid biosynthetic gene of Neurospora crassa. *Molecular and Cellular Biology*, 10(10), 5064–5070. doi:10.1128/mcb.10.10.5064
Schmidt, A. D., Heinekamp, T., Matuschek, M., Liebmann, B., Bollschweiler, C., & Brakhage, A. A. (2005). Analysis of mating-dependent transcription of *Blakeslea trispora* carotenoid biosynthesis genes carB and carRA by quantitative real-time PCR. *Applied Microbiology and Biotechnology*, 67(4), 549–555.
Schwartz, S. H., Qin, X., & Loewen, M. C. (2004). The biochemical characterization of two carotenoid cleavage enzymes from Arabidopsis indicates that a carotenoid-derived compound inhibits lateral branching. *Journal of Biological Chemistry*, 279, 46940–46945.
Schwartz, S. H., Tan, B. C., Gage, D. A., Zeevaart, J. A. D., & McCarty, D. R. (1997). Specific oxidative cleavage of carotenoids by VP14 of maize. *Science*, 276(5320), 1872–1874.
Shete, V., & Quadro, L. (2013). Mammalian metabolism of β-carotene: Gaps in knowledge. *Nutrients*, 5(12), 4849–4868. doi:10.3390/nu5124849
Shinozawa, A., Otake, R., Takezawa, D., Umezawa, T., Komatsu, K., Tanaka, K., ... Sakata, Y. (2019). SnRK2 protein kinases represent an ancient system in plants for adaptation to a terrestrial environment. *Communications Biology*, 2, 30. doi:10.1038/s42003-019-0281-1
Shumbe, L., D'Alessandro, S., Shao, N., Chevalier, A., Ksas, B., Bock, R., & Havaux, M. (2017). Methylene blue sensitivity 1 (MBS1) is required for acclimation of Arabidopsis to singlet oxygen and acts downstream of β-cyclocitral. *Plant Cell Environ*, 40(2), 216–226. doi:10.1111/pce.12856
Sicilia, T., Bub, A., Rechkemmer, G., Kraemer, K., Hoppe, P. P., & Kulling, S. E. (2005). Novel lycopene metabolites are detectable in plasma of preruminant calves after lycopene supplementation. *The Journal of Nutrition*, 135, 2616–2621.

Spiegler, E., Kim, Y. K., Hoyos, B., Narayanasamy, S., Jiang, H., Savio, N., ... Quadro, L. (2018). β-apo-10'-carotenoids support normal embryonic development during vitamin A deficiency. *Scientific Reports*, 8(1), 8834. doi:10.1038/s41598-018-27071-3

Spudich, J. L. (2006). The multitalented microbial sensory rhodopsins. *Trends in Microbiology*, 14(11), 480–487.

Strobel, I., Breitenbach, J., Scheckhuber, C., Osiewacz, H., & Sandmann, G. (2009). Carotenoids and carotenogenic genes in Podospora anserina: Engineering of the carotenoid composition extends the life span of the mycelium. *Current Genetics*, 55(2), 175–184.

Sun, J., Li, H., & Yuan, Q. (2012). Metabolic regulation of trisporic acid on *Blakeslea trispora* revealed by a GC-MS-based metabolomic approach. *PLos One*, 7(9), e46110.

Sun, J., Narayanasamy, S., Curley, Jr., R. W., & Harrison, E. H. (2014). Beta-apo-13-carotenone regulates retinoid X receptor transcriptional activity through tetramerization of the receptor. *Journal Biology Chemistry*, 289, 33118–33124.

Sussmilch, F. C., Brodribb, T. J., & McAdam, S. A. (2017). Up-regulation of NCED3 and ABA biosynthesis occur within minutes of a decrease in leaf turgor but AHK1 is not required. *Journal of Experimental Botany*, 68(11), 2913–2918. doi:10.1093/jxb/erx124

Sutter, R. P., Capage, D. A., Harrison, T. L., & Keen, W. A. (1973). Trisporic acid biosynthesis in separate plus and minus cultures of *Blakeslea trispora*: Identification by Mucor assay of two mating-type-specific components. *The Journal of Bacteriology*, 114(3), 1074–1082.

Tan, B. C., Joseph, L. M., Deng, W. T., Liu, L., Li, Q. B., Cline, K., & McCarty, D. R. (2003). Molecular characterization of the Arabidopsis 9-cis epoxycarotenoid dioxygenase gene family. *The Plant Journal*, 35, 44–56.

Tanumihardjo, S. A., Russell, R. M., Stephensen, C. B., Gannon, B. M., Craft, N. E., Haskell, M. J., ... Raiten, D. J. (2016). Biomarkers of nutrition for development (BOND) – Vitamin A review. *The Journal of Nutrition*, 146(9), 1816S–1848S. doi:10.3945/jn.115.229708

Tatman, D., & Mo, H. (2002). Volatile isoprenoid constituents of fruits, vegetables and herbs cumulatively suppress the proliferation of murine B16 melanoma and human HL-60 leukemia cells. *Cancer Letters*, 175, 129–139.

Teixeira da Silva, J. A., Zeng, S., Godoy-Hernández, G., Rivera-Madrid, R., & Dobránszki, J. (2018). *Bixa orellana* L. (achiote) tissue culture: A review. *In Vitro Cellular & Developmental Biology – Plant*, 55, 231–241. doi:10.1007/s11627-019-09969-3

Thewes, S., Prado-Cabrero, A., Prado, M. M., Tudzynski, B., & Avalos, J. (2005). Characterization of a gene in the car cluster of Fusarium fujikuroi that codes for a protein of the carotenoid oxygenase family. *Molecular Genetics and Genomics*, 274(3), 217–228.

Timoneda, J., Rodríguez-Fernández, L., Zaragozá, R., Marín, M. P., Cabezuelo, M. T., Torres, L., ... Barber, T. (2018). Vitamin A deficiency and the lung. *Nutrients*, 10(9), 1132. doi:10.3390/nu10091132

Tsin, A., Betts-Obregon, B., & Grigsby, J. (2018). Visual cycle proteins: Structure, function, and roles in human retinal disease. *Journal of Biological Chemistry*, 293(34), 13016–13021.

Ueda, H., & Kusaba, M. (2015). Strigolactone regulates leaf senescence in concert with ethylene in Arabidopsis. *Plant Physiology*, 169(1), 138–147. doi:10.1104/pp.15.00325

Vallabhaneni, R., Bradbury, L. M., & Wurtzel, E. T. (2010). The carotenoid dioxygenase gene family in maize, sorghum, and rice. *Archives of Biochemistry and Biophysics*, 504(1), 104–111. doi:10.1016/j.abb.2010.07.019

Van Den Ende, H. (1967). Sexual factor of the mucorales. *Nature*, 215(5097), 211–212.

Van Rongen, M., Bennett, T., Ticchiarelli, F., & Leyser, O. (2019). Connective auxin transport contributes to strigolactone-mediated shoot branching control independent of the transcription factor BRC1. *PLoS Genetics*, 15(3), e1008023. doi:10.1371/journal.pgen.1008023

Vedavathy, S. (2003). Annatto and its potential in international market. *Natural Product Radiance*, 2(2), 72–75.

Vilar, D., Vilar, M. S., de Lima e Moura, T. F., Raffin, F. N., de Oliveira, M. R., Franco, C. F., ... Barbosa-Filho, J. M. (2014). Traditional uses, chemical constituents, and biological activities of *Bixa orellana L.*: A review. *The Scientific World Journal*, 2014, 857292. doi:10.1155/2014/857292

Vogel, J. T., Tan, B. C., McCarty, D. R., & Klee, H. J. (2008). The carotenoid cleavage dioxygenase 1 enzyme has broad substrate specificity, cleaving multiple carotenoids at two different bond positions. *Journal of Biological Chemistry*, 283, 11364–11373.

Wang, J. Y., Haider, I., Jamil, M., Fiorilli, V., Saito, Y., Mi, J., ... Al-Babili, S. (2019). The apocarotenoid metabolite zaxinone regulates growth and strigolactone biosynthesis in rice. *Nature Communications*, 10(1), 810. doi:10.1038/s41467-019-08461-1

Wang, X., Kuang, T., & He, Y. (2010). Conservation between higher plants and the moss Physcomitrella patens in response to the phytohormone abscisic acid: A proteomics analysis. *BMC Plant Biol*, 10, 192. doi:10.1186/1471-2229-10-192

Waschuk, S. A., Bezerra, A. G., Shi, L., & Brown, L. S. (2005). Leptosphaeria rhodopsin: Bacteriorhodopsin-like proton pump from a eukaryote. *Proceedings of the National Academy of Sciences of the United States of America*, 102(19), 6879–6883.

Wieland, B., Feil, C., Gloria-Maercker, E., Thumm, G., Lechner, M., Bravo, J. M., ... Götz, F. (1994). Genetic and biochemical analyses of the biosynthesis of the yellow carotenoid 4,4'-diaponeurosporene of Staphylococcus aureus. *Journal of Bacteriology*, 176(24), 7719–7726. doi:10.1128/jb.176.24.7719-7726.1994

Wirth, J. P., Petry, N., Tanumihardjo, S. A., Rogers, L. M., McLean, E., Greig, A., ... Rohner, F. (2017). Vitamin A supplementation programs and country-level evidence of vitamin A deficiency. *Nutrients*, 9(3), 190. doi:10.3390/nu9030190

Wiysonge, C. S., Ndze, V. N., Kongnyuy, E. J., & Shey, M. S. (2017). Vitamin A supplements for reducing mother-to-child HIV transmission. *The Cochrane Database of Systematic Reviews*, 9(9), CD003648. doi:10.1002/14651858.CD003648.pub4

Wolf, L., Rizzini, L., Stracke, R., Ulm, R., & Rensing, S. A. (2010). The molecular and physiological responses of Physcomitrella patens to ultraviolet-B radiation. *Plant Physiology*, 153(3), 1123–1134. doi:10.1104/pp.110.154658

Xue-Xuan, X., Hong-Bo, S., Yuan-Yuan, M., Gang, X., Jun-Na, S., Dong-Gang, G., & Cheng-Jiang, R. (2010). Biotechnological implications from abscisic acid (ABA) roles in cold stress and leaf senescence as an important signal for improving plant sustainable survival under abiotic-stressed conditions. *Critical Reviews in Biotechnology*, 30(3), 222–230.

Yamaguchi-Shinozaki, K., & Shinozaki, K. (2006). Transcriptional regulatory networks in cellular responses and tolerance to dehydration and cold stresses. *Annual Review of Plant Biology*, 57, 781–803. doi:10.1146/annurev.arplant.57.032905.105444

Yan, W., Jang, G. F., Haeseleer, F., Esumi, N., Chang, J., Kerrigan, M., ... Zack, D. J. (2001). Cloning and characterization of a human β,β-carotene-15, 15′-dioxygenase that is highly expressed in the retinal pigment epithelium. *Genomics*, 72(2), 193–202.

Yao, D., Wang, Y., Li, Q., Ouyang, X., Li, Y., Wang, C., ... Xiao, Y. (2018). Specific upregulation of a cotton phytoene synthase gene produces golden cottonseeds with enhanced provitamin A. *Scientific Reports*, 8(1), 1348. doi:10.1038/s41598-018-19866-1

Zalokar, M. (1957). Isolation of an acidic pigment in Neurospora. *Archives of Biochemistry and Biophysics*, 70(2), 568–571.

Zhang, X., Liao, S., Cao, F., Zhao, L., Pei, J., & Tang, F. (2018). Cloning and characterization of enoate reductase with high β-ionone to dihydro-β-ionone bioconversion productivity. *BMC Biotechnology*, 18(1), 26. doi:10.1186/s12896-018-0438-x

Zhao, M., Li, Q., Chen, Z., Lv, Q., Bao, F., Wang, X., & He, Y. (2018). Regulatory mechanism of ABA and ABI3 on vegetative development in the moss *Physcomitrella patens*. *International Journal of Molecular Sciences*, 19(9), 2728. doi:10.3390/ijms19092728

Zhong, M., Kawaguchi, R., Kassai, M., & Sun, H. (2012). Retina, retinol, retinal and the natural history of vitamin A as a light sensor. *Nutrients*, 4(12), 2069–2096. doi:10.3390/nu4122069

Ziouzenkova, O., Orashanu, G., Sukhova, G., Lau, E., Berger, J. P., Tang, G., ... Plutzky, J. (2007). Asymmetric cleavage of {beta}-carotene yields a transcriptional repressor of RXR and PPAR responses. *Molecular Endocrinology*, 21(1), 77–88.

3 Apocarotenoid Molecules Are Continuously Being Discovered in All Taxa

Renata Rivera-Madrid and Siva Ramamoorthy

CONTENTS

INTRODUCTION

Carotenoids belong to the isoprenoid family and are found in nature as numerous coloured compounds. A structural characteristic of these molecules is a conjugated double bond (Auldridge et al., 2006b). In plants, carotenoids share a common C_{40}-skeleton, but differ in the number of their conjugated double bonds, stereo-configuration, oxygen content, and end-group structures. During biosynthesis, they suffer various chemical changes by desaturation, cycling or oxidation processes, or are otherwise modified by downstream enzymes (Cazzonelli and Pogson, 2010), giving rise to different kinds of carotenoids that are not the end products of their pathway. A broad diversity of carotenoid derivatives produced by metabolic processes are known as apocarotenoid (Walter et al., 2010; Walter and Strack, 2011) (Figure 3.1). These molecules are essential for the primary and secondary metabolism of living organisms, and contribute to the health of the animal kingdom, including humans. These compounds are generally catalyzed by one enzyme family known as carotenoid cleavage oxygenases (CCOs); Dioxygenases enzymes are usually found in plants, and monooxygenases in mammals (Walter and Strack, 2011; Serra, 2015). Carotenoid derivates can also be generated spontaneously through other mechanisms such as oxidative stress, photooxidation or other unidentified mechanisms acting on a specific double bond in a carotenoid precursor (Schwartz et al., 1997; Wang et al., 2019). Interestingly, apocarotenoids themselves act as substrates for several CCOs and are frequently modified by different enzymes including cytochrome P450 enzymes (Wang et al., 2019). The broad diversity of enzymatic cleavages carried out by CCOs and the mechanisms mentioned above generate a myriad of apocarotenoid compounds in nature. Carotenoids can be cleaved by CCOs in different sites according to the enzyme type. The resulting molecules are

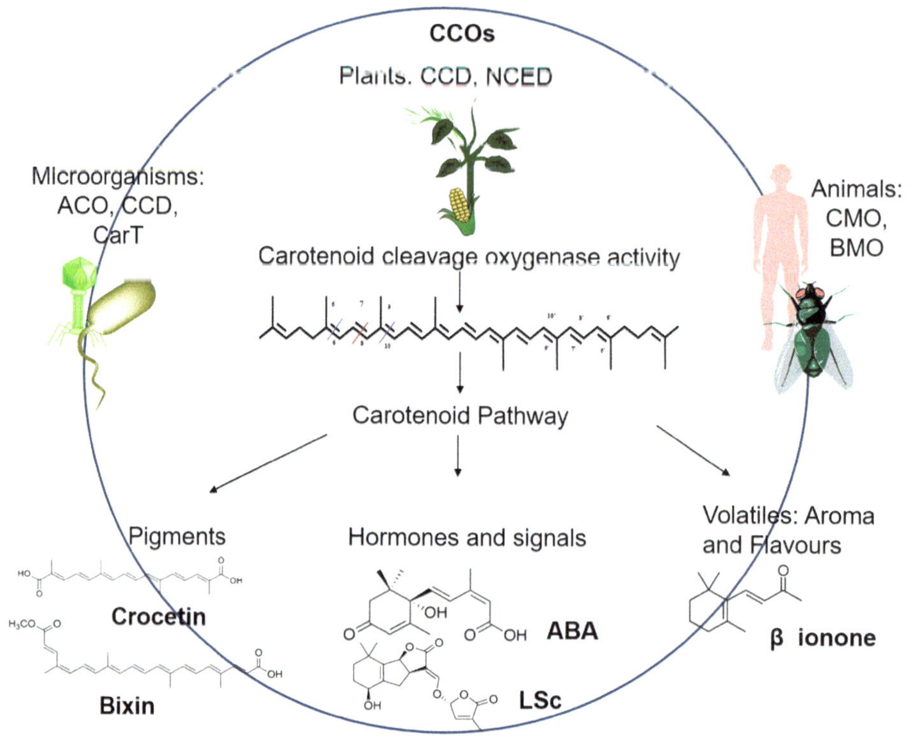

FIGURE 3.1 CCOs are classified by taxa and named according to the organism of origin: thus, CCOs from microorganisms have been called ACO, CCD, NSC CarT, CarX, CarS, TSP; those found in plants have been named CCDs, NCEDs, ZASs; animal CCOs have received the names of CMO and BMO. These enzymes produce different types of apocarotenoids, such as pigments, hormones, signals, aromas and flavours.

classified as pigments, hormones, signals, volatiles, and aromas (Giuliano et al., 2003; Auldridge et al., 2006b; Nisar et al., 2015) (Figure 3.1).

APOCAROTENOID PIGMENTS

Carotenoids comprise a broad group of mostly coloured molecules on which the oxidative cleavage of CCOs act to produce numerous derivatives known as apocarotenoids. Some of these apocarotenoid have colour, such as bixin, a red-orange monomethylester of the C_{24} dicarboxylic acid norbixin that is accumulated mainly in the seeds of *Bixa orellana* (Rivera-Madrid et al., 2016) and has commercial value. This pigment is industrially used to colour food, pharmacy and makeup products, among others. Bixin biosynthesis was first studied in *Escherichia coli* heterologous system (Bouvier et al., 2003a). However, researchers have recently proposed that more genes are involved in an alternative biosynthesis pathway, suggesting its synthesis is far more complex (Rodríguez-Ávila et al., 2011; Cárdenas-Conejo et al., 2015; Rivera-Madrid et al., 2016; Carballo-Uicab et al., 2019).

Another commercially important compound is *crocin*, found in the stigma of saffron (*Crocus sativus*). Crocin is a C_{20} polyene derivative generated by the oxidative cleavage of β-carotene and zeaxanthin (Rubio et al., 2008; Ahrazem et al., 2018). It is used as a spice to flavour and colour food products. In addition, this compound has been demonstrated to protect humans from some diseases, including heart and blood disorders, cancer, and renal ischaemia (Abdullaev, 2002; Ochiai et al., 2007; Ahrazem et al., 2010). Thus, it has attracted the attention of several research groups that have elucidated the mechanisms of its biosynthesis pathway (Bouvier et al., 2003b; Rubio et al., 2008; Ahrazem et al., 2010; Frusciante et al., 2014; Demurtas et al., 2019). There are other less known, but no less significant, apocarotenoid pigments such as those from the roots of *Escobedia scabrifolia*, an endemic plant in tropical America that accumulates azafrin in its root, a yellow-orange C_{27} carotenoic acid exhibiting a diol substitution at the cyclohexane ring. Its yellow pigment is used to colour textiles and it also has similar applications as crocin from saffron. Additionally, azafrin has medicinal use for treating diseases such as jaundice, hepatitis and hyperlipidemia, among others (Muriel Ruiz et al., 2015).

Other interesting pigments, such as red β-citraurin and orange β-citraurinene, are found in citrus peel. Both C_{30} apocarotenoids are rarely found in other plants (Curl, 1965; Leuenberger and Stewart, 1976; Alquézar et al., 2008; Zheng et al., 2019; Luan et al., 2020). β-citraurin is generated by oxidative cleavage of two common plant carotenoids – β-cryptoxanthin and zeaxanthin – catalyzed by dioxygenase 4 (CCD4), as reported by recent studies in mandarins and clementines (Alquézar et al., 2008; Ma et al., 2013; Zheng et al., 2019; Luan et al., 2020). The enhanced expression of CCD4b may contribute to the development of red-coloured peels in mandarin. The difference in the chemical structure between β-citraurinene and β-citraurin is a methyl group in the former and an aldehyde group in the latter, at the 80 position (Zheng et al., 2019).

Another well-characterized yellow pigment is mycorradicin, a C_{14} polyenic acid (10,10′-diapocarotene-10,10′-dioic acid) found in the colonized roots of mycorrhizal fungi (Klingner et al., 1995). This pigment has also been found in low concentrations in other plants in the absence of colonized roots (Walter et al., 2007). Mycorradicin is stored as globules in root chromoplasts and its accumulation leads to changes in root plastid morphology (Scannerini and Bonfante-Fasolo, 1977; Fiorilli et al., 2019; see more details in chapter 10).

Retinal and retinoic acid derivatives are visual pigments (vitamin A) or signalling photosensory compounds in animals (Wald, 1968), green algae and Halobacterium (Foster et al., 1984). Retinoids, derived from vitamin A, are found in animals (Giguere et al., 1987) and have important clinical applications. However, this compound has now been found in fungous host that also synthesize retinal and neurosporaxanthin (Bibikov et al., 1993).

Apocarotenoids from cyanobacteria, such as rhodopsin, are photoprotective molecules and accessory pigments in thylakoid membranes. Certain archaea and eubacteria also produce retinal, a chromophore for rhodopsins that functions as a photosensory receptor in its natural environment, suggesting that it transduces a signal from the receptor that may signalize through a soluble cytoplasmic protein, analogous to higher animal visual pigments (Prado-Cabrero et al., 2007; Walter and Strack, 2011; Ahrazem et al., 2016). This topic is discussed more extensively in the Chapter 2, by Dzib-Cauich et al. Some fungi are excellent carotenoid producers of retinal and have been adopted as biotechnological carotenoid sources (Jung et al., 2003; Balashov and Lanyi, 2007; Ahrazem et al., 2016).

Microorganisms have a potential use as hosts to produce apocarotenoid pigments. So far, many microorganisms have been genetically modified by introducing genes from the carotenoid synthesis pathway for its production. Many of the carotenoid and apocarotene enzymes have been characterized. The identification of these compounds has revealed the multiple functions they play in the growth and development of plants, mainly those related to communication between roots and their hosts.

HORMONES AND SIGNALS

Hormones and signalling molecules, which include metabolites derived from carotenoids, mediate the effects of environmental, developmental, stress and genetic factors on living organisms. They act as chemical signal to coordinate growth and developmental processes, overcome environmental changes, and communicate with symbiotic organisms. Hormones are involved in shoot and root architecture, contribute to nutrients availability, and participate in the dehydration processes of symbiotic mycorrhizal (Ahrazem et al., 2016). The most known families of carotenoid derivatives hormones are abscisic acid (ABA), strigolactones (SLs), and zagynones (Walter et al., 2010), regulating plant growth and development, and acting against pathogens (Foster et al., 1984; Ton et al., 2009; Walter and Strack, 2011; Wang et al., 2019).

ABA

The apocarotenoid ABA C_{15} is a phytohormone that has been thoroughly studied. It has an important role in plant response to environmental stress. It is also found in some algae and several phytopathogenic fungi (Zeevaart and Creelman, 1988). ABA is involved in tolerance to desiccation processes and is present in dormant seeds; it also enhances the biosynthesis of LEA proteins (abundant in the late stages of embryogenesis) and soluble carbohydrates (Ma et al., 2013). ABA also induces stomata closure via its biochemical effect on guard cell metabolism. Its significance increased with the evolution of the signalling network, contributing to the increasing complexity of land plants (Zeevaart and Creelman, 1988; Schwartz et al., 1997; Peleg and Blumwald, 2011).

NCED type of CCO enzyme catalyze the first step in ABA biosynthesis by cleaving 9-cis-violaxanthin or 9′-cis-neoxanthin, forming the ABA precursor C_{15} xanthoxin, which is further metabolized to ABA by the short chain dehydrogenase reductase (SDR) and abscisic aldehyde oxidase (AAO) (Brunetti et al., 2019). VP14 was isolated from an ABA-deficient viviparous mutant in maize (Schwartz et al., 1997; Nambara and Marion-Poll, 2005). The derived amino acid sequence of VP14 shows significant sequence similarity lignostilbene dioxygenases (LSDs), a CCO found in *Pseudomonas paucimobilis* lignostilbene dioxygenases (LSDs) (Schwartz et al., 1997). ABA biosynthetic mutants have also been identified in a wide range of plants (Schwartz et al., 1997). ABA seems to play a role in a transduction signal; however, this function needs further study. A Mitogen-activated protein kinase (MAPK) from maize regulates ABA homeostasis and seedlings drought sensitivity (Schwartz et al., 1997; Walter et al., 2010).

Evidence from different host plants indicates that ABA may have a direct role in mycorrhizal root colonization. ABA defective mutants of *Solanum lycopersicum* (tomato) have been found to show lower levels of arbuscular mycorrhizal fungus (AMF) colonization and arbuscule formation and functionality, which may be associated with an increase of ethylene (Zheng et al., 2019; Zhu et al., 2020).

ABA promotes AM (arbuscular mycorrhizal) colonization in *Medicago truncatula*, but only at low concentrations; this positive effect is mediated by the protein phosphatase 2A (PP2A) which is activated during the AM symbiosis and also under ABA treatment (Herrera-Medina et al., 2007; Martín-Rodríguez et al., 2011). Higher colonization and arbuscule level were also found in mycorrhizal roots of *Solanum tuberosum* pre-treated with ABA (Charpentier et al., 2014), suggesting the positive influence of ABA on mycorrhizal formation (Ludwig-Müller, 2010; Mercy et al., 2017; see more details in chapter 9).

Strigolactones

Strigolactones (SLs), a class of terpenoid hormones, were discovered approximately 40 years ago. They were thought to be seed germination stimulants, especially for Striga species (López-Ráez et al., 2010). Although there has been a significant advance in the study of SL, there is still much to be learned (Borghi et al., 2016; Bythell-Douglas et al., 2017).

Strigolactones are also hormones that determine plant architecture, mediate mycorrhization, and facilitate infestation by root parasitic weeds. In addition, SLs take part in leaf senescence and are involved in biotic and abiotic stress (Bythell-Douglas et al., 2017; Machin et al., 2019). SLs are rhizospheric signals released by roots into the soil by the action of specific SL transporters; they facilitate recruitment of AM fungi and initiate the symbiosis function of many flowering plants (Xie et al., 2010; Al-Babili and Bouwmeester, 2015; Decker et al., 2017; Saeed et al., 2017; Wang et al., 2019). This process occurs under conditions of phosphate-deficit in the rhizosphere (Borghi et al., 2016; see Walter's chapter for more details). SLs are used by the seeds of parasites plants such as

Striga spp. which needs a germination signal ensuring host disposal (Bythell-Douglas et al., 2017; Wang et al., 2019). This topic is discussed more extensively in chapter 10.

Plants, parasites and fungi strigolactone receptors differ substantially, so probably not all strigolactones are bioactive in a given system (Bouwmeester et al., 2007; Umehara et al., 2008; Felemban et al., 2019; Wang et al., 2019). Even within plants, there could be major differences in the way diverse strigolactones' signals are perceived. This means that strigolactones can be characterized based on the exact receptor it uses (Machin et al., 2019). However, a better definition could be based on their biosynthesis in plants themselves (Machin et al., 2019). Advances have been made in the understanding of how SLs and other hormone signalling pathways influence plant developmental and environmental responses. The involvement of SLs in regulating the responses to drought, salinity, and nutrient deficiency stress, as well as tolerance to cold has been demonstrated. However, the downstream targets of SLs signalling pathways are not fully understood, and further research is required to identify the components of the SLs signalling pathways and unveil their functions (Jia et al., 2018) (See more details in chapter 11).

SLs have been shown to crosstalk with other hormone signalling pathways, including auxin, gibberellic acid (GA), abscisic acid (ABA), cytokinin (CK), and ethylene (ET), under normal and abiotic stress conditions (Yang et al., 2019; Zhu et al., 2019).

Zaxinone and D'orenone

Zaxinone, a member of the carotenoid-derived signalling molecules family, is a hydroxylated form of β-apo-13-carotenone (Yang et al., 2019; Zhu et al., 2019), that regulates the architecture, growth and development of rice and the biotic interactions with rice metabolites (Wang et al., 2019). It also controls the level of SLs – important in the response to biotic and abiotic stress – and it is a key component of the communication process that takes place between beneficial symbiotic partners and harmful parasitic invaders, such as *Striga*. Further study of zaxinone may help us to better understand the evolution of land plant and the plant-AM fungi symbiosis that occurs in most land plants (Alder et al., 2008; Felemban et al., 2019; Fiorilli et al., 2019; Wang et al., 2019).

Studies on a rice *zas* mutant showed lower root zaxinone content, severely retarded root and shoot growth, and higher SLs levels. In wild-type plants under conditions of low phosphate supply, the exogenous application of zaxinone corrected root defects, promoted root growth and reduced SLs biosynthesis and exudation, underscoring the crucial role of zaxinone in rice development and growth (Felemban et al., 2019; Fiorilli et al., 2019; Wang et al., 2019). However, despite a higher level of SLs, mutant rice (*Oryza sativa,* mutant) displayed a lower level of AM colonization as compared to wild-type plants, although arbuscule morphology was unaltered. Hence, our understanding of the complex communication process between zaxinone and SLs during mycorrhization is quite limited. Since zaxinone controls SLs biosynthesis and root parasitc weeds require SLs released by the host to germinate (Felemban et al., 2019; Fiorilli et al., 2019; Wang et al., 2019), a better understanding of zaxione biology will allow us to use it to control *Striga* and other infestations by root parasitic weeds (Al-Babili and Bouwmeester, 2015), hence decreasing the high loss in crop yield that now occur in warm and temperate zones (Felemban et al., 2019; Fiorilli et al., 2019; Wang et al., 2019).

D'orenone is another apocarotenoid compound which has recently proved to affect ectomycorrhizal formation. The possible mechanism is the modulation of auxin metabolism in both partners (Parker, 2009). Its chemical structure is similar to zaxinone so they possibly have similar functions in the regulation of plant development or in the interaction with microorganisms (Felemban et al., 2019; Fiorilli et al., 2019; Wang et al., 2019).

β-Cyclocitral

β-cyclocitral is a small apocarotenoide molecule produced by the cleaving action of enzymes from CCD1 and CCD4 subgroups on β-carotene (Rodrigo et al., 2013; Dickinson et al., 2019). It can also be formed non-enzymatically by the oxidative action of singlet oxygen ($^1O_2$) (Wagner et al., 2016; Fiorilli et al., 2019). It has been studied in Arabidopsis, rice, and tomato, and proven to regulate the behaviour of root stem cells promoting root stem cell divisions to enhance growth and branching. Hence, it has important uses in agriculture. Under salt stress conditions, β-cyclocitral also improves rice shoot and root growth (Ramel et al., 2012; Dickinson et al., 2019; Felemban et al., 2019; Luan et al., 2020).

Gao et al. (2019) showed that the amount of β-cyclocitral (and other carotenoid-derived volatiles) was proportionally related to the level of expression of LOX 13S-lipoxygenase; TomLocX antisense transgenic tomato had decreased levels of apocarotenoids. The AB part in both enzyme families can be modified through methylation, hydroxylation, epoxidation, or ketolation, giving rise to the diversity of SLs (Fiorilli et al., 2019). Further research on the synthesis and role of β-cyclocytral and other phytohormones which could be playing an important role in crop yield by acting on root development and growth (Al-Babili and Bouwmeester, 2015). In addition, β-cyclocitral confers Arabidopsis resistance to oxidative stress (Dickinson et al., 2019) by acting as another messenger, taking the $^1O_2$ stress signal to the nucleus, thus altering the transcription of $^1O_2$-regulated genes by means of a methylene blue sensitivity protein (D'Alessandro et al., 2018). The resulting transcriptome reprograming process increases plant tolerance to photo-oxidative stress and starts the process of acclimation to high light conditions (Ramel et al., 2012; Shumbe et al., 2014). In such an environment, the content of glycosylated β-cyclocitral increases significantly, which might be a mechanism to deactivate this signalling molecule (D'Alessandro et al., 2018).

Recently, a study underpinned the biological importance of β-cyclocitral by revealing its role as root growth regulator (Ramel et al., 2012; Al-Babili and Bouwmeester, 2015; D'Alessandro et al., 2018). Apocarotenoid signals are required for root branching and lateral roots emergence and probably some function remain unknown.

VOLATILES: AROMA AND FLAVOURS

Numerous flavours and fragrances are possibly the product of the combination of carotenoids chemical structures formed during the different degradation pathways (Dickinson et al., 2019). The type of chain (linear or rings) of the carotene molecule gives flavour (geranyl acetone or pseudoionone) or aroma (cyclohexene derivatives, ionone and damascone isomers) (Serra, 2015). Most of the flavour apocarotenoids are produced naturally in very small quantities so bioconversions of natural precursors using biocatalysis or by fermentation processes, would be more useful (Serra, 2015). The amount and diffusion of a given apocarotenoid in nature depends on the abundance of the carotenoid from which it is derived, and on the chemical stability of the apocarotenoid molecule. For instance, β-ionone, dihydroactinidiolide, and ketoisophorone are the degradation products of different abundant carotenoids and are stable compounds (Brenna et al., 2003). Ionone and damascone isomers are relevant in flavour and fragrance industries. Megastigmatrienone and megastigmadienone isomers are constituents of the aroma of different fruits, rums, and tobacco (Serra, 2015). Actinidiolide has only been found in the essential oil of *Actinidia polygama;* dihydroa ctinidiolide and *cis*-tetrahydroactinidiolide are flavour components in tea, tobacco, wine, apricot, algae, melon and cactaceae (Ito et al., 1997; Serra, 2015). A kind of riesling acetal was identified in Riesling wine and quince brandy. These compounds are widely distributed in plants and algae and trace amount have even been found in ocean water. Using them as potential volatile markers could help determine the origin of sea salt (Serra, 2015).

Apocarotenoids in cyanobacteria can give undesired flavours to water and aquaculture fish. Apocarotenoids include derivatives of β-ionone, which generally have pleasant odours and flavours, are found in many flowers, and are widely used by the industry. Isomeric forms of ionones have notably different odours (Serra, 2015). Plants release volatile organic compounds to interact with their environment, attracting pollinators, discouraging herbivore and defending them from pathogens (Serra, 2015).

During flower blossoming and fruit ripening, carotenoids are broken down to scent substances of less than 15 carbon atoms (Walter et al., 2010).

C_{14} polyene cleavage products are usually undetectable and the metabolic fate of the primary C_{14} dialdehyde cleavage product remains unknown. Derivatives of C_{13} cyclohexene moieties, on the contrary, are frequently found in many flowers and fruits as well as in mycorrhizal roots (Al-Babili and Bouwmeester, 2015; D'Alessandro et al., 2018). They contribute to the scent of roses (Eugster and Märki-Fischer, 1991; Bouvier et al., 2005; Auldridge et al., 2006a; Strack and Fester, 2006; Walter et al., 2007), tomato (Huang et al., 2009) and wine (Mendes-Pinto, 2009). Glycosylation is common in vegetal species that form volatiles, apparently serving as a general storage principle of aromatic compounds (Maicas and Mateo, 2005).

Apocarotenoids Synthesis

Apocarotenoids are produced from carotenoids by the enzyme activity of CCOs that cleave the double bonds of the carotenoid structure (Delaux et al., 2012). Since the nomenclature of oxygenase enzymes is controversial, we use the term CCO in this chapter to includes all types of carotenoid cleavage enzymes. The first CCO was studied in *Microcystis* PPC 7806; β-carotene 7,8-7′,8′ oxygenase, β-cyclocitral, and crocetindial were identified in this organism (Simkin et al., 2004). The first plant CCO was isolated from *Zea mays*, VP14; its characterization helped to better understand the mechanisms of action of these enzymes in plants as well as in animals (Jüttner and Höflacher, 1985). The analysis of maize VP14/NCED1 allowed further advances on cleavage specificity by using modelling and/or site-directed mutagenesis approaches on other plant CCDs (Schwartz et al., 1997; Tan et al., 2003).

CCOs are characterized by the presence of an oxygenase with a non-heme iron character and are evolutionary related (Hou et al., 2016). The protein structure of the enzyme contains a Fe^{+} ion, which is coordinated to four conserved histidine residues in the active site. All CCOs have the 4-histidine–iron arrangement, but their amino acid sequences differ (Walter et al., 2010). These enzymes are monooxygenases or dioxygenases. Animal monooxygenases are called CMOs or BMOs (von Lintig and Vogt, 2000; Redmond et al., 2001; von Lintig, 2010; Walter and Strack, 2011). In plants CCOs are regularly named CCDs. So far, the CCDs family of enzymes have been classified into six subfamilies according to their cleaving position or substrate of preference: CCD1, CCD2, CCD4, CCD7, CCD8, and NECDs (Hessel et al., 2007) (Figure 3.2). A new CCD subfamily has been recently identified in rice, ZAS (Zaxinone Synthase); its enzymatic activity converts 3-OH-all-trans-β-apo-10′-carotenal (apo-10′-zeaxanthinal) into zaxinone (3-OH-all-trans-apo-13-carotenone), a regulatory metabolite required for normal rice growth (Auldridge et al., 2006a; Kloer and Schulz, 2006; Walter and Strack, 2011; Frusciante et al., 2014). Biochemical characterization has established that most CCDs have wide substrate promiscuity but are highly specific for the position of the double bond that they cleave (Wang et al., 2019; Figures 3.2 and 3.3).

Studies to understand the role of CCOs have been based on their expression in *E. coli*. A CCO is purified and incubated with different carotenoid substrates to characterize its enzyme activity and specificity. Some CCO enzymes exhibit high specificity, cleaving at a certain position of the polyene chain, while others are less specific (Auldridge et al., 2006b; Walter et al., 2010; Walter and Strack, 2011). However, further *in vivo* biochemical research is needed to define the specificity of the CCOs.

Bacteria and Fungi CCOs

Microorganisms also have different types of CCO: ACO, NSC, CCD, CarT, CarX, TSP3 (Ahrazem et al., 2016) (Figures 3.1 and 3.3). The CCO in *Novosphingobium aromaticivorans* is know as ACO (NACOX1). It is an apo-carotenoid 13,14-dioxygenase, identified as such because of its substrate specificity, cleavage mechanism, and resulting products. It acts exclusively on carotenoids with one

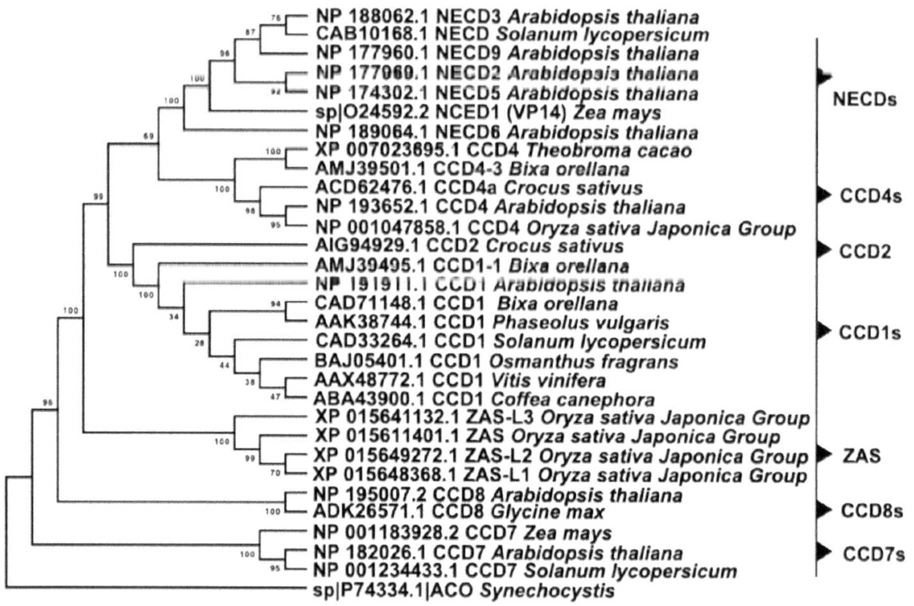

FIGURE 3.2 The phylogenetic tree was inferred using the maximum-likelihood method based on the Jones–Taylor–Thornton (JTT) substitution model and was gamma distributed with Invariant sites (G + I) in MEGA6. Numbers near the branch points represent the bootstrap value produced by 1,000 replications. Phylogenetic trees were rooted with Synechocystis apocarotenoid cleavage oxygenase (ACO) (P74334).

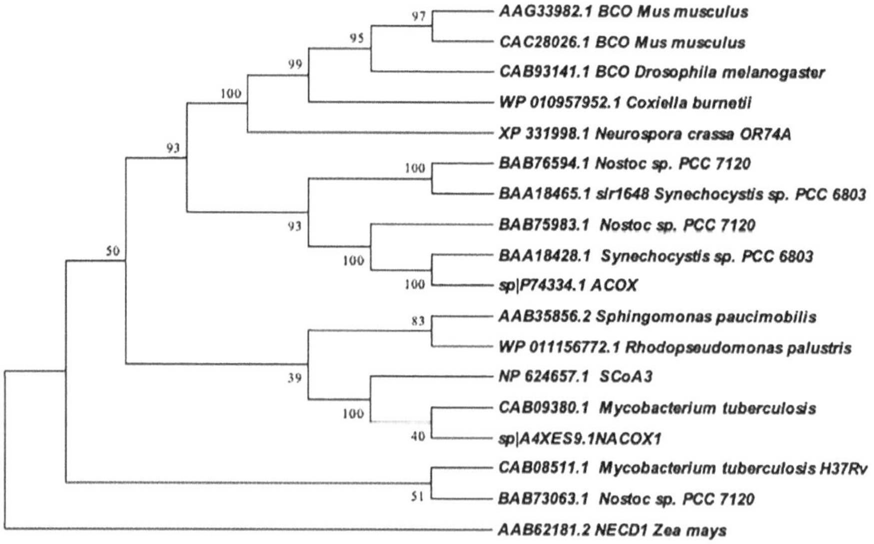

FIGURE 3.3 The phylogenetic tree was inferred using the maximum-likelihood method based on the Jones–Taylor–Thornton (JTT) substitution model and was gamma distributed with Invariant sites (G + I) in MEGA6. Numbers near the branch points represent the bootstrap value produced by 1,000 replications. Phylogenetic trees were rooted with NECD1 *Zea mays*.

β-ionone ring, including β-apo-80-carotenal, β-apo-40-carotenal, and c-carotene, converting β-apo-80-carotenal to β-apo-13-carotenone (Walter et al., 2010; Ahrazem et al., 2016).

In *M. tuberculosis,* MtCCO cleaves β-carotene at the 15,15′ double bond to form C_{20}-retinal, and at the 13,14 double bond to form β-apo-14′-carotenal (C_{22}) and β-apo-13-carotenone (C_{18}) (Kim et al., 2012; Ahrazem et al., 2016).

Lignostilbene-α,β-dioxygenase from *Pseudomonas paucimobilis* is crucial in the metabolic pathway of the Cα-Cβ cleavage of dimeric lignin model compounds, and is significantly homologous to the bacterial and cyanobacterial CCDs (Scherzinger et al., 2010).

In cyanobacteria, apocarotenoid oxygenase (SynACO, or Diox1) from *Synechocystis* sp. converts β-apo-carotenals (instead of β-carotene) into retinal (β-apo-15′-carotenal) at the 15,15′ position (Kamoda and Saburi, 1993; Liang et al., 2018). Retinal was identified as a chromophore of the light-activated photoreceptor rhodopsin in the cyanobacterium Nostoc sp. strain PCC 7120, in which three enzymes have been identified (NSC1, NSC2, and NSC3) (15–17) (Figure 3.3). *In vitro*, NSC1 cleaves C9-C10 and C9′-C10′ in bicyclic carotenoids and C9-C10 and C7-C8 in monocyclic carotenoids. NSC2, cleaves monocyclic or acyclic (linear) carotenoids at C15-C15′ *in vitro*, producing retinal, but *in vivo* its action on β-carotene is minimal. NSC3, does not cleave full-length carotenoids but cleaves β-apo-8β-carotenal at C9-C10 double bonds *in vitro*. 9,10-apocarotenoid cleavage specificity has not been previously described (Ruch et al., 2005; Marasco et al., 2006; Scherzinger et al., 2006; Scherzinger et al., 2010), but the preferred substrates have not yet been identified.

Cyanobacterial CCDs in different *E. coli* carotenoid backgrounds proved to have different bleaching efficiencies. NSC1 and NSC2 expressed in cells producing lycopene, β,β-carotene, zeaxanthin, torulene, or C_{30} dialdehyde yielded less pigment accumulation. NSC1, NSC2, and SYO (CCO from *Synechocystis*) showed distinctive bleaching with a range of linear and cyclic carotenoid substrates. In contrast, NSC3 and SYC1 produced no observable bleaching.

The first CarX (also known as Ccol) retinal-synthesizing enzyme was reported in fungi. In ascomycete *Fusarium fujikuroi*, it cleaves mono- and bi-cyclic carotenoids and apocarotenoids – but not acyclic substrates – at the C15-C15′ bond. Thus, CarX is involved in retinal biosynthesis, but not in the synthesis of the acidic C_{35} apocarotenoid neurosporaxanthin as previously suspected. A different cleavage enzyme, CarT is involved in neurosporaxanthin biosynthesis. It cleaves monocyclic torulene producing β-apo-45-carotenal. In *Neurospora crassa*, two putative CCOs were identified by sequence mining from protein databases; NcCAO2 turned out to catalyze torulene cleavage.

CarX cleaves β-carotene, γ-carotene, torulene, and β-apo-8′-carotenal (at least one β-ionone ring in the substrates) at the C15-C15′ double bond to produce retinal (Scherzinger et al., 2006). CarX from the basidiomycete *Ustilago maydis* catalyzes the symmetrical cleavage of β-carotene, producing two molecules of retinal (Estrada et al., 2009).

In *Neurospora crassa*, two putative CCOs were identified by sequence mining from protein databases, NcCAO1 acts in a similar way as CarX (Saelices et al., 2007), cleaving β-carotene into retinal (Prado-Cabrero et al., 2007). NcCAO2 turned out to catalyze torulene cleavage. A different cleavage enzyme, CarT, is involved in neurosporaxanthin biosynthesis. It cleaves monocyclic torulene producing β-apo-45-carotenal.

In *P. blakesleeanus*, the initial reaction in the apocarotenoid pathway is carried out by TSP3 (also known as CarS), cleaving β-carotene at its 11′-12′ double bond to form β-apo-12′-carotenal (C_{25}), which is then cleaved at the 13,14 double bond by AcaA to produce β-apo-13-carotenone (C_{18}, the precursor for trisporic acids biosynthesis) (Medina et al., 2011; Tagua et al., 2012; Liang et al., 2018).

In *M. circinelloides*, CCDs convert β-carotene into the precursors of three families of apocarotenoids, both *in vitro* and in the *E. coli* heterologous *in vivo* system. CarS, cleaves the C_{40} β-carotene into the C_{15} precursor of cyclofarnesoids and a C_{25} apocarotenal; the later is converted by AcaA, into the C_{18} precursor of trisporoids and the C_7 precursor of methylhexanoids (Alcalde et al., 2019). CCOs have been identified in *Rhizopus oryzae* and *Blakeslea trispora* (TSP3 and TSP4, respectively). *In vitro*, TSP3 is a β-carotene-cleaving enzyme (Liang et al., 2018) (Figure 3.3).

Plant CCDs

In plants, CCO genes and enzymes have been characterized in a growing number of species, including several crops (Simkin et al., 2004a; Ibdah et al., 2006; García-Limones et al., 2008; Huang et al., 2009; Ilg et al., 2010) such as petunia (Simkin et al., 2004), tomato (*Solanum lycopersicum* L (Simkin et al., 2004), saffron (*Crocus sativus*) (Bouvier et al., Camara 2003b; Rubio et al., 2008), grapes (*Vitis vinifera* L.) (Mathieu et al., 2005), melon (*Cucumis melo* L.) (Ibdah et al., 2006), citrus fruit (Kato et al., 2006), rice (Ilg et al., 2010), nectarine (*Prunus persica* L.) (Baldermann et al., 2005), rose (*Rosa damascena Mill*) (Huang et al., 2009), *Medicago truncatula* var. (Floss et al., 2008), and *Bixa orellana* L. (Rodríguez-Ávila et al., 2011; Cárdenas-Conejo et al., 2015; Rivera-Madrid et al., 2016; Carballo-Uicab et al., 2019) (Figure 3.2).

Concerning the NCEDs (9-cis-epoxy-carotenoid dioxigenases) family enzymes in plants, they are involved in the biosynthesis of ABA. NCED1 of *Zea mays* L. was the first biochemically characterized carotenoid dioxygenase enzyme (VP14) and catalyzes the first step for ABA biosynthesis by cutting at the 11,12-11′,12′ positions of the 9-cis-violaxanthin or 9-cis-neoxanthin double bond (Schwartz et al., 1997; Tan et al., 1997). Enzymes CCD1, CCD 2, CCD 4, CCD 7, and CCD 8 cleave different carotenoids at one or both ends and are involved in the production of different metabolites (Simkin et al., 2004; Vogel et al., 2008; Walter and Strack, 2011). Arabidopsis has nine CCO members (Tan et al., 2003). CCD1s and CCD2 are identified by their subcellular location in the cytosol. The other CCDs have transit peptides that target plastid or plastoglobule (Tan et al., 2003; Rubio et al., 2008; Frusciante et al., 2014). These enzymes are involved in the production of volatiles substances, such as β-ionone (Tan et al., 2003; Schwartz et al., 2004).

Recombinant CCD1 enzymes can also cleave the 5,6-5′,6′ and 7,8-7′,8′ double bonds under certain buffer conditions and on specific substrates (Vogel et al., 2008; Ilg et al., 2010), but it is not yet clear whether this activity is also relevant *in planta*. A revised overview of the *in planta* role of CCD1 from an RNAi approach found that CCD1 cleavage activity on 9,10 and/or 9′,10′ is required to form C_{13} apocarotenoid scent and aroma compounds (Simkin et al., 2004; Simkin et al., 2004; Mathieu et al., 2005; Ibdah et al., 2006; Kato et al., 2006; García-Limones et al., 2008; Huang et al., 2009). CCD1 enzymes can accept a wide range of C_{40} carotenoid substrates *in vitro* (Vogel et al., 2008; Huang et al., 2009; Ilg et al., 2010). *Solanum lycopersicum* produces two CCD1 isoforms, but the difference between them appears to rest on the differential expression of the corresponding genes, rather than on differential enzyme characteristics (Simkin et al., 2004). Studies using functional expression in carotenoid accumulating *E. coli* strains have demonstrated that CCD1 enzymes from different plant species produce volatiles 6-methyl-5-hepten-2-one (C_8), geranial (C_{10}), and a series of C_{13} cyclohexenones including α- and β-ionone. CCD1 cleaves C_{40} carotenoids or apo-10′-carotenoids to form simultaneously different dialdehyde products, such as rosafluene-dialdehyde and C_{13}-ionone(s) (Vogel et al., 2008; Ilg et al., 2010; Ilg et al., 2014). CCD2, closely related to the CCD1 family (Figure 3.2) is the key enzyme involved in the synthesis of saffron apocarotenoids *Crocus* species (Frusciante et al., 2014). It can sequentially cleave the 7,8 and 7′,8′ double bonds adjacent to a 3-hydroxy-β-ionone ring, converting zeaxanthin into crocetin dialdehyde (Frusciante et al., 2014). CCD2 have also been identified in other plants (Ahrazem et al., 2016). CCD4 cleave bicyclic all-*trans*-carotenoids such as all-trans-β-carotene at the C9, C10 or C9′, C10′ double bond producing apo-10′-carotenoids (C_{27}), such as β-apo-10′-carotenal, and the corresponding C_{13} cyclohexenone product, β-ionone (Bruno et al., 2015; Bruno and Al-Babili, 2016; Fiorilli et al., 2019).

The *Citrus* pigment citraurin (3-OH-β-apo-8′-carotenoid; C_{30}) is produced by other types of CCD4 by cleaving the C7, C8-C7′, C8′ double bond in hydroxylated bicyclic carotenoids (Rodrigo et al., 2013).

CCD7 and CCD8 synthesize strigolactone (SL) sequentially converting 9-cis-β-carotene produced by the carotene isomerase DWARF27 (D27) into the SL precursor carlactone, substrate of cytochrome P450 enzymes from the 711 clade, such as the rice carlactone oxidase (OsCO), that forms 4-deoxyorobanchol (Alder et al., 2012; Abe et al., 2014; Bruno et al., 2014; Zhang et al.,

2014; Bruno et al., 2015; Bruno and Al-Babili, 2016; Abuauf et al., 2018; Zhang et al., 2018) CCD7 also participates in the production of C_{13} cyclohexenones (Floss et al., 2008; Walter et al., 2015). ZASs were recently identified as a CCD subfamily converting 3-OH-all-trans-β-apo-10′-carotenal (apo-10′-zeaxanthinal) into zaxinone (3-OH-all-trans-apo-13-carotenone), a regulatory metabolite required for normal rice growth (Wang et al., 2019).

CCD7 and CCD8 asymmetrically cleave carotenoids and apocarotenoids and take part in the synthesis of strigolactones (Schwartz et al., 2004; Snowden et al., 2005; Sergeant et al., 2009) (Figure 3.2). The specific role of both enzymes the CCD7 and the CCD8 classes defined so far is in the concerted production of an apocarotenoid shoot branching inhibitor hormone, strigolactones (Walter et al., 2010; Hou et al., 2016). The protein structure of the enzyme has a 4-histidine–iron arrangement in the active site, as is common to all CCOs; however, their amino acid sequences are rather heterogeneous. The iron is embedded in a 7-bladed β-propeller chain arrangement, topped by a dome of six large loops (Hou et al., 2016).

Several studies point out the possible involvement of two types of CCDs in apocarotenoids production through a sequential two-steps cleavage: a C_{40} carotenoid is cleaved by CCD7 or possibly CCD4, producing a C_{27} apocarotenoid and a C_{13} cyclohexenone; the C_{27} apocarotenoid is then converted by CCD1 into rosafluene-dialdehyde (C_{14}), the precursor of mycorradicin, and C_{13} cyclohexenone (Floss et al., 2008; Walter et al., 2010; Hou et al., 2016). In bixin synthesis, the sequential participations of a CCD4 and a CCD1 to obtain bixin aldehyde has also been proposed (Rodríguez-Ávila et al., 2011; Carballo-Uicab et al., 2019).

CCD7 and CCD8 act sequentially to produce strigolactone, providing the 9-cis-configured C_{27} intermediate in SL biosynthesis; CCD7 is also a candidate enzyme in the synthesis of the C_{27} precursor for the CCD8-mediated formation of the C_{18}-ketone β-apo-13-carotenone, the precursor for strigolactone biosynthesis and an acyclic dialdehyde at the 13,14 double bond (Alder et al., 2008; Vogel et al., 2008; Walter et al., 2010; Kohlen et al., 2011; Seto et al., 2014). CCD7 is a stereospecific enzyme that solely cleaves 9-cis-configured carotenoids, forming accordingly configured C_{27}-apocarotenoids (Bruno et al., 2014). Therefore, the involvement of CCD7 in the latter two metabolic processes implies a *cis* to *trans* isomerization of the formed C_{27}-apocarotenoids, preparing them to be converted into β-apo-13-carotenone and C_{13} and C_{14} mycorrhizal apocarotenoids (Walter et al., 2010). Then CCD8 converts β-apo-10′-carotenal into C_{18} β-apo-13-carotenone.

Canonical CCD7 and CCD8 seem to be absent in green algae and most Chlorophyta; *C. reinhardtii* does not contain strigolactones (Delaux et al., 2012). While *Marchantia polymorpha* (liverwort) and Charophyte green algae (such as *Nitella hyalina*, *Nitella pseudoflabellata* and *Chara corallina*) seem to lack CCD8 but produce strigolactones, suggesting the presence of another strigolactones biosynthesis pathway (Delaux et al., 2012).

The study of the synthesis and function of apocarotenoids is increasingly surprising. They act as pigments, flavours, aromas, hormones and signalling molecules. Due to their chemical and biochemical reactions, signalling apocarotenoids play an important role in plants and in the symbiotic plant–microorganism relationship. New products derived from carotenoids are continuously being discovered, as well as catalytic enzymes involved in their production. Research perspectives on the biogenesis and function of these molecules are becoming very attractive and exciting in all taxa.

Finally, the catalytic mechanisms of different CCOs are still unknown, and efforts are still needed to study the newly identified CCOs involved in the metabolic conversions of carotenoids and apocarotenoids. Moreover, determining the biological functions of new apocarotenoids in all living organisms will be a future challenge.

ACKNOWLEDGEMENTS

Author thanks to Consejo Nacional de Ciencia y Tecnología (CONACYT) for their support with the project: Fronteras de la Ciencia No. 2016-01-1716. DC. Author thanks to Victor Carballo-Uicab for help with the designing the figures.

REFERENCES

Abdullaev F. 2002. Cancer chemopreventive and tumoricidal properties of saffron (*Crocus sativus* L.). *Exp Biol Med.* 227(1):20–25.

Abe S, Sado A, Tanaka K, Kisugi T, Asami K, Ota S, Il H, Yoneyama K. 2014. Carlactone is converted to carlactonoic acid by MAX1 in Arabidopsis and its methyl ester can directly interact with AtD14 *in vitro*. *Proc Natl Acad Sci U S A.* 111(50):18084–18093.

Abuauf H, Haider I, Jia K, Ablazov A, Mi J, Blilou I, Al-Babili S. 2018. Plant Science isomerase and is induced by auxin, *abscisic acid and phosphate deficiency. Plant Sci.* 277(June):33–42.

Ahrazem O, Argandoña J, Fiore A, Aguado C, Luján R, Rubio-Moraga Á, Marro M, Araujo-Andrade C, Loza-Alvarez P, Diretto G, Gómez-Gómez L. 2018. Transcriptome analysis in tissue sectors with contrasting crocins accumulation provides novel insights into apocarotenoid biosynthesis and regulation during chromoplast biogenesis. *Sci Rep.* 8(1):1–17.

Ahrazem O, Gómez-Gómez L, Rodrigo MJ, Avalos J, Limón MC. 2016. Carotenoid cleavage oxygenases from microbes and photosynthetic organisms: Features and functions. *Int J Mol Sci.* 17(11):1781.

Ahrazem O, Rubio-Moraga A, López RC, Gómez-Gómez L. 2010. The expression of a chromoplast-specific lycopene beta cyclase gene is involved in the high production of saffron's apocarotenoid precursors. *J Exp Bot.* 61(1):105–119.

Al-Babili S, Bouwmeester HJ. 2015. Strigolactones, a novel carotenoid-derived plant hormone. *Annu Rev Plant Biol.* 66(1):161–186.

Alcalde E, Cerdá-Olmedo E, Al-Babili S. 2019. Apocarotenoids produced from β-carotene by dioxygenases from *Mucor circinelloides. Microbiology.* 165(4):433–438.

Alder A, Holdermann I, Beyer P, Al-Babili S. 2008. Carotenoid oxygenases involved in plant branching catalyse a highly specific conserved apocarotenoid cleavage reaction. *Biochem J.* 416(2):289–296.

Alder A, Jamil M, Marzorati M, Bruno M, Vermathen M, Bigler P, Ghisla S, Bouwmeester H, Beyer P, Al-Babili S. 2012. The path from β-carotene to carlactone, a strigolactone-like plant hormone. *Science.* 335(6074):1348–1351.

Alquézar B, Jesús M, Lorenzo R. 2008. Carotenoid biosynthesis and their regulation in *Citrus fruits*.

Auldridge ME, Block A, Vogel JT, Dabney-Smith C, Mila I, Bouzayen M, Magallanes-Lundback M, DellaPenna D, McCarty DR, Klee HJ. 2006a. Characterization of three members of the Arabidopsis carotenoid cleavage dioxygenase family demonstrates the divergent roles of this multifunctional enzyme family. *Plant J.* 45(6):982–993.

Auldridge ME, McCarty DR, Klee HJ. 2006b. Plant carotenoid cleavage oxygenases and their apocarotenoid products. *Curr Opin Plant Biol.* 9(3):315–321.

Balashov SP, Lanyi JK. 2007. Xanthorhodopsin: Proton pump with a carotenoid antenna. *Cell Mol Life Sci.* 64(18):2323–2328.

Baldermann S, Naim M, Fleischmann P. 2005. Enzymatic carotenoid degradation and aroma formation in nectarines (*Prunus persica*). *Food Res Int.* 38(8–9):833–836.

Bibikov SI, Grishanin RN, Kaulen AD, Marwan W, Oesterhelt D, Skulachev VP. 1993. Bacteriorhodopsin is involved in halobacterial photoreception. *Proc Natl Acad Sci U S A.* 90(20):9446–9450.

Borghi L, Liu GW, Emonet A, Kretzschmar T, Martinoia E. 2016. The importance of strigolactone transport regulation for symbiotic signaling and shoot branching. *Planta.* 243(6):1351–1360.

Bouvier F, Dogbo O, Camara B. 2003a. Biosynthesis of the food and cosmetic plant pigment bixin (annatto). *Science.* 300(5628):2089–2091.

Bouvier F, Isner JC, Dogbo O, Camara B. 2005. Oxidative tailoring of carotenoids: A prospect towards novel functions in plants. *Trends Plant Sci.* 10(4):187–194.

Bouvier F, Suire C, Camara B. 2003b. Oxidative remodeling of chromoplast carotenoids: identification of the carotenoid dioxygenase. *Society.* 15(January):47–62.

Bouwmeester HJ, Roux C, López-Ráez JA, Bécard G. 2007. Rhizosphere communication of plants, parasitic plants and AM fungi. *Trends Plant Sci.* 12(5):224–230.

Brenna E, Fuganti C, Serra S. 2003. Enantioselective perception of chiral odorants. *Tetrahedron: Asymmetry.* 14(1):1–42.

Brunetti C, Sebastiani F, Tattini M. 2019. Review: ABA, flavonols, and the evolvability of land plants. *Plant Sci.* 280(December 2018):448–454.

Bruno M, Al-Babili S. 2016. On the substrate specificity of the rice strigolactone biosynthesis enzyme DWARF27. *Planta.* 243(6):1429–1440.

Bruno M, Beyer P, Al-Babili S. 2015. The potato carotenoid cleavage dioxygenase 4 catalyzes a single cleavage of β-ionone ring-containing carotenes and non-epoxidated xanthophylls. *Arch Biochem Biophys.* 572:126–133.

Bruno M, Hofmann M, Vermathen M, Alder A, Beyer P, Al-Babili S. 2014. On the substrate- and stereospecificity of the plant carotenoid cleavage dioxygenase 7. *FEBS Lett* 588(9):1802–1807.

Bythell-Douglas R, Rothfels CJ, Stevenson DWD, Graham SW, Wong GKS, Nelson DC, Bennett T. 2017. Evolution of strigolactone receptors by gradual neo-functionalization of KAI2 paralogues. *BMC Biol.* 15(1):1–21.

Carballo-Uicab VM, Cárdenas-Conejo Y, Vallejo-Cardona AA, Aguilar-Espinosa M, Rodríguez-Campos J, Serrano-Posada H, Narváez-Zapata JA, Vázquez-Flota F, Rivera-Madrid R. 2019. Isolation and functional characterization of two dioxygenases putatively involved in bixin biosynthesis in annatto (*Bixa orellana* L.). *PeerJ.* 2019(6).

Cárdenas-Conejo Y, Carballo-Uicab V, Lieberman M, Aguilar-Espinosa M, Comai L, Rivera-Madrid R. 2015. De novo transcriptome sequencing in *Bixa orellana* to identify genes involved in methylerythritol phosphate, carotenoid and bixin biosynthesis. *BMC Genomics.* 16(1):877.

Cazzonelli CI, Pogson BJ. 2010. Source to sink: Regulation of carotenoid biosynthesis in plants. *Trends Plant Sci.* 15(5):266–274.

Charpentier M, Sun J, Wen J, Mysore KS, Oldroyd GED. 2014. Abscisic acid promotion of arbuscular mycorrhizal colonization requires a component of the PROTEIN PHOSPHATASE 2A complex. *Plant Physiol.* 166(4):2077–2090.

Curl AL. 1965. The occurrence of beta-citraurin and of beta-apo-8′-carotenal in the peels of California tangerines and oranges. *J Food Sci.* 30(1):13–16.

D'alessandro S, Ksas B, Havaux M. 2018. Decoding β-cyclocitral-mediated retrograde signaling reveals the role of a detoxification response in plant tolerance to photooxidative stress. *Plant Cell.* 30(10):2495–2511.

Decker EL, Alder A, Hunn S, Ferguson J, Lehtonen MT, Scheler B, Kerres KL, Wiedemann G, Safavi-Rizi V, Nordzieke S, et al. 2017. Strigolactone biosynthesis is evolutionarily conserved, regulated by phosphate starvation and contributes to resistance against phytopathogenic fungi in a moss, *Physcomitrella patens. New Phytol.* 216(2):455–468.

Delaux P, Xie X, Timme RE, Puech-Pages V, Dunand C, Lecompte E, Delwiche CF, Yoneyama K, Bécard G, Séjalon-Delmas N. 2012. Origin of strigolactones in the green lineage. *New Phytol.* 195(4):857–871.

Demurtas OC, Francisco R, Diretto G, Ferrante P, Frusciante S, Pietrella M, Aprea G, Borghi L, Feeney M, Frigerio L, et al. 2019. ABCC transporters mediate the vacuolar accumulation of crocins in saffron stigmas. Plant Cell. *31(November):tpc.*00193.2019.

Dickinson AJ, Lehner K, Mi J, Jia KP, Mijar M, Dinneny J, Al-Babili S, Benfey PN. 2019. β-Cyclocitral is a conserved root growth regulator. *Proc Natl Acad Sci U S A.* 116(21):10563–10567.

Estrada AF, Brefort T, Mengel C, Díaz-Sánchez V, Alder A, Al-Babili S, Avalos J. 2009. Ustilago maydis accumulates β-carotene at levels determined by a retinal-forming carotenoid oxygenase. *Fungal Genet Biol.* 46(10):803–813.

Eugster CH, Märki-Fischer E. 1991. The chemistry of rose pigments. *Angew Chemie Int Ed English.* 30(6):654–672.

Felemban A, Braguy J, Zurbriggen MD, Al-Babili S. 2019. Apocarotenoids Involved in Plant Development and Stress Response. *Front Plant Sci.* 10(September):1–16.

Fiorilli V, Wang JY, Bonfante P, Lanfranco L, Al-Babili S. 2019. Apocarotenoids: Old and new mediators of the arbuscular mycorrhizal symbiosis. *Front Plant Sci.* 10(September):1–9.

Floss DS, Schliemann W, Schmidt J, Strack D, Walter MH. 2008. RNA interference-mediated repression of MtCCD1 in mycorrhizal roots of Medicago truncatula causes accumulation of C27 apocarotenoids, shedding light on the functional role of CCD1. *Plant Physiol.* 148(3):1267–1282.

Foster KW, Saranak J, Patel N, Zarilli G, Okabe M, Kline T, Nakanishi K. 1984. A rhodopsin is the functional photoreceptor for phototaxis in the unicellular eukaryote Chlamydomonas. *Nature.* 311(5988):756–759.

Frusciante S, Diretto G, Bruno M, Ferrante P, Pietrella M, Prado-Cabrero A, Rubio-Moraga A, Beyer P, Gomez-Gomez L, Al-Babili S, Giuliano G. 2014. Novel carotenoid cleavage dioxygenase catalyzes the first dedicated step in saffron crocin biosynthesis. *Proc Natl Acad Sci U S A.* 111(33):12246–12251.

García-Limones C, Schnäbele K, Blanco-Portales R, Luz Bellido M, Caballero JL, Schwab W, Muñoz-Blanco J. 2008. Functional characterization of FaCCD1: A carotenoid cleavage dioxygenase from strawberry involved in lutein degradation during fruit ripening. *J Agric Food Chem.* 56(19):9277–9285.

Giguere V, Ong ES, Segui P, Evans RM. 1987. Identification of a receptor for the morphogen retinoic acid. *Nature.* 330(6149):624–629.

Giuliano G, Rosati C, Bramley PM. 2003. To dye or not to dye: Biochemistry of annatto unveiled. *Trends Biotechnol.* 21(12):513–516.

Herrera-Medina MJ, Steinkellner S, Vierheilig H, Ocampo Bote JA, García Garrido JM. 2007. Abscisic acid determines arbuscule development and functionality in the tomato arbuscular mycorrhiza. *New Phytol.* 175(3):554–564.

Hessel S, Eichinger A, Isken A, Amengual J, Hunzelmann S, Hoeller U, Elste V, Hunziker W, Goralczyk R, Oberhauser V, et al. 2007. CMO1 deficiency abolishes vitamin a production from β-carotene and alters lipid metabolism in mice. *J Biol Chem.* 282(46):33553–33561.
Hou X, Rivers J, León P, McQuinn RP, Pogson BJ. 2016. Synthesis and function of apocarotenoid signals in plants. *Trends Plant Sci.* 21(9):792–803.
Huang FC, Molnár P, Schwab W. 2009. Cloning and functional characterization of carotenoid cleavage dioxygenase 4 genes. *J Exp Bot.* 60(11):3011–3022.
Ibdah M, Azulay Y, Portnoy V, Wasserman B, Bar E, Meir A, Burger Y, Hirschberg J, Schaffer AA, Katzir N. 2006. Functional characterization of CmCCD1, a carotenoid cleavage dioxygenase from melon. *Phytochemistry.* 67(15):1579–1589.
Ilg A, Bruno M, Beyer P, Al-Babili S. 2014. Tomato carotenoid cleavage dioxygenases 1A and 1B: Relaxed double bond specificity leads to a plenitude of dialdehydes, mono-apocarotenoids and isoprenoid volatiles. *FEBS Open Bio.* 4:584–593.
Ilg A, Yu Q, Schaub P, Beyer P, Al-Babili S. 2010. Overexpression of the rice carotenoid cleavage dioxygenase 1 gene in Golden Rice endosperm suggests apocarotenoids as substrates in planta. *Planta.* 232(3): 691–699.
Ito N, Etoh T, Hagiwara H, Kato M. 1997. Novel synthesis of degradation products of carotenoids, megastigmatrienone analogues and blumenol. *A J Chem Soc – Perkin Trans 1.* 3(10):1571–1579.
Jia KP, Baz L, Al-Babili S. 2018. From carotenoids to strigolactones. *J Exp Bot.* 69(9):2189–2204.
Jung KH, Trivedi VD, Spudich JL. 2003. Demonstration of a sensory rhodopsin in eubacteria. *Mol Microbiol.* 47(6):1513–1522.
Jüttner F, Höflacher B. 1985. Evidence of β-carotene 7,8(7′,8′) oxygenase (β-cyclocitral, crocetindial generating) in Microcystis. *Arch Microbiol.* 141(4):337–343.
Kamoda S, Saburi Y. 1993. Cloning, expression, and sequence analysis of a lignostilbene- α,β-dioxygenase gene from pseudomonas paucimobilis TMY1009. *Biosci Biotechnol Biochem.* 57(6):926–930.
Kato M, Matsumoto H, Ikoma Y, Okuda H, Yano M. 2006. The role of carotenoid cleavage dioxygenases in the regulation of carotenoid profiles during maturation in citrus fruit. *J Exp Bot.* 57(10):2153–2164.
Kim YS, Seo ES, Oh DK. 2012. Characterization of an apo-carotenoid 13,14-dioxygenase from Novosphingobium aromaticivorans that converts β-apo-8'-carotenal to β-apo-13-carotenone. *Biotechnol Lett.* 34(10):1851–1856.
Klingner A, Hundeshagen B, Kernebeck H, Bothe H. 1995. Localization of the yellow pigment formed in roots of gramineous plants colonized by arbuscular fungi. *Protoplasma.* 185(1–2):50–57.
Kloer DP, Schulz GE. 2006. Structural and biological aspects of carotenoid cleavage. *Cell Mol Life Sci.* 63(19–20):2291–2303.
Kohlen W, Charnikhova T, Liu Q, Bours R, Domagalska MA, Beguerie S, Verstappen F, Leyser O, Bouwmeester H, Ruyter-Spira Carolien, Laboratory. 2011. Strigolactones are transported through the xylem and play a key role in shoot architectural response to phosphate deficiency in nonarbuscular mycorrhizal. *Plant Physiol.* 155(2):974–987.
Leuenberger U, Stewart I. 1976. β-citraurinene, a new C30-citrus carotenoid. *Phytochemistry.* 15(1):227–229.
Liang MH, Zhu J, Jiang JG. 2018. Carotenoids biosynthesis and cleavage related genes from bacteria to plants. *Crit Rev Food Sci Nutr.* 58(14):2314–2333.
López-Ráez JA, Kohlen W, Charnikhova T, Mulder P, Undas AK, Sergeant MJ, Verstappen F, Bugg TDH, Thompson AJ, Ruyter-Spira C, Bouwmeester H. 2010. Does abscisic acid affect strigolactone biosynthesis? *New Phytol.* 187(2):343–354.
Luan Y, Wang S, Wang R, Xu C. 2020. Accumulation of red apocarotenoid β -citraurin in peel of a spontaneous mutant of huyou (*Citrus changshanensis*) and the effects of storage temperature and ethylene application. *Food Chem.* 309(October 2019):125705.
Ludwig-Müller J. 2010. Hormonal responses in host plants triggered by arbuscular mycorrhizal fungi. In: *Arbuscular mycorrhizas Physiol Funct.* Springer; p. 169–190.
Ma G, Zhang L, Matsuta A, Matsutani K, Yamawaki K, Yahata M, Wahyudi A, Motohashi R, Kato M. 2013. Enzymatic formation of β-citraurin from β-cryptoxanthin and zeaxanthin by carotenoid cleavage dioxygenase4 in the flavedo of citrus fruit. *Plant Physiol.* 163(2):682–695.
Machin DC, Hamon-Josse M, Bennett T. 2019. Fellowship of the rings: A saga of strigolactones and other small signals. *New Phytol.* 225(2):621–636.
Maicas S, Mateo JJ. 2005. Hydrolysis of terpenyl glycosides in grape juice and other fruit juices: A review. *Appl Microbiol Biotechnol.* 67(3):322–335.
Marasco EK, Vay K, Schmidt-Dannert C. 2006. Identification of carotenoid cleavage dioxygenases from Nostoc sp. PCC 7120 with different cleavage activities. *J Biol Chem.* 281(42):31583–31593.

Martín-Rodríguez JÁ, León-Morcillo R, Vierheilig H, Ocampo JA, Ludwig-Müller J, García-Garrido JM. 2011. Ethylene-dependent/ethylene-independent ABA regulation of tomato plants colonized by arbuscular mycorrhiza fungi. *New Phytol.* 190(1):193–205.

Mathieu S, Terrier N, Procureur J, Bigey F, Günata Z. 2005. A carotenoid cleavage dioxygenase from *Vitis vinifera* L.: Functional characterization and expression during grape berry development in relation to C13-norisoprenoid accumulation. *J Exp Bot.* 56(420):2721–2731.

Medina HR, Cerdá-Olmedo E, Al-Babili S. 2011. Cleavage oxygenases for the biosynthesis of trisporoids and other apocarotenoids in Phycomyces. *Mol Microbiol.* 82(1):199–208.

Mendes-Pinto MM. 2009. Carotenoid breakdown products the—norisoprenoids—in wine aroma. *Arch Biochem Biophys.* 483(2):236–245.

Mercy L, Lucic-Mercy E, Nogales A, Poghosyan A, Schneider C, Arnholdt-Schmitt B. 2017. A functional approach towards understanding the role of the mitochondrial respiratory Chain in an endomycorrhizal symbiosis. *Front Plant Sci.* 8(March):1–27.

Muriel Ruiz SB, Cardona-medina E, Arias-ruiz E, Gómez-Gómez A. 2015. Indagaciones acerca del azafrán de raiz (Escobedia grandiflora (l. F.) Kuntze) en Antioquia - Colombia : Una especie olvidada. 85:9–12.

Nambara E, Marion-Poll A. 2005. Abscisic acid biosynthesis and catabolism. *Annu Rev Plant Biol.* 56(1):165–185.

Nisar N, Li L, Lu S, Khin NC, Pogson BJ. 2015. Carotenoid metabolism in plants. *Mol Plant.* 8(1):68–82.

Ochiai T, Shimeno H, Mishima K ichi, Iwasaki K, Fujiwara M, Tanaka H, Shoyama Y, Toda A, Eyanagi R, Soeda S. 2007. Protective effects of carotenoids from saffron on neuronal injury *in vitro* and *in vivo*. *Biochim Biophys Acta - Gen Subj.* 1770(4):578–584.

Parker C. 2009. Observations on the current status of orobanche and striga problems worldwide. *Pest Manag Sci.* 65(5):453–459.

Peleg Z, Blumwald E. 2011. Hormone balance and abiotic stress tolerance in crop plants. *Curr Opin Plant Biol* 14(3):290–295.

Prado-Cabrero A, Scherzinger D, Avalos J, Al-Babili S. 2007. Retinal biosynthesis in fungi: Characterization of the carotenoid oxygenase CarX from *Fusarium fujikuroi. Eukaryot Cell.* 6(4):650–657.

Ramel F, Birtic S, Ginies C, Soubigou-Taconnat L, Triantaphylidès C, Havaux M. 2012. Carotenoid oxidation products are stress signals that mediate gene responses to singlet oxygen in plants. *Proc Natl Acad Sci U S A.* 109(14):5535–5540.

Redmond TM, Gentleman S, Duncan T, Yu S, Wiggert B, Gantt E, Cunningham FX. 2001. Identification, expression, and substrate specificity of a mammalian β-carotene 15, 15′-dioxygenase. *J Biol Chem.* 276(9):6560–6565.

Rivera-Madrid R, Aguilar-Espinosa M, Cárdenas-Conejo Y, Garza-Caligaris LE. 2016. Carotenoid derivates in achiote (*Bixa orellana*) seeds: Synthesis and health promoting properties. *Front Plant Sci.* 7(September):1406.

Rodrigo MJ, Alquézar B, Alós E, Medina V, Carmona L, Bruno M, Al-Babili S, Zacarías L. 2013. A novel carotenoid cleavage activity involved in the biosynthesis of Citrus fruit-specific apocarotenoid pigments. *J Exp Bot.* 64(14):4461–4478.

Rodríguez-Ávila NL, Narváez-Zapata JA, Ramírez-Benítez JE, Aguilar-Espinosa ML, Rivera-Madrid R. 2011. Identification and expression pattern of a new carotenoid cleavage dioxygenase gene member from *Bixa orellana. J Exp Bot.* 62(15).

Rubio A, Rambla JL, Santaella M, Gómez MD, Orzaez D, Granell A, Gómez-Gómez L. 2008. Cytosolic and plastoglobule-targeted carotenoid dioxygenases from *Crocus sativus* are both involved in β-ionone release. *J Biol Chem.* 283(36):24816–24825.

Ruch S, Beyer P, Ernst H, Al-Babili S. 2005. Retinal biosynthesis in Eubacteria: *In vitro* characterization of a novel carotenoid oxygenase from Synechocystis sp. PCC 6803. *Mol Microbiol.* 55(4):1015–1024.

Saeed W, Naseem S, Ali Z. 2017. Strigolactones biosynthesis and their role in abiotic stress resilience in plants: A critical review. *Front Plant Sci.* 8(August):1–13.

Saelices L, Youssar L, Holdermann I, Al-Babili S, Avalos J. 2007. Identification of the gene responsible for torulene cleavage in the Neurospora carotenoid pathway. *Mol Genet Genomics.* 278(5):527–537.

Scannerini S, Bonfante-Fasolo P. 1977. Unusual plasties in an endomycorrhizal root. *Can J Bot.* 55(18):2471–2474.

Scherzinger D, Ruch S, Kloer DP, Wilde A, Al-Babili S. 2006. Retinal is formed from apo-carotenoids in Nostoc sp. PCC7120: *In vitro* characterization of an apo-carotenoid oxygenase. *Biochem J.* 398(3):361–369.

Scherzinger D, Scheffer E, Bär C, Ernst H, Al-Babili S. 2010. The *Mycobacterium tuberculosis* ORF Rv0654 encodes a carotenoid oxygenase mediating central and excentric cleavage of conventional and aromatic carotenoids. *FEBS J.* 277(22):4662–4673.

Schwartz SH, Qin X, Loewen MC. 2004. The biochemical characterization of two carotenoid cleavage enzymes from Arabidopsis indicates that a carotenoid-derived compound inhibits lateral branching. *J Biol Chem.* 279(45):46940–46945.

Schwartz SH, Tan BC, Gage DA, Zeevaart JAD, McCarty DR. 1997. Specific oxidative cleavage of carotenoids by VP14 of maize. *Science.* 276(5320):1872–1874.

Sergeant MJ, Li J, Fox C, Brookbank N, Rea D, Bugg TDH, Thompson AJ. 2009. Selective inhibition of carotenoid cleavage dioxygenases. 284(8):5257–5264.

Serra S. 2015. Recent advances in the synthesis of carotenoid-derived flavours and fragrances. *Molecules.* 20(7):12817–12840.

Seto Y, Sado A, Asami K, Hanada A, Umehara M, Akiyama K. 2014. Carlactone is an endogenous biosynthetic precursor for strigolactones. 111(4):1–6.

Shumbe L, Bott R, Havaux M. 2014. Dihydroactinidiolide, a high light-induced β-carotene derivative that can regulate gene expression and photoacclimation in Arabidopsis. *Mol Plant.* 7(7):1248–1251.

Simkin AJ, Schwartz SH, Auldridge M, Taylor MG, Klee HJ. 2004a. The tomato carotenoid cleavage dioxygenase 1 genes contribute to the formation of the flavor volatiles β-ionone, pseudoionone, and geranylacetone. *Plant J.* 40(6):882–892.

Simkin AJ, Underwood BA, Auldridge M, Loucas HM, Shibuya K, Schmelz E, Clark DG, Klee HJ. 2004b. Circadian regulation of the PhCCD1 carotenoid cleavage dioxygenase controls emission of β-ionone, a fragrance volatile of petunia flowers. *Plant Physiol.* 136(3):3504–3514.

Snowden KC, Simkin AJ, Janssen BJ, Templeton KR, Loucas HM, Simons JL, Karunairetnam S, Gleave AP, Clark DG, Klee HJ. 2005. The decreased apical dominance1 / petunia hybrida *CAROTENOID CLEAVAGE DIOXYGENASE*8 gene affects branch production and plays a role in leaf senescence, root growth, and flower development. 17(March):746–759.

Strack D, Fester T. 2006. Isoprenoid metabolism and plastid reorganization in arbuscular mycorrhizal roots. *New Phytol.* 172(1):22–34.

Tagua VG, Medina HR, Martín-Domínguez R, Eslava AP, Corrochano LM, Cerdá-Olmedo E, Idnurm A. 2012. A gene for carotene cleavage required for pheromone biosynthesis and carotene regulation in the fungus *Phycomyces blakesleeanus. Fungal Genet Biol.* 49(5):398–404.

Tan BC, Joseph LM, Deng WT, Liu L, Li QB, Cline K, McCarty DR. 2003. Molecular characterization of the Arabidopsis 9-cis epoxycarotenoid dioxygenase gene family. *Plant J.* 35(1):44–56.

Tan BC, Schwartz SH, Zeevaart JAD, McCarty DR. 1997. Genetic control of abscisic acid biosynthesis in maize. *Proc Natl Acad Sci.* 94(22):12235–12240.

Ton J, Flors V, Mauch-Mani B. 2009. The multifaceted role of ABA in disease resistance. *Trends Plant Sci.* 14(6):310–317.

Umehara M, Hanada A, Yoshida S, Akiyama K, Arite T, Takeda-Kamiya N, Magome H, Kamiya Y, Shirasu K, Yoneyama K, et al. 2008. Inhibition of shoot branching by new terpenoid plant hormones. *Nature.* 455(7210):195–200.

Vogel JT, Tan B-C, McCarty DR, Klee HJ. 2008. The carotenoid cleavage dioxygenase 1 enzyme has broad substrate specificity, cleaving multiple carotenoids at two different bond positions. *J Biol Chem.* 283(17):11364–11373.

von Lintig J. 2010. Colors with functions: Elucidating the biochemical and molecular basis of carotenoid metabolism. *Annu Rev Nutr.* 30(1):35–56.

von Lintig J, Vogt K. 2000. Filling the gap in vitamin a research molecular identification of an enzyme cleaving β-carotene to retinal. *J Biol Chem.* 275(16):11915–11920.

Wagner K, Krause K, David A, Kai M, Jung EM, Sammer D, Kniemeyer O, Boland W, Kothe E. 2016. Influence of zygomycete-derived D'orenone on IAA signalling in Tricholoma-spruce ectomycorrhiza. *Environ Microbiol.* 18(8):2470–2480.

Wald G. 1968. The molecular basis of visual excitation. *Nature.*

Walter MH, Floß DS, Hans J, Fester T, Strack D. 2007. Apocarotenoid biosynthesis in arbuscular mycorrhizal roots: Contributions from methylerythritol phosphate pathway isogenes and tools for its manipulation. *Phytochemistry.* 68(1):130–138.

Walter MH, Floss DS, Strack D. 2010. Apocarotenoids: Hormones, mycorrhizal metabolites and aroma volatiles. *Planta.* 232(1):1–17.

Walter MH, Stauder R, Tissier A. 2015. *Plant Science Evolution of root-specific carotenoid precursor pathways for apocarotenoid signal biogenesis.* 233:1–10.

Walter MH, Strack D. 2011. Carotenoids and their cleavage products: Biosynthesis and functions. *Nat Prod Rep.* 28(4):663–692.

Wang JY, Haider I, Jamil M, Fiorilli V, Saito Y, Mi J, Baz L, Kountche BA, Jia KP, Guo X, et al. 2019. The apocarotenoid metabolite zaxinone regulates growth and strigolactone biosynthesis in rice. *Nat Commun.* 10(1):1–9.

Xie X, Yoneyama Kaori, Yoneyama Koichi. 2010. The strigolactone story. *Annu Rev Phytopathol.* 48:93–117.

Yang T, Lian Y, Wang C. 2019. Comparing and contrasting the multiple roles of butenolide plant growth regulators: Strigolactones and karrikins in plant development and adaptation to abiotic stresses. *Int J Mol Sci.* 20(24):6–11.

Zeevaart JAD, Creelman RA. 1988. Metabolism and Physiology of Abscisic Acid. *Annu Rev Plant Physiol Plant Mol Biol.* 39(1):439–473.

Zhang C, Chen X, Lindley ND, Too HP. 2018. A 'plug-n-play' modular metabolic system for the production of apocarotenoids. *Biotechnol Bioeng.* 115(1):174–183.

Zhang Y, Dijk ADJ Van, Scaffidi A, Flematti GR, Hofmann M, Charnikhova T, Verstappen F, Hepworth J, Krol S Van Der, Leyser O. 2014. Rice cytochrome P450 MAX1 homologs catalyz distinct steps in strigolactone biosynthesis. *Nat Chem Biol.* 10(12):1028–1033.

Zheng X, Zhu K, Sun Q, Zhang W, Wang X, Cao H, Tan M, Xie Z, Zeng Y, Ye J, et al. 2019. Natural variation in CCD4 promoter underpins species-specific evolution of red coloration in citrus peel. *Mol Plant* 12(9):1294–1307.

Zhu A, Lao C, Wang Z, Chen Y, Bai C. 2019. Characterization of crocetin-monoglucuronide as a neuron-protective metabolite of crocin-1. *Mol Nutr Food Res.* 63(13):1–9.

Zhu D, Chang Y, Pei T, Zhang X, Liu L, Li Y, Zhuang J, Yang H, Qin F. 2020. MAPK-like protein 1 positively regulates maize seedling drought sensitivity by suppressing ABA biosynthesis. *Plant J.* 102(4):747–760

4 Unravelling Apocarotenoid Biosynthesis Pathways: Transcriptomic Tools for Gene Discovery

David Chan-Rodriguez and Renata Rivera-Madrid

CONTENTS

INTRODUCTION

To decipher the genetic complexity of plant secondary metabolite pathways, also known as specialized metabolites, researchers had employed a variety of approaches, often involving genetic, biochemical, and molecular tools. Model organisms such as *Arabidopsis thaliana* and *Oryza sativa* paved the way to unravelling a myriad of fundamental biological processes, due to the vast genetic and genomic resources available to the scientific community. Such efforts to generate these tools lies in the biological amenability of both systems. The eudicot *Arabidopsis thaliana* has been adopted as the favourite model organism for plant research for decades because of its short life cycle, small size, simple growth requirements, small genome (~135 Mb), self-fertilization, and high seed production (Somerville and Koornneef, 2002; Woodward and Bartel, 2018). Researchers also have access to the vast functional genomic toolbox through The Arabidopsis Information Resource (TAIR) web-based platform. TAIR holds genomic, transcriptomic, and metabolomic data, as well as T-DNA mutant collection and seed stocks for forward and reverse genetic approaches. Although Arabidopsis significantly contributed to breakthroughs in our understanding of plant biology, this model organism fell short of describing fundamental features present in crops. Rice is closely related to the cereal crops barley, wheat, and maize (Jackson, 2016), which makes it suitable as a model system for crop grasses.

The release of the Arabidopsis and rice genome provided us a deeper understanding of plant metabolism, including secondary metabolites' biosynthesis and their functions (Kliebenstein, 2004; D'Auria and Gershenzon, 2005; Tohge et al., 2018; Wang et al., 2018). However, to figure out all the steps of a metabolic pathway is not a trivial task. This task becomes dauting when studying biosynthetic pathways in non-model plants, which lack many, if not all, the genetic and genomic tools available for Arabidopsis and rice. Many of the secondary metabolites treasured by the pharmaceutical industry are synthesized by non-model plants.

Plants synthesize a myriad of secondary metabolites to face the challenges imposed by their surroundings. Often, these compounds are produced in a tissue- and developmental-specific manner (Yang et al., 2012). Thus, scientists had paid close attention to these hotspots, searching for clues regarding the biosynthetic pathways. Researchers have identified metabolic-related genes by surveying what genes are expressed through transcriptome profiling. One of the first approaches to capture a snapshot of the transcript present at specific tissues and developmental stages was generating Expression Sequences Tags (EST) collections. Similarly, microarray technology has been paramount for tissue-specific transcriptome profiling and identifying candidate genes for functional analysis. The development of the next generation sequencing (NGS) technologies rapidly increased the number of genes discovered, which previously were out of reach using former transcriptomic approaches.

Terpenoids – also known as isoprenoids – are molecules widely distributed across all kingdoms of life, playing diverse functions in development and environmental stress responses (Cheng et al., 2007; Vranová et al., 2012). Many terpenoids are extremely valuable in the pharmaceutical and food industry (Pichersky and Raguso, 2018). The building blocks of all isoprenoids are isopentyl pyrophosphate (IPP) and dimethylallyl pyrophosphate (DMAPP). IPP and DMAPP are derived by the mevalonate (MAV) and methylerythitrol (MEP) pathways (Rohmer et al., 1996). Plants, some non-photosynthetic fungi, and bacteria produce carotenoids, which are pigments belonging to the group of isoprenoids molecules (Nisar et al., 2015; Hou et al., 2016). However, the functions of carotenoids go beyond conferring just colouring fruits and flowers. Carotenoids also function as photosynthesis pigment accessories, photoprotective agents, hormone precursors and signalling molecules (Hou et al., 2016). Apocarotenoids – molecules derived from the oxidative cleavages of carotenoids – have caught much interest because of their roles in animal and plant development, as well as for their economic value (Beltran and Stange, 2016; Hou et al., 2016). For example, the plant hormones abscisic acid (ABA) and strigolactone (SL) – products of β-carotene oxidative cleavage – are essential in development and biotic and abiotic stress responses (Finkelstein 2013; Zwanenburg et al., 2016). Similarly, the apocarotenoid β-cyclocitral has been suggested to be involved in oxidative stress response in plants (Ramel et al., 2012). In animals, vitamin A and its derivates (namely retinoids) – apocarotenoids arise from the oxidative cleavage of β-carotene – are vital in embryonic development, immunity, and cellular differentiation (Tang and Gudas, 2011; von Lintig, 2012). The apocarotenoids bixin and saffron are highly valuable in the food industry.

Apocarotenoid can be derived from the enzymatic or non-enzymatic cleavage of carotenoids. The carotenoid cleavage dioxygenase (CCD) or reactive oxygen species (ROS) are responsible for apocarotenoid synthesis. The viviparous *vp14* maize mutant played an essential role in cloning and characterizing the first CCD enzyme (Schwartz et al., 1997) and set the way to identifying CCD in other plants. *Vp14* encodes a 9-cis-epoxycarotenoid dioxygenase, which oxidatively cleaves 9-cis-neoxanthin and 9-cis-violaxanthin. This reaction is considered the first committed step in the ABA biosynthetic pathway of ABA (Schwartz et al., 1997).

The essential roles of apocarotenoids in human health and plant physiology, as well as their industrial applications, are the driving force to study their functions and biosynthesis. Table 4.1 summarizes the approach used for gene discovery in some apocarotenoid pathways. In this chapter, we will discuss the different transcriptomic approaches applied to the investigate apocarotenoid pathways.

EXPRESSION SEQUENCES TAG (EST)

Nowadays, genome sequencing and transcriptome analysis have become an achievable and relatively straightforward goal thanks to next-generation sequencing technologies. 30 years ago, the scientific scenario was different. Back then, genome sequencing was a complex, lengthy, and expensive task involving multiple cloning-based strategies and relying on Sanger sequencing. EST collections provided a feasible alternative to whole-genome sequencing because of the inexpensive and efficient

TABLE 4.1
Approaches for gene discovery of apocarotenoid-related genes

Apocarotenoid	Organism	Gene	Approach	Reference
Bixin	*Bixa orellana*	*BoCCD1(4)* *BoCCD4(5)* *BoALDH(20)* *BoSABATH(12)*	RNA-seq	Cárdenas-Conejo et al. (2015)
	Vaccinium corymbosum	*VcBADH1* *VcBADH2* *VcBADH3* *VcnBMT1* *VcnBMT2*	RNA-seq	Gupta et al. (2015)
Saffron	*Crocus sativus*	*CsCCD2*	RNA-seq	Frusciante et al. (2014)
		CsULT1	EST analysis	Ashraf, Jain, and Vishwakarma (2015)
Abscisic acid (ABA)	*Arabidopsis thaliana*	*AtCCD1*	EST analysis	Schwartz, Qin, and Zeevaart (2001)
β-ionone	*Lycopersicon esculentum*, cv. M82	*LeCC1a* and *LeCC1b*	EST analysis	Simkin et al. (2004)
	Cucumis melo	*CuCC1*	EST analysis	Ibdah et al. (2006)
	Vitis vinifera	*VvCC1*	EST analysis	Lashbrooke et al. (2013)
Mycorradicin	*Zea mays*	*ZmCC1*	EST analysis	Sun et al. (2008)
Vitamin A	*Mus musculus*	*β-CD*	EST analysis	Redmond et al. (2001)
		BCO2	EST analysis	Kiefer et al. (2001)

In the Gene column, numbers in parenthesis () indicate the quantity of genes identified.

nature of this approach. ESTs are partial sequences of mRNA obtained by single-pass sequencing of randomly picked clones from cDNA libraries (Parkinson and Blaxter, 2009). In the nineties and early two-thousands, the analysis of EST collections was a widely adopted tool for gene discovery, mapping, and structure prediction (Bouchez and Höfte, 1998; Gruber, 2007). In the quest to fill up the gaps in metabolic pathways, the screening of EST collection led to the identification of genes involved in primary and secondary metabolism. Over the years, many research groups dedicated to generating EST sequences from many organisms built up a vast repository of sequencing data that is publicly available in the National Center of Biotechnology Information (NCBI https://www.ncbi.nlm.nih.gov/).

In plant biology research, EST collections have proved to be an invaluable resource to study apocarotenoid metabolism and biosynthesis. For instance, a homology-sequence search in Arabidopsis EST collections, using the maize VP14 as query, allowed the identification of the Arabidopsis CCDs (Schwartz et al., 2001; Tan et al., 2003). Moreover, Arabidopsis ESTs facilitated identifying and cloning of CCDs in other plants. Because of its importance as a model plant, the Arabidopsis research community developed EST collections from different tissue such as roots, leaves, inflorescence, whole seedlings, and seeds (Höfte et al., 1993; Newman et al., 1994; White et al., 2000). Other EST collections were also developed from other model and crop plants like rice, maize, and brassica (Uchimiya et al., 1992; Keith et al., 1993; Park et al., 1993; Sasaki et al., 1994).

Non-model organisms also benefited from the inexpensive nature of ESTs. Saffron is one of the most commercially valuable spices worldwide. The apocarotenoids crocetin, picrocrocin, and safranal provide the colour, flavour, and aroma, respectively (Caballero-Ortega et al., 2007). These apocarotenoids – products of the oxidative cleavage of zeaxanthin – accumulate tenaciously in the stigmas of *Crocus sativus* (Bouvier et al., 2003b). Therefore, stigma was the obvious choice to look for transcripts associated with saffron-apocarotenoid biosynthesis. An EST collection derived from mature dried stigmas from *C. sativus* was developed to explore saffron-apocarotenoids biosynthesis and other mechanisms related to stigma development and Iridaceae family (D'Agostino et al., 2007). This EST database has been useful not only to study biosynthetic genes but also to identify transcriptional regulators as demonstrated by Ashraf et al. (Ashraf et al., 2015). This group identified and cloned the transcription factor CsULT1 (Ashraf et al., 2015), taking advantage of this genomic resource. Transient overexpression of *CsULT1* in *Crocus* calli increased the transcript levels of genes involved in the MEP pathway, carotenoid, and saffron-related apocarotenoid biosynthetic pathway.

Bixin is another commercially valuable apocarotenoid to the food industry. *Bixa orellana* accumulates large quantities of bixin in the seeds. Thus, this tissue has been subjected to transcriptome analysis. A bixin biosynthetic pathway was proposed by (Bouvier et al., 2003a), suggesting a three-step pathway and lycopene as a precursor. The genes proposed to be involved in bixin biosynthesis are lycopene cleavage dioxygenase (*BoLCD*), bixin aldehyde dehydrogenase (*BoBALDH*), norbixin carboxyl methyltransferase (*BonBMT*) (Bouvier et al., 2003a). EST collections from *B. orellana* have provided valuable insight into the bixin biosynthesis and carotenoid metabolism (Jako et al., 2002; Soares et al., 2011).

In fruits, volatile compounds play an important role in conferring flavour and aroma. For example, β-ionone is a volatile apocarotenoid, present in aroma of fresh tomatoes – although at low concentrations – and contributes greatly to tomatoes' flavour (Elizabeth et al., 2000). By mining an EST tomato (*Lycopersicon esculentum*) library searching for *AtCCD1* homologs, *LeCCD1a*, and *LeCCD1b* were identified (Simkin et al., 2004). Functional analysis of both genes in a heterologous system and *in vitro* showed their abroad carotenoid-substrate specify, and that they cleave β-carotene to yield β-ionone. Analysis of knock-down plants for either gene revealed a significant reduction of β-ionone in transgenic plants compared to wild-type controls. These results suggested that both *LeCC1a* and *LeCC1b* are involved in the biosynthesis β-ionone. Studies in melon (*Cucumis melo* L.) (Ibdah et al., 2006) and grapes (*Vitis vinifera*) (Lashbrooke et al., 2013) – searching in EST libraries – showed that *CuCCD1* and *VvCCD1* oxidatively cleave β-carotene to generate β-ionone.

In mammals, vitamin A and its derivates (namely retinoids) are essential apocarotenoid molecules for development (Zile, 2001), cellular differentiation and immunity (Tang and Gudas, 2011; von Lintig, 2012). Mammals cannot synthesize retinoids; thus, they must obtain the precursor from plant or animal sources as provitamin and preassemble vitamin A (Toti et al., 2018), respectively. β-carotene – provitamin A carotenoid – is metabolized in mammalian tissues to produce retinoids. β-carotene is cleaved by carotenoid cleavage dioxygenase enzyme (CCD). The mouse β-carotene 15,15′-dioxygenase (β-CD) is the first CCD identified in mammals (Redmond et al., 2001). Redmond et al. (2001) screened a mouse kidney EST database for sequence similarities to the fruit fly (*Drosophila melanogaster*) retinal pigment epithelium-expressed protein RPE65, maize VP14, and the bacterial lignostilbene α, β-dioxygenase. β-CD catalyses the central cleavage of β-carotene to generate all-*trans*-retinal, a key, and necessary step on vitamin A biosynthesis. EST database analysis also revealed another carotenoid dioxygenase, B-diox-II (also known as BCO2) which cleaves β carotenoid to give rise to β-apo-10′-carotenal and β-ionone (Kiefer et al., 2001). Mice *BCO2* mutants show impairment in carotenoid metabolism and mitochondria oxidative stress (Amengual et al., 2010), which ultimately leads to apoptosis (Lobo et al., 2012).

The index of expressed genes contained in EST libraries has helped research on secondary metabolism pathways moving forward. Scientists have been able to identify key players in the biosynthesis pathway of important apocarotenoids such as ABA, bixin, saffron-apocarotenoids, and retinoids through homology-based screening of EST collections. This genomic resource allowed the cloning and characterization of CCDs from Arabidopsis and non-model plants. Even though the large-scale single-pass sequencing approach allows generating data at a low-cost and makes the study of non-model organisms feasible, it comes with some limitations that hampered gene discovery (Alba et al., 2004). For example, the EST collections do not represent the entire transcript population from where they are derived. Low abundant or rare transcripts are likely to be under-represented; in consequence, researchers had to take additional normalization or subtraction steps (Bonaldo et al., 1996). Also, the quality of the short sequence (~200–800 bp) is low. EST captures partial sequences from either the 5′ or/and 3′ end by sequencing only once, which makes this approach prone to errors.

TRANSCRIPTOME PROFILING BY MICROARRAYS

Microarray technology – also known as DNA chip – gave scientists the ability to study gene expression at the genome level (Schena et al., 1995) by analysing multiple genes in parallel. Before microarrays, we were only able to study a few genes at the time. DNA microarrays provided a tool to explore transcriptome in a global way (Brown and Botstein, 1999). The basis of microarray technology lies in the hybridization of complementary sequences. In brief, microarrays employ thousands of known DNA sequences (probes) printed to a specific location at a surface or matrix. The samples of interest are labelled with a fluorescent dye during cDNA synthesis; hence, if a gene is expressed, it will hybridize to a complementary DNA probe, giving a fluorescent signal. The intensity of the fluorescent signal is proportionate to the amount of transcript present in the sample. Therefore, one could compare the expression levels of two samples using a different fluorescent dye in each sample (Shalon et al., 1996). Unlike EST gene expression studies, microarray analysis allowed to quantify transcript abundance from specific tissues in a reliable way. For the very first time, microarrays allowed researchers to look at global gene expression patterns and how different external or environmental conditions, as well as mutations at specific genes, impact the whole organism.

The function of many apocarotenoids is still a question up in the air. Some authors have proposed some apocarotenoids functioning as signal molecules (Estavillo et al., 2013). β-cyclocitral has been suggested to play a role as a stress signal molecule (Ramel et al., 2012; Estavillo et al., 2013; Havaux 2014; Hou et al., 2016; D'Alessandro et al., 2019) as well as root growth regulator (Dickinson et al., 2019). Reactive oxygen species (ROS) are by-products of photosynthesis, which is exacerbated by environmental stress such as excess of light (D'Alessandro et al., 2018; D'Alessandro and Havaux, 2019). The singlet oxygen ($_1O^2$) – generated by triplet-excited chlorophyll transferring electrons to O_2 (Krieger-Liszkay et al., 2008; Triantaphylidès and Havaux, 2009) – is the major player on photooxidative damage to plant leaves (Triantaphylidès and Havaux, 2009). The β-carotene molecules from photosystem II are oxidized by $_1O^2$ generating various products; one of these is β-cyclocitral (Krieger-Liszkay et al., 2008; González-Pérez et al., 2011; D'Alessandro and Havaux, 2019). Ramel et al. (2012) performed experiments in Arabidopsis, aiming to understand the role of β-cyclocitral in oxidative stress signalling (Ramel et al., 2012). Their results revealed that β-cyclocitral levels increased under high-light stress and suggested this apocarotenoid as intermediate in the $_1O^2$ signalling pathway. Microarrays experiments showed that β-cyclocitral increases the transcript levels of genes involved in defence mechanisms.

Microarrays also helped to reveal the role of some CCDs from crop plants. For example, transcriptome profiling of two varieties of potato (*Solanum tuberosum*) – differing on the carotenoid content in the tuber (Ducreux et al., 2008) – showed that *CCD4* expression was higher in the low-carotenoid variety compared to the high-carotenoid variety (Ducreux et al., 2008). The results of the microarray analysis guided the researchers to explore the function of the potato CCD4 further. Campbell et al. (2010) performed microarrays analysis on *CCD4* RNA interference (RNAi) knockdown plants to determine whether CCD4 was responsible for the carotenoid-content trait (Campbell et al., 2010). The RNAi experiment confirmed that the downregulation of *CCD4* results in higher carotenoid content in tubers compared to control (Campbell et al., 2010).

Microarrays have been a valuable tool to explore the apocarotenoid-mediated signal pathways. In human health, retinoids have been reported to reduce and prevent some types of cancers (Soprano et al., 2004; Sharoni et al., 2012). Also, retinoids have been reported to inhibit second primary cancer growth (Hong et al., 1990; Muto et al., 1996). Experiments in human bronchial epithelian cells suggested that retinoids are linked to cell cycle control (Langenfeld et al., 1997). To elucidate the biological mechanism and signalling pathway of retinoids, Ma et al. (2003) looked for genes differentially expressed between retinoid sensitive and resistant human bronchial epithelial cell lines. The research group identified three novel retinoid targets: placental bone morphogenic protein (PLAB), polyamine oxidase isoform1 (PAOh1), and E74-like factor 3 (ELF3) (Ma et al., 2003).

Moreover, transcriptome profiling of pancreatic tissue from cancer patients showed abnormal expression of genes involved in the retinoid signalling pathway (Segara et al., 2005). Here, the author provided evidence that a retinoid-inducible transcription factor HOXB2 is expressed at higher levels in cancer tissue compared to control. Abnormal expression of *HOX* genes has been linked to the development of kidney, breast, and colon cancers (Cillo et al., 1992; De Vita et al., 1993; Naora et al., 2001). Other studies in retinoid-receptors knock out teratocarcinoma stem cell lines (Su and Gudas, 2008a, 2008b) and murine embryonic stem cell shed to light on retinoids mechanism of cell proliferation and differentiation (Kashyap et al., 2013).

While microarrays offered the opportunity to survey gene expression at the whole-genome level, some limitations hindered capturing all the transcripts present in a given sample (Black et al., 2013; Bumgarner, 2013). Some drawbacks of microarray technology are the requirements of prior genomic information, cross-hybridization, and restricted range of detection. Microarrays require sequence information – EST or genome annotation – to design the probes. Hence, we were limited to obtain expression data from known genes, hampering the discovery of novel genes. As a result, a substantial portion of all the transcripts present in the sample were left out from the analysis (Black et al., 2013; Lowe et al., 2017). For research on non-model organisms, microarray analysis was out of reach since many of them lacked genomic resources.

Another limitation is cross-hybridization. Cross-hybridization can occur when more than a gene shares high sequence similarity with a specific probe. Consequently, unspecific hybridization creates background signal (Marioni et al., 2008).

Furthermore, the dynamic range of microarrays is limited. The dynamic range depends on fluorescence detention and saturation. As a consequence, genes expressed at high levels can cause signal saturation, while genes expressed at low levels result in poor resolution (Black et al., 2013; Lowe et al., 2017).

NEXT GENERATION SEQUENCING (NGS)

The search for cost-effective, faster, and more accurate sequencing methods stimulated technological advances leading to the next generation sequencing platforms (NGS). Compared to the traditional Sanger sequencing method, which processes one DNA fragment at the time, NGS technology allows high throughput by parallel sequencing (Metzker, 2010). As better sequencing methods are developed, researchers have more options at hand to explore the genome and transcriptome of any organism. Some of the commercially available NGS platforms are illumina, PacBio (Pacific Bioscience), and Oxford nanopore, each one differing in the length and number of reads output (Lu et al., 2016). For instance, illumina platform generates short reads (up to 300 bp) while PacBio (~20 Kb) and Oxford nanopore (up to 150Kb) give long reads (Eid et al., 2009; Goodwin et al., 2016; Miten et al., 2016; Li and Harkess, 2018).

Transcriptome profiling through NGS platforms is a method known as RNA-sequencing (RNA-seq). RNA-seq overcame the drawbacks of microarrays-based transcriptomic studies such as prior knowledge of gene sequences and cross-hybridization. More importantly, this approach allows capturing the entire transcriptome revealing not only novel transcripts, but also alternative splicing, and non-coding RNAs. RNA-seq opened the door to explore the secondary metabolism of commercially important non-model plants (Unamba et al., 2015).

Research in *B. orellana* and *C. sativus* were favoured by high-throughput sequencing, since neither plants have complete genome sequences, and the only gene expression data came from EST libraries (Jako et al., 2002; D'Agostino et al., 2007; Soares et al., 2011). For *B. orellana*, the EST libraries only contain partial sequences of transcripts from developing seed coats and whole-developing seeds. RNA-seq analysis of immature and mature seeds and leaves of *B. orellana* provided further insight into the MEP pathway and identified new candidates of the bixin biosynthetic pathway (Cárdenas-Conejo et al., 2015). Interestingly, the proposed bixin biosynthetic genes (Bouvier et al., 2003a) did not appear in the RNA-seq data, suggesting that other genes are responsible for bixin synthesis. Carballo-Uicab et al. (2019) functionally characterized some of these candidates, BoCCD1-1 and BoCC4-3, in a heterologous system demonstrating that both CCD cleave lycopene to produce bixin aldehyde, an intermedia metabolite in the bixin biosynthesis (Carballo-Uicab et al., 2019). RNA-seq analysis in blueberry (*Vaccinium corymbosum*) uncovered the expression of bixin biosynthetic genes in green fruits (Gupta et al., 2015). Moreover, CCD genes expression has been detected in the grapevine tree (*Vitis vinifera*) (Leng et al., 2017).

C. sativus share same rocky road as *B. orellana*; the limited genomic resources available for both organisms have imposed a bit of a challenge not only to search biosynthetic genes, but also to explore transcriptional networks and apocarotenoid transporters. RNA-seq-based transcriptomics has greatly contributed to revealing critical components in the saffron biosynthetic pathway. Transcriptome analysis of different stigma developmental stages showed that *CCD2* expression was high during crocetin and crocin accumulation. This gene, a novel CCD class, cleaves zeaxanthin, producing crocetin dialdehyde and crocetin (Frusciante et al., 2014). Even though the saffron-apocarotenoids pathways have received attention, and many steps have been figured out, there still are gaps to be filled. Baba et al. (2015) analysed flowers and stigma to provide the whole picture of genes expression aiming to discover genes and transcription factors regulating carotenoid and apocarotenoid biosynthesis (Baba et al., 2015). Transcriptome profiling of corm, tepal, leaf, stigma, and stamen revealed the tissue-specific expression pattern of saffron-apocarotenoid biosynthetic genes (Mukesh et al., 2016). The expression of genes on apocarotenoid biosynthesis was higher in stigma compared to that of other tissues. Moreover, RNA-seq data showed transcription factors such as WRKY, MYB, bHLH, MADS, C2C2-YABBY, and SNF in stigma (Mukesh et al., 2016).

Another study analysed the transcriptome of tepal sector of the flowers *Crocus sieberi* (Ahrazem et al., 2018). In *Crocus* species, the characteristic yellow colour of the flower is due to crocin. The flowers of *C. sieberi* have lilac, and yellow sectors and crocin accumulate in the yellow sector (Rubio Moraga et al., 2013) as well as in the stigmas. The RNA-seq data revealed information about the apocarotenoid biosynthesis regulation and chromoplast biogenesis.

FUTURE DIRECTIONS

We are experiencing rapid progress in sequencing technologies, which, in turn, has been translated into making genomic research affordable and accessible to many laboratories around the world. Next-generation sequencing platforms have contributed significantly to obtain whole-genome sequences from many plant and animal organisms, in a cost-effective manner. These platforms have also favoured transcriptome analysis in a way predecessor technology could not achieve. Also, researchers continue to innovate genome-scale gene expression survey protocols to obtain a more accurate and precise landscape of transcripts at tissue and cell levels. Since NGS platforms have been widely adopted for sequencing projects, we have generated an unprecedented amount of genomic and transcriptome data. All of the raw RNA-seq data are deposited into the publicly available NCBI databases Refseq (https://www.ncbi.nlm.nih.gov/refseq/about/). For plant biologists, a user-friendly web tool is available containing not only RNA-seq, but also small RNAs, chromatin immunoprecipitation and degradome sequencing data from many plant species (https://mpss.danforthcenter.org/) (Nakano et al., 2020). Therefore, data mining those resources represents an opportunity to gain more in-depth insight into the metabolic pathways and in their regulation.

Data mining of RNA-seq databases targeting transcription factors can uncover regulatory networks in apocarotenoid pathways. The transcriptional regulation of apocarotenoid biosynthesis – such as bixin, crocin, and β-ionone – is still a mystery. Similarly, data mining can also target apocarotenoid transporters. After synthesis, some metabolites are sequestered into the vacuole to avoid feedback regulation and toxic effects. Demurtas et al. (2018) proposed that the crocin biosynthesis takes place in different cell compartments with the final product – crocin – ending up stored in the vacuole (Demurtas et al., 2018). Data mining of *C. sativus* RNA-seq helped to find candidate genes for tonoplast transporters – CsABCC4a and CsABCC2 – that co-expressed with saffron-apocarotenoid biosynthetic genes (Demurtas et al., 2019).

The number of genomes available is rising since the NGS platform came into the scene, becoming the predominant approach for sequencing. The genome sequencing of *B. orellana* and *C. sativus* will provide the genome structure and the tools pinpoint the location of the biosynthetic genes. We will be able to perform comparative genomics analysis. Another interesting aspect we will be able to explore is the promoter regions of the apocarotenoid-related genes.

ACKNOWLEDGEMENTS

The authors thank Consejo Nacional de Ciencia y Tecnología (CONACYT) for their support with the project: Fronteras de la Ciencia No. 2016-01-1716. DC, postdoctoral position (FC No. 2016-01-1716) and Laboratory technical, respectively, were financially supported by CONACYT.

REFERENCES

Ahrazem, Oussama, Javier Argandoña, Alessia Fiore, Carolina Aguado, Rafael Luján, Ángela Rubio-Moraga, Mónica Marro, et al. 2018. 'Transcriptome Analysis in Tissue Sectors with Contrasting Crocins Accumulation Provides Novel Insights into Apocarotenoid Biosynthesis and Regulation during Chromoplast Biogenesis.' *Scientific Reports* 8 (1): 2843. https://doi.org/10.1038/s41598-018-21225-z.

Alba, Rob, Zhangjun Fei, Paxton Payton, Yang Liu, Shanna L Moore, Paul Debbie, Jonathan Cohn, et al. 2004. 'ESTs, CDNA Microarrays, and Gene Expression Profiling: Tools for Dissecting Plant Physiology and Development.' *The Plant Journal* 39 (5): 697–714. https://doi.org/10.1111/j.1365-313X.2004.02178.x.

Amengual, Jaume, Glenn P Lobo, Marcin Golczak, Hua Nan M Li, Tatyana Klimova, Charles L Hoppel, Adrian Wyss, Krzysztof Palczewski, and Johannes von Lintig. 2010. 'A Mitochondrial Enzyme Degrades Carotenoids and Protects against Oxidative Stress.' *The FASEB Journal* 25 (3): 948–959. https://doi.org/10.1096/fj.10-173906.

Ashraf, Nasheeman, Deepti Jain, and Ram A Vishwakarma. 2015. 'Identification, Cloning and Characterization of an Ultrapetala Transcription Factor CsULT1 from Crocus: A Novel Regulator of Apocarotenoid Biosynthesis.' *BMC Plant Biology* 15 (1): 25. https://doi.org/10.1186/s12870-015-0423-7.

Baba, Shoib Ahmad, Tabasum Mohiuddin, Swaraj Basu, Mohit Kumar Swarnkar, Aubid Hussain Malik, Zahoor Ahmed Wani, Nazia Abbas, Anil Kumar Singh, and Nasheeman Ashraf. 2015. 'Comprehensive Transcriptome Analysis of Crocus Sativus for Discovery and Expression of Genes Involved in Apocarotenoid Biosynthesis.' *BMC Genomics* 16 (1): 698. https://doi.org/10.1186/s12864-015-1894-5.

Beltran, Juan Camilo Moreno, and Claudia Stange. 2016. 'Apocarotenoids: A New Carotenoid-Derived Pathway BT – Carotenoids in Nature: Biosynthesis, Regulation and Function.' In *Carotenoids in Nature. Subcellular Biochemistry*, vol 79, edited by Claudia Stange, 239–272. Cham: Springer International Publishing. https://doi.org/10.1007/978-3-319-39126-7_9.

Black, Michael B, Bethany B Parks, Linda Pluta, Tzu-Ming Chu, Bruce C Allen, Russell D Wolfinger, and Russell S Thomas. 2013. 'Comparison of Microarrays and RNA-Seq for Gene Expression Analyses of Dose-Response Experiments.' *Toxicological Sciences* 137 (2): 385–403. https://doi.org/10.1093/toxsci/kft249.

Bonaldo, MF, G Lennon, and MB Soares. 1996. 'Normalization and Subtraction: Two Approaches to Facilitate Gene Discovery.' *Genome Research* 6 (9): 791–806. http://genome.cshlp.org/content/6/9/791.abstract.

Bouchez, David, and Herman Höfte. 1998. 'Functional Genomics in Plants.' *Plant Physiology* 118 (3): 725 LP–732. https://doi.org/10.1104/pp.118.3.725.

Bouvier, Florence, Odette Dogbo, and Bilal Camara. 2003a. 'Biosynthesis of the Food and Cosmetic Plant Pigment Bixin (Annatto).' *Science* 300 (5628): 2089. https://doi.org/10.1126/science.1085162.

Bouvier, Florence, Claude Suire, Jérôme Mutterer, and Bilal Camara. 2003b. 'Oxidative Remodeling of Chromoplast Carotenoids.' *The Plant Cell* 15 (1): 47. https://doi.org/10.1105/tpc.006536.

Brown, Patrick O, and David Botstein. 1999. 'Exploring the New World of the Genome with DNA Microarrays.' *Nature Genetics* 21 (1): 33–37. https://doi.org/10.1038/4462.

Bumgarner, Roger. 2013. 'Overview of DNA Microarrays: Types, Applications, and Their Future.' *Current Protocols in Molecular Biology* 101 (1): 22.1.1–22.1.11. https://doi.org/10.1002/0471142727.mb2201s101.

Caballero-Ortega, Heriberto, Rogelio Pereda-Miranda, and Fikrat I Abdullaev. 2007. 'HPLC Quantification of Major Active Components from 11 Different Saffron (*Crocus sativus L.*) Sources.' *Food Chemistry* 100 (3): 1126–1131. https://doi.org/10.1016/j.foodchem.2005.11.020.

Campbell, Raymond, Laurence JM Ducreux, Wayne L Morris, Jenny A Morris, Jeffrey C Suttle, Gavin Ramsay, Glenn J Bryan, Pete E Hedley, and Mark A Taylor. 2010. 'The Metabolic and Developmental Roles of Carotenoid Cleavage Dioxygenase4 from Potato.' *Plant Physiology* 154 (2): 656. https://doi.org/10.1104/pp.110.158733.

Carballo-Uicab, Víctor Manuel, Yair Cárdenas-Conejo, Alba Adriana Vallejo-Cardona, Margarita Aguilar-Espinosa, Jacobo Rodríguez-Campos, Hugo Serrano-Posada, José Alberto Narváez-Zapata, Felipe Vázquez-Flota, and Renata Rivera-Madrid. 2019. 'Isolation and Functional Characterization of Two Dioxygenases Putatively Involved in Bixin Biosynthesis in Annatto (*Bixa orellana L.*).' Edited by Axel Tiessen. *PeerJ* 7: e7064. https://doi.org/10.7717/peerj.7064.

Cárdenas-Conejo, Yair, Víctor Carballo-Uicab, Meric Lieberman, Margarita Aguilar-Espinosa, Luca Comai, and Renata Rivera-Madrid. 2015. 'De Novo Transcriptome Sequencing in Bixa Orellana to Identify Genes Involved in Methylerythritol Phosphate, Carotenoid and Bixin Biosynthesis.' *BMC Genomics* 16 (1): 877. https://doi.org/10.1186/s12864-015-2065-4.

Cheng, Ai-Xia, Yong-Gen Lou, Ying-Bo Mao, Shan Lu, Ling-Jian Wang, and Xiao-Ya Chen. 2007. 'Plant Terpenoids: Biosynthesis and Ecological Functions.' *Journal of Integrative Plant Biology* 49 (2): 179–186. https://doi.org/10.1111/j.1744-7909.2007.00395.x.

Cillo, C, P Barba, G Freschi, G Bucciarelli, MC Magli, and E Boncinelli. 1992. 'HOX Gene Expression in Normal and Neoplastic Human Kidney.' *International Journal of Cancer* 51 (6): 892–897. https://doi.org/10.1002/ijc.2910510610.

D'Alessandro, Stefano, and Michel Havaux. 2019. 'Sensing β-Carotene Oxidation in Photosystem II to Master Plant Stress Tolerance.' *New Phytologist* 223 (4): 1776–1783. https://doi.org/10.1111/nph.15924.

D'Alessandro, Stefano, Brigitte Ksas, and Michel Havaux. 2018. 'Decoding β-Cyclocitral-Mediated Retrograde Signaling Reveals the Role of a Detoxification Response in Plant Tolerance to Photooxidative Stress.' *The Plant Cell* 30 (10): 2495 LP–2511. https://doi.org/10.1105/tpc.18.00578.

D'Alessandro, Stefano, Yusuke Mizokami, Bertrand Légeret, and Michel Havaux. 2019. 'The Apocarotenoid β-Cyclocitric Acid Elicits Drought Tolerance in Plants.' *IScience* 19: 461–473. https://doi.org/10.1016/j.isci.2019.08.003.

D'Agostino, Nunzio, Daniele Pizzichini, Maria Luisa Chiusano, and Giovanni Giuliano. 2007. 'An EST Database from Saffron Stigmas.' *BMC Plant Biology* 7 (1): 53. https://doi.org/10.1186/1471-2229-7-53.

D'Auria, John C, and Jonathan Gershenzon. 2005. 'The Secondary Metabolism of Arabidopsis Thaliana: Growing like a Weed.' *Current Opinion in Plant Biology* 8 (3): 308–316. https://doi.org/10.1016/j.pbi.2005.03.012.

Demurtas, Olivia Costantina, Rita de Brito Francisco, Gianfranco Diretto, Paola Ferrante, Sarah Frusciante, Marco Pietrella, Giuseppe Aprea, et al. 2019. 'ABCC Transporters Mediate the Vacuolar Accumulation of Crocins in Saffron Stigmas.' *The Plant Cell* 31 (11): 2789 LP–2804. https://doi.org/10.1105/tpc.19.00193.

Demurtas, Olivia Costantina, Sarah Frusciante, Paola Ferrante, Gianfranco Diretto, Noraddin Hosseinpour Azad, Marco Pietrella, Giuseppe Aprea, et al. 2018. 'Candidate Enzymes for Saffron Crocin Biosynthesis Are Localized in Multiple Cellular Compartments.' *Plant Physiology* 177 (3): 990 LP – 1006. https://doi.org/10.1104/pp.17.01815.

Dickinson, Alexandra J, Kevin Lehner, Jianing Mi, Kun-Peng Jia, Medhavinee Mijar, José Dinneny, Salim Al-Babili, and Philip N Benfey. 2019. 'β-Cyclocitral Is a Conserved Root Growth Regulator.' *Proceedings of the National Academy of Sciences* 116 (21): 10563–10567. https://doi.org/10.1073/pnas.1821445116.

Ducreux, Laurence JM, Wayne L Morris, Ian M Prosser, Jenny A Morris, Michael H Beale, Frank Wright, Tom Shepherd, Glenn J Bryan, Pete E Hedley, and Mark A Taylor. 2008. 'Expression Profiling of Potato Germplasm Differentiated in Quality Traits Leads to the Identification of Candidate Flavour and Texture Genes.' *Journal of Experimental Botany* 59 (15): 4219–4231. https://doi.org/10.1093/jxb/ern264.

Eid, John, Adrian Fehr, Jeremy Gray, Khai Luong, John Lyle, Geoff Otto, Paul Peluso, et al. 2009. 'Real-Time DNA Sequencing from Single Polymerase Molecules.' *Science* 323 (5910): 133. https://doi.org/10.1126/science.1162986.

Elizabeth, A Baldwin, W Scott John, K Shewmaker Christine, and Schuch Wolfgang. 2000. 'Flavor Trivia and Tomato Aroma: Biochemistry and Possible Mechanisms for Control of Important Aroma Components.' *HortScience HortSci* 35 (6): 1013–1022. https://doi.org/10.21273/HORTSCI.35.6.1013.

Estavillo, Gonzalo, Kai Chan, Su Yin Phua, and Barry Pogson. 2013. 'Reconsidering the Nature and Mode of Action of Metabolite Retrograde Signals from the Chloroplast.' *Frontiers in Plant Science* 3: 300. https://www.frontiersin.org/article/10.3389/fpls.2012.00300.

Finkelstein, Ruth. 2013. 'Abscisic Acid Synthesis and Response.' *The Arabidopsis Book* 11 (*November*): e0166–e0166. https://doi.org/10.1199/tab.0166.

Frusciante, Sarah, Gianfranco Diretto, Mark Bruno, Paola Ferrante, Marco Pietrella, Alfonso Prado-Cabrero, Angela Rubio-Moraga, et al. 2014. 'Novel Carotenoid Cleavage Dioxygenase Catalyzes the First Dedicated Step in Saffron Crocin Biosynthesis.' *Proceedings of the National Academy of Sciences* 111 (33): 12246. https://doi.org/10.1073/pnas.1404629111.

González-Pérez, Sergio, Jorge Gutiérrez, Francisco García-García, Daniel Osuna, Joaquín Dopazo, Óscar Lorenzo, José L Revuelta, and Juan B Arellano. 2011. 'Early Transcriptional Defense Responses in Arabidopsis Cell Suspension Culture under High-Light Conditions.' *Plant Physiology* 156 (3): 1439 LP 1456. https://doi.org/10.1104/pp.111.177766.

Goodwin, Sara, John D McPherson, and W Richard McCombie. 2016. 'Coming of Age: Ten Years of Next-Generation Sequencing Technologies.' *Nature Reviews Genetics* 17: 333. https://doi.org/10.1038/nrg.2016.49.

Gruber, Arthur. 2007. 'Expressed Sequence Tags.' In *Bioinformatics*, edited by PH Dear, 141–167. USA: Scion Publishing.

Gupta, Vikas, April D Estrada, Ivory Blakley, Rob Reid, Ketan Patel, Mason D Meyer, Stig Uggerhøj Andersen, Allan F Brown, Mary Ann Lila, and Ann E Loraine. 2015. 'RNA-Seq Analysis and Annotation of a Draft Blueberry Genome Assembly Identifies Candidate Genes Involved in Fruit Ripening, Biosynthesis of Bioactive Compounds, and Stage-Specific Alternative Splicing.' *GigaScience* 4 (1). https://doi.org/10.1186/s13742-015-0046-9.

Havaux, Michel. 2014. 'Carotenoid Oxidation Products as Stress Signals in Plants.' *The Plant Journal* 79 (4): 597–606. https://doi.org/10.1111/tpj.12386.

Höfte, Herman, Thierry Desprez, Joëlle Amselem, Hélène Chiapello, Michel Caboche, Annick Moisan, Marie-Françoise Jourjon, et al. 1993. 'An Inventory of 1152 Expressed Sequence Tags Obtained by Partial Sequencing of CDNAs from Arabidopsis Thaliana†.' *The Plant Journal* 4 (6): 1051–1061. https://doi.org/10.1046/j.1365-313X.1993.04061051.x.

Hong, Waun KI, Scott M Lippman, Loretta M Itri, Daniel D Karp, Jin S Lee, Robert M Byers, Stimson P Schantz, et al. 1990. 'Prevention of Second Primary Tumors with Isotretinoin in Squamous-Cell Carcinoma of the Head and Neck.' *New England Journal of Medicine* 323 (12): 795–801. https://doi.org/10.1056/NEJM199009203231205.

Hou, Xin, John Rivers, Patricia León, Ryan P McQuinn, and Barry J Pogson. 2016. 'Synthesis and Function of Apocarotenoid Signals in Plants.' *Trends in Plant Science* 21 (9): 792–803. https://doi.org/10.1016/j.tplants.2016.06.001.

Ibdah, Mwafaq, Yaniv Azulay, Vitaly Portnoy, Boris Wasserman, Einat Bar, Ayala Meir, Yossi Burger, et al. 2006. 'Functional Characterization of CmCCD1, a Carotenoid Cleavage Dioxygenase from Melon.' *Phytochemistry* 67 (15): 1579–1589. https://doi.org/10.1016/j.phytochem.2006.02.009.

Jackson, Scott A. 2016. 'Rice: The First Crop Genome.' *Rice* 9 (1): 14. https://doi.org/10.1186/s12284-016-0087-4.

Miten, Jain, Hugh E Olsen, Benedict Paten, and Mark Akeson. 2016. 'The Oxford Nanopore MinION: Delivery of Nanopore Sequencing to the Genomics Community.' *Genome Biology* 17 (1): 239. https://doi.org/10.1186/s13059-016-1103-0.

Mukesh, Jain, Prabhakar Lal Srivastava, Mohit Verma, Rajesh Ghangal, and Rohini Garg. 2016. 'De Novo Transcriptome Assembly and Comprehensive Expression Profiling in Crocus Sativus to Gain Insights into Apocarotenoid Biosynthesis.' *Scientific Reports* 6: 22456. https://doi.org/10.1038/srep22456; https://www.nature.com/articles/srep22456#supplementary-information.

Jako, Colette, Catherine Coutu, Ingeborg Roewer, Darwin W Reed, Lawrence E Pelcher, and Patrick S Covello. 2002. 'Probing Carotenoid Biosynthesis in Developing Seed Coats of Bixa Orellana (Bixaceae) through Expressed Sequence Tag Analysis.' *Plant Science* 163 (1): 141–145. https://doi.org/10.1016/S0168-9452(02)00083-3.

Kashyap, Vasundhra, Kristian B Laursen, Fabienne Brenet, Agnes J Viale, Joseph M Scandura, and Lorraine J Gudas. 2013. 'RARγ Is Essential for Retinoic Acid Induced Chromatin Remodeling and Transcriptional Activation in Embryonic Stem Cells.' *Journal of Cell Science* 126 (4): 999–1008. https://doi.org/10.1242/jcs.119701.

Keith, CS, DO Hoang, BM Barrett, B Feigelman, MC Nelson, H Thai, and C Baysdorfer. 1993. 'Partial Sequence Analysis of 130 Randomly Selected Maize CDNA Clones.' *Plant Physiology* 101 (1): 329. https://doi.org/10.1104/pp.101.1.329.

Kiefer, Cornelia, Susanne Hessel, Johanna M Lampert, Klaus Vogt, Markus O Lederer, Dietmar E Breithaupt, and Johannes von Lintig. 2001. 'Identification and Characterization of a Mammalian Enzyme Catalyzing

the Asymmetric Oxidative Cleavage of Provitamin A.' *Journal of Biological Chemistry* 276 (17): 14110–14116. http://www.jbc.org/content/276/17/14110.abstract.
Kliebenstein, D J. 2004. 'Secondary Metabolites and Plant/Environment Interactions: A View through Arabidopsis Thaliana Tinged Glasses.' *Plant, Cell & Environment* 27 (6): 675–684. https://doi.org/10.1111/j.1365-3040.2004.01180.x.
Krieger-Liszkay, Anja, Christian Fufezan, and Achim Trebst. 2008. 'Singlet Oxygen Production in Photosystem II and Related Protection Mechanism.' *Photosynthesis Research* 98 (1): 551–564. https://doi.org/10.1007/s11120-008-9349-3.
Langenfeld, John, Hiroaki Kiyokawa, David Sekula, Jay Boyle, and Ethan Dmitrovsky. 1997. 'Posttranslational Regulation of Cyclin D1 by Retinoic Acid: A Chemoprevention Mechanism.' *Proceedings of the National Academy of Sciences* 94 (22): 12070–12074. https://doi.org/10.1073/pnas.94.22.12070.
Lashbrooke, Justin G, Philip R Young, Samantha J Dockrall, Krishnan Vasanth, and Melané A Vivier. 2013. 'Functional Characterisation of Three Members of the Vitis Vinifera L. Carotenoid Cleavage Dioxygenase Gene Family.' *BMC Plant Biology* 13 (1): 156. https://doi.org/10.1186/1471-2229-13-156.
Leng, Xiangpeng, Peipei Wang, Chen Wang, Xudong Zhu, Xiaopeng Li, Hongyan Li, Qian Mu, Ao Li, Zhongjie Liu, and Jinggui Fang. 2017. 'Genome-Wide Identification and Characterization of Genes Involved in Carotenoid Metabolic in Three Stages of Grapevine Fruit Development.' *Scientific Reports* 7 (1): 4216. https://doi.org/10.1038/s41598-017-04004-0.
Li, Fay-Wei, and Alex Harkess. 2018. 'A Guide to Sequence Your Favorite Plant Genomes.' *Applications in Plant Sciences* 6 (3): e1030. https://doi.org/10.1002/aps3.1030.
Lintig, Johannes von. 2012. 'Provitamin A Metabolism and Functions in Mammalian Biology.' *The American Journal of Clinical Nutrition* 96 (5): 1234S-1244S. https://doi.org/10.3945/ajcn.112.034629.
Lobo, Glenn P, Andrea Isken, Sylvia Hoff, Darwin Babino, and Johannes von Lintig. 2012. 'BCDO2 Acts as a Carotenoid Scavenger and Gatekeeper for the Mitochondrial Apoptotic Pathway.' *Development* 139 (16): 2966. https://doi.org/10.1242/dev.079632.
Lowe, Rohan, Neil Shirley, Mark Bleackley, Stephen Dolan, and Thomas Shafee. 2017. 'Transcriptomics Technologies.' *PLOS Computational Biology* 13 (5): e1005457. https://doi.org/10.1371/journal.pcbi.1005457.
Lu, Hengyun, Francesca Giordano, and Zemin Ning. 2016. 'Oxford Nanopore MinION Sequencing and Genome Assembly.' *Genomics, Proteomics & Bioinformatics* 14 (5): 265–279. https://doi.org/10.1016/j.gpb.2016.05.004.
Ma, Yan, Petra H Koza-Taylor, Debra A DiMattia, Lynn Hames, Haoning Fu, Konstantin H Dragnev, Tom Turi, Jean S Beebe, Sarah J Freemantle, and Ethan Dmitrovsky. 2003. 'Microarray Analysis Uncovers Retinoid Targets in Human Bronchial Epithelial Cells.' *Oncogene* 22 (31): 4924–4932. https://doi.org/10.1038/sj.onc.1206728.
Marioni, John C, Christopher E Mason, Shrikant M Mane, Matthew Stephens, and Yoav Gilad. 2008. 'RNA-Seq: An Assessment of Technical Reproducibility and Comparison with Gene Expression Arrays.' *Genome Research* 18 (9): 1509–1517. http://genome.cshlp.org/content/18/9/1509.abstract.
Metzker, Michael L. 2010. 'Sequencing Technologies — The Next Generation.' *Nature Reviews Genetics* 11 (1): 31–46. https://doi.org/10.1038/nrg2626.
Muto, Yasutoshi, Hisataka Moriwaki, Mitsuo Ninomiya, Sadashi Adachi, Akiko Saito, Ken Takeshi Takasaki, Takuji Tanaka, et al. 1996. 'Prevention of Second Primary Tumors by an Acyclic Retinoid, Polyprenoic Acid, in Patients with Hepatocellular Carcinoma.' *New England Journal of Medicine* 334 (24): 1561–1568. https://doi.org/10.1056/NEJM199606133342402.
Nakano, Mayumi, Kevin McCormick, Caghan Demirci, Feray Demirci, Sai Guna Ranjan Gurazada, Deepti Ramachandruni, Ayush Dusia, Joshua A Rothhaupt, and Blake C Meyers. 2020. 'Next-Generation Sequence Databases: RNA and Genomic Informatics Resources for Plants.' *Plant Physiology* 182 (1): 136 LP–146. https://doi.org/10.1104/pp.19.00957.
Naora, Honami, YanQin Yang, Fredrick J Montz, Jeffrey D Seidman, Robert J Kurman, and Richard BS Roden. 2001. 'A Serologically Identified Tumor Antigen Encoded by a Homeobox Gene Promotes Growth of Ovarian Epithelial Cells.' *Proceedings of the National Academy of Sciences* 98 (7): 4060. https://doi.org/10.1073/pnas.071594398.
Newman, T, FJ de Bruijn, P Green, K Keegstra, H Kende, L McIntosh, J Ohlrogge, et al. 1994. 'Genes Galore: A Summary of Methods for Accessing Results from Large-Scale Partial Sequencing of Anonymous Arabidopsis CDNA Clones.' *Plant Physiology* 106 (4): 1241. https://doi.org/10.1104/pp.106.4.1241.
Nisar, Nazia, Li Li, Shan Lu, Nay Chi Khin, and Barry J. Pogson. 2015. 'Carotenoid Metabolism in Plants.' *Molecular Plant* 8 (1): 68–82. https://doi.org/10.1016/j.molp.2014.12.007.

Park, YS, JM Kwak, OY Kwon, YS Kim, DS Lee, MJ Cho, HH Lee, and HG Nam. 1993. 'Generation of Expressed Sequence Tags of Random Root CDNA Clones of Brassica Napus by Single-Run Partial Sequencing.' *Plant Physiology* 103 (2): 359. https://doi.org/10.1104/pp.103.2.359.

Parkinson, John, and Mark Blaxter. 2009. 'Expressed Sequence Tags: An Overview.' In *Expressed Sequence Tags (ESTs): Generation and Analysis*, edited by John Parkinson, 1–12. Totowa, NJ: Humana Press. https://doi.org/10.1007/978-1-60327-136-3_1.

Pichersky, Eran, and Robert A Raguso. 2018. 'Why Do Plants Produce so Many Terpenoid Compounds?' *New Phytologist* 220 (3): 692–702. https://doi.org/10.1111/nph.14178.

Ramel, Fanny, Simona Birtic, Christian Ginies, Ludivine Soubigou-Taconnat, Christian Triantaphylidès, and Michel Havaux. 2012. 'Carotenoid Oxidation Products Are Stress Signals That Mediate Gene Responses to Singlet Oxygen in Plants.' *Proceedings of the National Academy of Sciences* 109 (14): 5535. https://doi.org/10.1073/pnas.1115982109.

Redmond, T Michael, Susan Gentleman, Todd Duncan, Shirley Yu, Barbara Wiggert, Elisabeth Gantt, and Francis X Cunningham. 2001. 'Identification, Expression, and Substrate Specificity of a Mammalian β-Carotene 15,15′-Dioxygenase.' *Journal of Biological Chemistry* 276 (9): 6560–6565. http://www.jbc.org/content/276/9/6560.abstract.

Rohmer, Michel, Myriam Seemann, Silke Horbach, Stephanie Bringer-Meyer, and Hermann Sahm. 1996. 'Glyceraldehyde 3-Phosphate and Pyruvate as Precursors of Isoprenic Units in an Alternative Non-Mevalonate Pathway for Terpenoid Biosynthesis.' *Journal of the American Chemical Society* 118 (11): 2564–2566. https://doi.org/10.1021/ja9538344.

Rubio Moraga, Angela, Oussama Ahrazem, José Luis Rambla, Antonio Granell, and Lourdes Gómez Gómez. 2013. 'Crocins with High Levels of Sugar Conjugation Contribute to the Yellow Colours of Early-Spring Flowering Crocus Tepals.' *PLOS ONE* 8 (9): e71946. https://doi.org/10.1371/journal.pone.0071946.

Sasaki, Takuji, Jianyu Song, Yasunori Koga-Ban, Eriko Matsui, Fang Fang, Hiromi Higo, Hideki Nagasaki, et al. 1994. 'Toward Cataloguing All Rice Genes: Large-Scale Sequencing of Randomly Chosen Rice CDNAs from a Callus CDNA Library.' *The Plant Journal* 6 (4): 615–624. https://doi.org/10.1046/j.1365-313X.1994.6040615.x.

Schena, Mark, Dari Shalon, Ronald W Davis, and Patrick O Brown. 1995. 'Quantitative Monitoring of Gene Expression Patterns with a Complementary DNA Microarray.' *Science* 270 (5235): 467. https://doi.org/10.1126/science.270.5235.467.

Schwartz, Steven H, Bao Cai Tan, Douglas A Gage, Jan AD Zeevaart, and Donald R McCarty. 1997. 'Specific Oxidative Cleavage of Carotenoids by VP14 of Maize.' *Science* 276 (5320): 1872. https://doi.org/10.1126/science.276.5320.1872.

Schwartz, Steven H, Xiaoqiong Qin, and Jan AD Zeevaart. 2001. 'Characterization of a Novel Carotenoid Cleavage Dioxygenase from Plants.' *Journal of Biological Chemistry* 276 (27): 25208–25211. http://www.jbc.org/content/276/27/25208.abstract.

Segara, Davendra, Andrew V Biankin, James G Kench, Catherine C Langusch, Amanda C Dawson, David A Skalicky, David C Gotley, Maxwell J Coleman, Robert L Sutherland, and Susan M Henshall. 2005. 'Expression of HOXB2, a Retinoic Acid Signaling Target in Pancreatic Cancer and Pancreatic Intraepithelial Neoplasia.' *Clinical Cancer Research* 11 (9): 3587. https://doi.org/10.1158/1078-0432.CCR-04-1813.

Shalon, Dari, Stephen J Smith, and Patrick O Brown. 1996. 'A DNA Microarray System for Analyzing Complex DNA Samples Using Two-Color Fluorescent Probe Hybridization.' *Genome Research* 6 (7): 639–645.

Sharoni, Yoav, Karin Linnewiel-Hermoni, Marina Khanin, Hagar Salman, Anna Veprik, Michael Danilenko, and Joseph Levy. 2012. 'Carotenoids and Apocarotenoids in Cellular Signaling Related to Cancer: A Review.' *Molecular Nutrition & Food Research* 56 (2): 259–269. https://doi.org/10.1002/mnfr.201100311.

Simkin, Andrew J, Steven H Schwartz, Michele Auldridge, Mark G Taylor, and Harry J Klee. 2004. 'The Tomato Carotenoid Cleavage Dioxygenase 1 Genes Contribute to the Formation of the Flavor Volatiles β-Ionone, Pseudoionone, and Geranylacetone.' *The Plant Journal* 40 (6): 882–892. https://doi.org/10.1111/j.1365-313X.2004.02263.x.

Soares, Virgínia LF, Simone M Rodrigues, Tahise M de Oliveira, Talisson O de Queiroz, Lívia S Lima, Braz T Hora-Júnior, Karina P Gramacho, et al. 2011. 'Unraveling New Genes Associated with Seed Development and Metabolism in Bixa Orellana L. by Expressed Sequence Tag (EST) Analysis.' *Molecular Biology Reports* 38 (2): 1329–1340. https://doi.org/10.1007/s11033-010-0234-8.

Somerville, Chris, and Maarten Koornneef. 2002. 'A Fortunate Choice: The History of Arabidopsis as a Model Plant.' *Nature Reviews Genetics* 3 (11): 883–889. https://doi.org/10.1038/nrg927.

Soprano, Dianne Robert, Pu Qin, and Kenneth J Soprano. 2004. 'Retinoic Acid Receptors and Cancers.' *Annual Review of Nutrition* 24 (1): 201–221. https://doi.org/10.1146/annurev.nutr.24.012003.132407.

Su, Dan, and Lorraine J Gudas. 2008a. 'Gene Expression Profiling Elucidates a Specific Role for RARγ in the Retinoic Acid-Induced Differentiation of F9 Teratocarcinoma Stem Cells.' *Biochemical Pharmacology* 75 (5): 1129–1160. https://doi.org/10.1016/j.bcp.2007.11.006.

Su, Dan, and Lorraine J Gudas. 2008b. 'Retinoic Acid Receptor; Activates Receptor Tyrosine Kinase Tie1 Gene Transcription through Transcription Factor GATA4 in F9 Stem Cells.' *Experimental Hematology* 36 (5): 624–641. https://doi.org/10.1016/j.exphem.2007.12.016.

Sun, Zhongkui, Joachim Hans, Michael H Walter, Radoslava Matusova, Jules Beekwilder, Francel WA Verstappen, Zhao Ming, et al. 2008. 'Cloning and Characterisation of a Maize Carotenoid Cleavage Dioxygenase (ZmCCD1) and Its Involvement in the Biosynthesis of Apocarotenoids with Various Roles in Mutualistic and Parasitic Interactions.' *Planta* 228 (5): 789. https://doi.org/10.1007/s00425-008-0781-6.

Tan, Bao-Cai, Leina M Joseph, Wen-Tao Deng, Lijuan Liu, Qin-Bao Li, Kenneth Cline, and Donald R McCarty. 2003. 'Molecular Characterization of the Arabidopsis 9-Cis Epoxycarotenoid Dioxygenase Gene Family.' *The Plant Journal* 35 (1): 44–56. https://doi.org/10.1046/j.1365-313X.2003.01786.x.

Tang, Xiao-Han, and Lorraine J Gudas. 2011. 'Retinoids, Retinoic Acid Receptors, and Cancer.' *Annual Review of Pathology: Mechanisms of Disease* 6 (1): 345–364. https://doi.org/10.1146/annurev-pathol-011110-130303.

Tohge, Takayuki, Monica Borghi, and Alisdair R Fernie. 2018. 'The Natural Variance of the Arabidopsis Floral Secondary Metabolites.' *Scientific Data* 5: 180051. https://doi.org/10.1038/sdata.2018.51.

Toti, Elisabetta, C-Y Oliver Chen, Maura Palmery, Débora Villaño Valencia, and Ilaria Peluso. 2018. 'Non-Provitamin A and Provitamin A Carotenoids as Immunomodulators: Recommended Dietary Allowance, Therapeutic Index, or Personalized Nutrition?' *Oxidative Medicine and Cellular Longevity* 2018 (May): 4637861. https://doi.org/10.1155/2018/4637861.

Triantaphylidès, Christian, and Michel Havaux. 2009. 'Singlet Oxygen in Plants: Production, Detoxification and Signaling.' *Trends in Plant Science* 14 (4): 219–228. https://doi.org/10.1016/j.tplants.2009.01.008.

Uchimiya, Hirofumi, Shin-ichiro Kidou, Tetsuo Shimazaki, Satoshi Aotsuka, Susumu Takamatsu, Ritsuko Nishi, Hiroshi Hashimoto, et al. 1992. 'Random Sequencing of CDNA Libraries Reveals a Variety of Expressed Genes in Cultured Cells of Rice (*Oryza sativa L.*).' *The Plant Journal* 2 (6): 1005–1009. https://doi.org/10.1111/j.1365-313X.1992.01005.x.

Unamba, Chibuikem IN, Akshay Nag, and Ram K Sharma. 2015. 'Next Generation Sequencing Technologies: The Doorway to the Unexplored Genomics of Non-Model Plants.' *Frontiers in Plant Science* 6 (1074). https://doi.org/10.3389/fpls.2015.01074.

Vita, Gabriella De, Pasquale Barba, Nicolas Odartchenko, Jean-Claude Givel, Giancarlo Freschi, Giorgio Bucciarelli, Maria Cristina Magli, Edoardo Boncinelli, and Clemente Cillo. 1993. 'Expression of Homeobox-Containing Genes in Primary and Metastatic Colorectal Cancer.' *European Journal of Cancer* 29 (6): 887–893. https://doi.org/10.1016/S0959-8049(05)80432-0.

Vranová, Eva, Diana Coman, and Wilhelm Gruissem. 2012. 'Structure and Dynamics of the Isoprenoid Pathway Network.' *Molecular Plant* 5 (2): 318–333. https://doi.org/10.1093/mp/sss015.

Wang, Weixuan, Yuying Li, Pengqin Dang, Siji Zhao, Daowan Lai, and Ligang Zhou. 2018. 'Rice Secondary Metabolites: Structures, Roles, Biosynthesis, and Metabolic Regulation.' *Molecules* 23 (12). https://doi.org/10.3390/molecules23123098.

White, Joseph A, Jim Todd, Tom Newman, Nicole Focks, Thomas Girke, Óscar Martínez de Ilárduya, Jan G Jaworski, John B Ohlrogge, and Christoph Benning. 2000. 'A New Set of Arabidopsis Expressed Sequence Tags from Developing Seeds. The Metabolic Pathway from Carbohydrates to Seed Oil.' *Plant Physiology* 124 (4): 1582. https://doi.org/10.1104/pp.124.4.1582.

Woodward, Andrew W, and Bonnie Bartel. 2018. 'Biology in Bloom: A Primer on the *Arabidopsis thaliana* Model System.' *Genetics* 208 (4): 1337. https://doi.org/10.1534/genetics.118.300755.

Yang, Chang-Qing, Xin Fang, Xiu-Ming Wu, Ying-Bo Mao, Ling-Jian Wang, and Xiao-Ya Chen. 2012. 'Transcriptional Regulation of Plant Secondary MetabolismF.' *Journal of Integrative Plant Biology* 54 (10): 703–712. https://doi.org/10.1111/j.1744-7909.2012.01161.x.

Zile, Maija H. 2001. 'Function of Vitamin A in Vertebrate Embryonic Development.' *The Journal of Nutrition* 131 (3): 705–708. https://doi.org/10.1093/jn/131.3.705.

Zwanenburg, Binne, Tomáš Pospíšil, and Sanja Ćavar Zeljković. 2016. 'Strigolactones: New Plant Hormones in Action.' *Planta* 243 (6): 1311–1326. https://doi.org/10.1007/s00425-015-2455-5.

5 Apocarotenoids: Natural Anti-Ageing Agents

Leepica Kapoor and Siva Ramamoorthy

CONTENTS

INTRODUCTION

Carotenoids are dynamic isoprenoids in terms of structure and function and are well known for their photoprotective role, and imparting red, yellow and orange hues to fruits, vegetables, flowers, and foliage. The enzymatic or non-enzymatic cleavage of carotenoid C_{40} hydrocarbon backbone in plants and in animal tissue (post carotenoid ingestion) at varied positions leads to structural and functional diversity of carotenoids in terms of apocarotenoids. In plants, carotenoids CCD (Carotenoid Cleavage Dioxygenase) genes are responsible for enzymatic cleavage of β carotene and form apocarotenoids such as carlactone which acts as a precursor of strigolactones. The CCD family also initiates oxidative cleavage of the polyene chain at the double bonds in β carotene structure generating a family of β apocarotenoids (Eroglu and Harrison, 2013). Apart from β carotene, CCD genes play a vital role in cleavage of other carotenoids like lycopene to generate apolycopenoids, bixin and norbixin. Zeaxanthin cleavage products include apocarotenoids like safranal, crocetin dialdehyde, crocin, picrocrocin and hydroxy beta cyclocitral. In addition, apocarotenoids like blumenin and mycorradicin are generated by CCD cleavage of lutein. The NCED genes cleave violaxanthin and neoxanthin and generate apocarotenoids such as xanthoxin and ABA (abscisic Acid) which is a plant growth regulator. In addition, non-enzymatic cleavage in plants rich in carotenoids, occurs as a result of response of carotenoids to environmental stimuli and intercellular antioxidant status resulting in formation of biologically active apocarotenoids (Walter and Strack, 2011).

In mammalian tissues, apocarotenoids such as retinoids can be generated by cleaving β carotene, which is a provitamin A (PVA) carotenoid resulting in formation of all-trans-retinal. This breakdown is mediated by the β-carotene 15,15′ dioxygenase enzyme (BCO1). Further breakdown of all-trans-retinal generates all-trans-retinoic acid, all-trans-retinol (vitamin A) and retinyl ester (Rodriguez-Concepcion et al., 2018). In addition, β-carotene-9′,10′-oxygenase (BCO2) has been reported to cleave other provitamin A carotenoids such as α-carotene, β-cryptoxanthin, and β apo carotenals to form retinal. Furthermore, BCO1 is also involved in cleavage of lycopene to form acycloretinal (Harrison and Quadro, 2018). The carotenoid cleavage products have been reported to possess bioactive properties with multi-fold functions and therapeutic efficacy. Apocarotenoids such as bixin, norbixin from *Bixa orellana* and safranal, crocin, crocetin, and picrocrocin from *Crocus sativus* (saffron) play a prominent role as an anti-ageing agent along with other apocarotenoids. Bixin is the first cis-apocarotenoid obtained from annatto seeds having tremendous industrial potential as an FDA approved food additive and food colourant as well as in cosmetics, with *Bixa orellana* tree being popularly named as the lipstick tree (Meléndez-Martínez, 2019). *Crocus sativus* has been documented in several trials for its therapeutic potential and antioxidant capacity due to the presence of apocarotenoids like crocin, crocetin and safranal (Bukhari et al., 2018). As per recent reports and epidemiological evidence, the therapeutic properties of these apocarotenoids mainly highlight their strong antioxidant properties, as their parent carotenoids, in scavenging of free radicals with specific underlying mechanisms (Rodriguez-Concepcion et al., 2018) thereby magnifying their role as an anti-ageing agent.

Ageing is a genetically programmed phenomenon leading to wear and tear of skin and internal organs. However, ageing could speed up by random damage caused by somatic mutation and escalated ROS (reactive oxygen species) (Liguori et al., 2018). Moreover, the chronic degenerative diseases related to ageing tend to accumulate the oxidized molecules and increase their concentration in cells. An increased accumulation of ROS in cells can lead to their exhaustion, and finally cell death. As these free radicals accumulate in the cells, ROS sensitive proteins involved in the ageing process are targeted, followed by the damage to epigenetic machinery which includes DNA methylation, noncoding RNAs, and histone modification, thereby ultimately causing irreversible injury to macromolecules like carbohydrates, lipids and proteins, and so accelerating ageing and ageing-related diseases. The chronic degeneration associated with ageing is mainly linked to prolonged exposure of skin and internal organs to reactive oxygen species, and manifest into photo-ageing, impaired metabolism of macromolecules, neurological disorders, inflammatory diseases, depression, and cardiovascular disease. This chapter focuses on the role of apocarotenoids as anti-ageing agents and their underlying mechanisms of action in prevention of ageing-related ailments. In addition, the therapeutic efficacy of β apocarotenoids, apolycopenoids along with industrially prominent apocarotenoids like bixin, crocin, crocetin and safranal with respect to degenerative diseases related to ageing has been discussed.

THE RONS FAMILY

The reactive oxygen and nitrogen species (RONS) are generated by all aerobic cellular mechanisms, and have detrimental effects on the phenomenon of ageing. In addition, they also participate in cellular signalling, immune defence, and energy extraction from molecules that are organic in nature. RONS like the superoxide anion could be generated during respiration within the body by NADPH (nicotinamide adenine dinucleotide phosphate) oxidase. The superoxide anions are acted upon by SOD (superoxide dismutase) and converted to hydrogen peroxide. Reactions such as Haber–Weiss and Fenton are involved in the formation of highly reactive hydroxyl radicals from hydrogen peroxide, that interact with phospholipids predominantly. Also, in neutrophils, in presence of chloride and MPO (myeloperoxidase) hydrogen peroxide conversion to hypochlorous acid is carried out, which is deleterious to proteins in the cells. In addition, nitric oxide is generated in the body as three isoforms of NOS (nitric oxide synthase) which includes neuronal NOS involved in signalling within the cells, epithelial NOS which affects vascular regulation, and inducible NOS activated by signalling from cytokine or other toxins. Furthermore, nitric oxide in reaction with oxygen forms peroxynitrite, which is a fairly reactive molecule (Salisbury and Bronas, 2015). Finally, the exposure of body to stimulants of RONS like pollution, metals, radiation, etc, can stimulate their production in the body and lead to oxidative changes in DNA, proteins and lipids.

The reactive oxygen species interact with amino acids by the Fenton and the Michael addition reactions to generate PC (protein carbonyl). Also, RNS (reactive nitrogen species) interact with residues of tyrosine to form NT (nitrotyrosine). The oxidative changes in low density lipoproteins which are cholesterol carriers may lead to cholesterol accumulation in body tissues (Trpkovic et al., 2015). The free radicals may also lead to lipid peroxidation of polyunsaturated fatty acids forming varied reactive aldehydes like MDA (malondialdehyde) and isoprostanes (Frijhoff et al., 2015). Also, due to the glycoxidation reaction of amino groups in proteins and carbonyl groups in carbohydrates, advanced glycation end products like hydroimidazolone, pentosidine and glucosepane are generated (Reynaert et al., 2016). Furthermore, several mutagenic lesions can occur due to oxidative changes in DNA (Jacob et al., 2013).

RONS MEDIATED AGEING THEORY

The oxidative stress ageing theory highlights the loss in functionality occurring due to free radical accretion due to RONS mediated injury to DNA, proteins and other macromolecules. The escalated RONS levels influence the mechanism of senescence in cells, which inhibits propagation of cells in response to their damage. These cells have been reported to attain SASP (senescence-associated secretory phenotype) which is irreversible in nature and involves secretions of factors like chemokines, interleukins, degenerative MMP (matrix metalloproteases) and components of ECM (extracellular matrix) (Pole et al., 2016). Therefore, RONS acts on various mechanisms of SASP and induce cellular senescence. The RONS initiate several SASP factors that initiate (Pole et al., 2016):

- Production of IL-1α which activates inflammation and increases NF-kB (Nuclear Factor kappa B) activity along with progression of tumour metastasis.
- Activation of MMP which facilitates the activity of deleterious enzymes like collagenase, elastase and hyaluronidase, which are involved in skin ageing of dermal fibroblasts. In addition, MMP also induces ageing-related chronic diseases such as atherosclerosis, Alzheimer's, and osteoarthritis.
- Oxidative stress related inhibition of activity of FOXO (Forkhead box) protein which induces insulin mediated protection from stress generated by free radicals.
- Reduction of calcium ATPase activity of sarco or endoplasmic reticulum which leads to senescence of cardiac system.
- Inhibition of SIRT1 (Sirtuin1); a regulator of lipid metabolism.

These SASP factors play a role in several ageing-related chronic pathological conditions like CVD (cardiovascular disease), CKD (chronic kidney disease), MD (macular degeneration), ND (neurodegenerative diseases) and cancer. Most of these degenerative diseases result due to oxidative damage to endocrine, immune and nervous systems along with activation of inflammatory pathways initiated by IL-1α, IL-6, IL-8. Majority of the diseases related to ageing share deleterious SASP profiles which involve MMPs. Therefore, oxidative stress, induced inflammation and ageing are closely linked leading to the theory of 'oxi-inflamm-ageing.' Therefore, ageing is a result of the vicious circle in which stress created by RONS and the resulting inflammation nourish each other and enhance age related chronic ailments (Miquel, 2009).

DEFENCE MECHANISMS: ANTIOXIDANTS

The reactive oxygens species have been documented to affect cellular signalling and maintain balance of input (UV, stress, cytokines, nutrients) and output (function) in cells, therefore, their concentrations are kept in a tight control by cellular defence mechanisms which involve several antioxidants (Meléndez-Martínez, 2019). The endogenous antioxidants include enzymes like catalase, glutathione peroxidase, glutathione-S-transferase, and glucose-6-phosphate dehydrogenase (Birben et al., 2012) In addition, non-enzymatic antioxidants like β-carotene, bilirubin and vitamin E are present in blood along with albumin and uric acid in the plasma (Wu et al., 2013). The antioxidants can also be administered exogenously in the form of vitamin C, vitamin E, drugs like acetylcysteine, stilbene derivatives (flavonoids, phenolic acids and resveratrol) zinc and selenium (Wu et al., 2013). Also, carotenoids and apocarotenoids play an indispensable role as potent antioxidants and scavengers of free radicals thereby slowing down the underlying mechanism of ageing. An imbalance in formation and neutralizing of the free radicals results in RONS mediated stress and its consecutive consequences in terms of ageing-related degenerative diseases.

APOCAROTENOIDS TARGET RNOS: UNDERLYING MECHANISMS

Apocarotenoids target the reactive oxygen species through well-defined mechanisms (Figure 5.1) and thereby delay the occurrence of chronic degenerative disorders which result due to chronic exposure of skin and body to the deleterious effects of reactive nitrogen and oxygen species.

Pro-oxidants and Antioxidants

Apocarotenoids can act as pro-oxidants or antioxidants based on the nature of ROS molecule and the existing cellular conditions thereby they exhibit vital functions against degenerative diseases like cancer, obesity, diabetes and inflammatory diseases like arthritis. Apocarotenoids may exhibit a pro-oxidant activity during their accumulation (aldehydes; epoxides and retinal) as a result of carotenoids' antioxidant activity, or during the down regulation of enzymatic carotenoid detoxification (Van Helden et al., 2009). Moreover, research trials on mice which were deficient in BCO2 demonstrate excess of mitochondrial carotenoid accumulation due to a compromised activity of BCO2, resulting in increased stress to oxidative mutilation and dysfunction of the mitochondria, ultimately leading to the apoptotic pathway (intrinsic) activation (Lobo et al., 2012). The pro-oxidant activity of apocarotenoids can prove to be useful under certain conditions as an escalated production of RONS induced by apocarotenoids would detain the cell cycle of damaged cells (Sosa et al., 2013). Furthermore, during progression of cancer, an increased intake of antioxidants may increase the rate of survival of damaged cells due to escalated resistance to oxidative stress leading to reduced apoptosis (Watson, 2013). Therefore, both pro-oxidant and antioxidant activities of apocarotenoids (Figure 5.1) are beneficial, and could act as a regulatory mechanism in cancer, being dependent on the stage of malignancy.

The antioxidant and anti-inflammatory capacity of apocarotenoids comes in handy against diseases like obesity and diabetes resulting from escalation of oxidative stress and adipose tissue inflammation (Marseglia et al., 2015). This stress induces standard inflammatory response like cytokines, which are proinflammatory, and chemokines, acting as signals for attracting macrophages. In addition, obese and inflamed adipocytes in tissues act as a precursor in the genesis of insulin resistance. However, this chain of events could be altered by apocarotenoids formed as a result of breakdown of fucoxanthin and lycopene thereby inhibiting the expression of inflammatory responses.

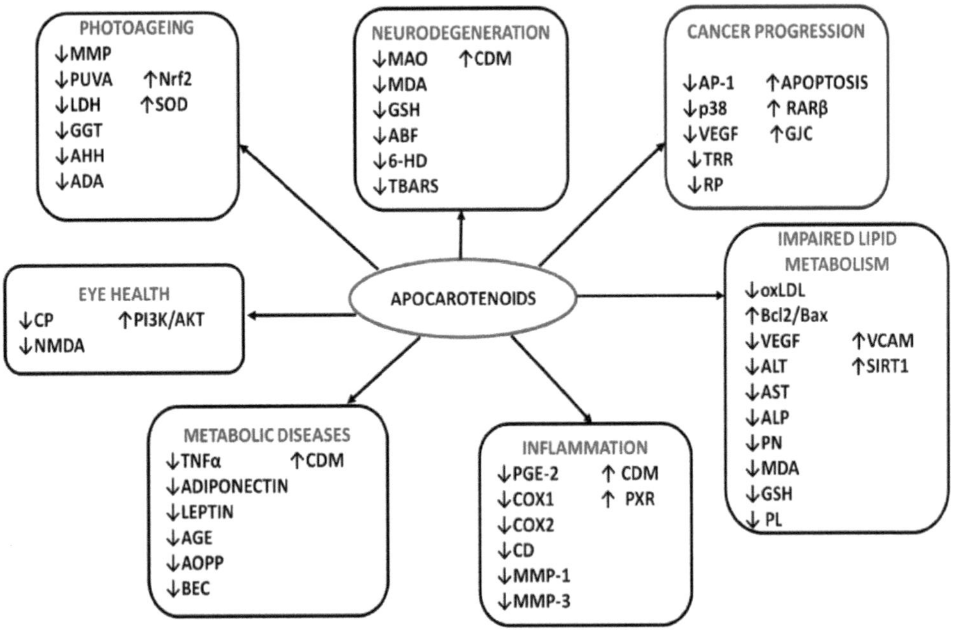

FIGURE 5.1 Anti-ageing mechanisms of apocarotenoids. The arrows pointing up (↑) refer to positive regulation, and the arrows pointing down (↓) refer to negative regulation of the factors associated with ageing by apocarotenoids (MMP: Matrix Metalloproteases, PUVA: Psoralen + UVA, LDH: Lactate Dehydrogenase, GGT: Gamma-Glutamyl Transferase, AHH: Aryl Hydrocarbon Hydroxylase and ADA: Adenosine Deaminase, Nrf2: nuclear factor erythroid 2 related factor 2, SOD: Super Oxide Dismutase, MAO: Monoamine Oxidase, MDA: malondialdehyde GSH: Glutathione, ABF: A-beta fibrillogenesis, 6-HD:6-Hydroxy Dopamine, TBARS: Thiobarbituric Acid, CDM: Cellular Defence Mechanisms, AP-1: Activator Protein-1, p38: Phosphorylated-P38 Mitogen activated Protein Kinase, VEGF (Vascular Endothelial Growth Factor), TRR (Thio Redoxin Reductase), RP: Regulatory Proteins (cell cycle) RARβ: Retinoid Acid Receptors, GJC: Gap Junction Communication, oxLDL: Oxidised Low Density Lipoprotein, Bcl2/Bax: BCl-2 associated X Protein, VCAM: Vascular Cell Adhesion Molecule, ALT: Alanine amino Transferase, SIRT1: Sirtuin 1, AST: Aspartate Transaminase, ALP: Alkaline Phosphatase, PL: Pancreatic Lipase, PN: Peroxynitrite, PGE-2: Prostaglandin E-2,COX1: Cyclo oxygenase 1, COX2: Cyclo oxygenase 2,CD: Cathepsin D, MMP-1: Matrix Metalloproteases 1,MMP-3: Matrix Metalloproteases3, PXR: human pregnane receptor, TNFα: Tumour Necrosis Factor, AGE: Advanced Glycation End products, AOPP: Advanced Oxidation Protein Products, BEC: Bovine Endothelial Cell, CP: Caspase Pathway, PI3K/AKT: phosphatidylinositol 3-kinase/Akt (Protein Kinase B), NMDA: N-methyl-D-aspartate.

Nrf2 pathway Activation

Apocarotenoids can indirectly evoke the cellular antioxidant activity by switching on the antioxidant response dependent on initiation of Nrf2 pathway (nuclear factor erythroid 2 related factor 2). Under a normal scenario, Nrf2 activity is connected to the protein Keap1 which is quite sensitive to redox ranking of cells. An increase in the RONS mediated stress breaks down the association between Nrf2 and Keap1 protein, resulting in activation of Nrf2, which in turn promotes cellular defence by increasing the expression of underlying genes involved. Apocarotenoids exhibit a vital role in inducing the intrinsic antioxidant capacity of the cells by activation of the Nrf2 pathway (Figure 5.1). In human bronchial cells (BEAS-2B), apocarotenoids, resulting from the breakdown of lycopene, e.g., apo-10′-lycopenoic acid, aldehyde carotenoid derivatives such as apo-10′-lycopenal and apo-10′lycopenol (lycopenoids) (Sharoni et al., 2012) have been documented to stimulate Nrf2 dependent antioxidant responses, and induction of cellular detoxifying enzymes (Lian and Wang, 2008).

NF-kB pathway Suppression

Apocarotenoids have been linked to suppression of NF-kB pathway (Figure 5.1) which stimulates inflammatory response and therefore affects cellular survival (Kaulmann and Bohn, 2014; Xia et al., 2014). NF-kB is bound to IkB and kept in an inactivated state. Stimuli like reactive oxygen species, radiation, metals, and antigens (bacterial or viral) can lead to breakdown of IkB protein, thereby making NF-kB active, which is transported to the nucleus, resulting in induction of genes related to activation of inflammatory response as cytokines. Apocarotenals that possess electrophilic functional groups are involved in suppression of NF-kB pathways by interacting with the cysteine remnants of NF-kB kinase and IkB Kinase (Kaulmann and Bohn, 2014).

INTERPLAY WITH TRANSCRIPTION FACTORS

The hydrophobic nature of apocarotenoids makes them potential candidates to interact with the NR (Nuclear Receptor) family transcription factor, because these transcription factors bind to ligands that have cell membrane permeability or intracellular ligands which are lipophilic in nature. In fact, Retinoic Acid (RA) ligands which are formed from β-carotene, which is a provitamin A carotenoid, activate specific transcription factors of the NR family, known as retinoid receptors. The retinoic acid receptors (RAR) are of two types: canonical RAR, which includes RAR α, β, γ; and retinoid X receptors, which includes RXR α, β, γ. Among the isomers of RA, 9-cis-RA isomer binds the receptors of RXR and RAR whereas atRA (all-trans-retinoic acid) isomer associates selectively with RAR receptors. Depending upon the activation by the ligand, the retinoid receptors alter and target definite gene transcription or obstruct the bioactivity of activator protein1 (AP-1) and CCAAT enhancer binding proteins (CEBP) (Bonet et al., 2012). Thereby, the role of apocarotenoids in prevention of hyperproliferative diseases is due to downregulation of the pro-proliferation activity of transcription factor AP1 (Bushue and Wan 2010). In addition, the RXRs are compelled to have a dimerization partnership with many other nuclear receptors like peroxisome proliferator-activated receptors, thyroid hormone receptor, and receptor linked to vitamin D apart from RAR (Aranda and Pascual, 2001). The interaction of retinoids with lipids and metabolism of energy is also mediated by their capacity in activation of nuclear receptors (Bonet et al., 2012).

Anti-ageing potential of Apocarotenoids

Apocarotenoids play a prominent role as an anti-ageing agent owing to their antioxidant properties and suppression of pro-senescence factors generated by the impact of free radicals and RONS family. Figure 5.1 depicts the role of apocarotenoids along with the underlying mechanisms in chronic diseases related to ageing such as neurodegeneration, photo-ageing, deteriorated eye health, impaired macromolecule metabolism, inflammation, and cancer progression, respectively. Apocarotenoid interaction with various transcription factors involved in these ailments have been described.

Skin and hair ageing: Photo-ageing

Apocarotenoids have been reported to protect skin against photo-ageing due to the prolonged exposure to free radicals and UV rays. Bixin is associated with inhibition of photo-ageing and greying of hair. Bixin has been linked to Nrf2 activation which is involved in antioxidant response at a cellular level thereby providing skin protection against highly reactive species, UV, skin cancer and reducing dermal hyperproliferation. In addition, bixin has also been observed to down regulate the activity of PUVA (psoralen + UVA) which induces greying of hair as documented in a mouse model *in vivo* (Rojo de la Vega et al., 2018). Among the apocarotenoids present in *Crocus sativus*, crocin, a potent hydrogen donor, has been reported to have highest antioxidant activity (higher than tocopherol). In addition, crocin has been reported to reverse cell injury due to oxidation, and increase the activity of antioxidants like superoxide dismutase (Ochiai et al., 2004). The antioxidant capacity of crocetin was found to be equivalent to Trolox and BHT (Kanakis et al., 2007) followed by safranal. Crocetin exhibits hydroxyl moieties on the carboxyl group, making it an efficient donor of H molecules, and thereby interacting actively with free radicals. Crocetin has been observed to decrease the concentration of enzymes which act as biomarkers of RONS mediated stress, such as lactate dehydrogenase gamma-glutamyl transferase, aryl hydrocarbon hydroxylase, and adenosine deaminase (Zheng et al., 2005). Furthermore, another apocarotenoid; safranal was reported for its potent inhibition of MMPs and demonstrated anti-elastase, anti-collagenase and anti-hyaluronidase activity (Madan and Nanda, 2018).

Apocarotenoids inhibit chronic oxidative stress related cancer progression

Chronic inflammation and carcinogenesis have been linked by various clinical trials. The proinflammatory cytokines like TNFα and RONS induce the activation of NF-kB which activates the genes connected with cellular proliferation and apoptosis. In addition, RONS mediated increase of *c*-fos and *c*-jun transcription factors increase the rate of angiogenesis. Pro-angiogenic factors include prostaglandins and growth factors like VEGF (vascular endothelial growth factor) which generate from macrophages and fibroblasts, and play a vital role in escalating the production of ROS, and increasing cancer risk (Khansari et al., 2009). RONS also interact with the DNA and introduce defects in the DNA such as arrest of transcription, and errors related to replication, which are associated with progression of cancer. Therefore, antioxidants like apocarotenoids affect the management of carcinogenesis, depending upon its stage of progression. Among the apocarotenoids, beta ionone has anti-carcinogenic properties, as it obstructs cancer cell proliferation, metastasis, and initiates apoptosis (Ansari and Emami, 2016). β apocarotenoids have been reported to be potent anti-ageing agents due to their anti-carcinogenic activity. Several trials have been conducted to study the biological functionality of β apocarotenoids (Eroglu and Harrison, 2013). The interaction between β-apo-14′-carotenoic acid present in bronchial epithelial of humans exhibiting no signs of morbidities along with benzo(a)pyrene, a carcinogenic agent present in the smoke of cigarettes, has been examined for its effect on RARβ expression levels (Prakash et al., 2004). Benzo(a)pyrene was found to downregulate the expression levels of RARβ, which were reversed by administration of β-apo-14′-carotenoic acid, thereby highlighting its role in transcriptionally activating RARβ. In sync with the antiproliferative activity, apo-10′-lycopenoic activates RARβ promoters, and thereby induces its expression (Lian et al., 2007).

β carotenoids have also been tested against their efficacy in inhibition of tumour cell growth in breast cancer in two types of cell lines which express RARs; MCF-7 (positive for oestrogen receptor), Hs578T and MDA-MB-231 (negative for oestrogen receptor). In MCF-7 it was observed that β-apo-12′-carotenoic acid and β-apo-14′-carotenoic acid demonstrated 50% inhibition in comparison to atRA at similar concentrations. In addition, β-apo-14′-carotenoic acid overpowered the activity of activator protein-1 and inhibited the growth of Hs578T cells (Tibaduiza et al., 2002).

Apolycopenoids, being the cleaved products of lycopene due to oxidation reactions, are bioactive molecules with specific mechanism of action (Lindshield et al., 2007; Erdman et al., 2009). Apolycopenoids, such as acycloretinoic acid, are known for their inactivation of prostate cancer cells and breast cancer cells MCF-7 (Ben-Dor et al., 2001) via Gap junction communication, and thereby they prevent the spreading of cancer cells, and induce their timely death (Kotake-Nara et al., 2002). Among the apolycopenoids, apo-10′-lycopenoic acid activity was found potent in the cell cycle arrest of lung cancer proliferation *in vitro* and *in vivo*. Bixin has tremendous research potential as an anti-ageing agent primarily due to its antioxidant properties as a result of its structure, which has double bonds that allow ease in transfer of electrons (Rivera-Madrid et al., 2016). Moreover, bixin is a scavenger of RONS generated from chemotherapeutic drugs like cisplatin (Rios et al., 2009). Apart from playing an important role in chemotherapy, bixin was found to have antimutagenic impact on UV exposed cells. The *cis bixin* configuration has been observed to possess cytotoxic activity by inhibiting the redox pathway of thioredoxin reductase (Tibodeau et al., 2010). All apocarotenoids of saffron have been reported to protect nucleic acid from harmful chemicals through various mechanisms, such as induction of apoptosis, inhibiting synthesis of DNA and RNA, inducing cytotoxicity, down regulating cell proliferation, and by inactivating the cell cycle regulatory proteins (Bukhari et al., 2018). Among the apocarotenoids in saffron, crocetin is linked to tumour growth inhibition by inhibiting p38 phosphorylation, and suppressing the activity of VEGF (vascular endothelial growth factor) involved in inducing tube formation in angiogenesis (Umigai et al., 2012). Safranal has been reported to prevent the growth of neuroblastoma (Samarghandian et al., 2014) and KB (oral squamous cell carcinoma) (Jabini et al., 2017) cell lines. Also, crocin was noted as a strong anticariogenic agent against cervical and breast cancer; common cancers in women above the age of 30 years (Mousavi et al., 2011).

Apocarotenoids inhibit RONS mediated neurological degeneration

The effects of dementia and impairment of cognition have a profound impact on the quality of life in the aged population. A recent study reported the relationship between biomarkers of RONS mediated inflammation such as malondialdehyde, protein carbonyl and poor cognition functionality (Baierle et al., 2015). In addition, stress granule formation is also activated by RONS (Chen and Liu, 2017). Under normal conditions these stress granules are degraded in the absence of stress-causing stimuli. However, these stress granules could be extremely stable due to mutagenic effects (Ramaswami et al., 2013). In the condition of impaired cognition, the stress remains constant, therefore these stress granules may affect neuronal functionality by impounding proteins like ribonucleoproteins, and silencing specific transcripts. Also, metals involved in reduction and oxidation reactions, such as zinc, iron, copper and aluminium, promote aggregation of amyloid (Tougu et al., 2008). These metals also positively effect Tau phosphorylation. A combined action of Tau protein, ROS and amyloid formation affect the NMDA (N-methyl-D-aspartate) receptors' activity, inducing escalated inflow of calcium ions in neurons after a synapse leading to a chain of events towards increased ROS and ultimately result in Alzheimer's disease due to synaptic dysfunction (Tönnies and Trushina, 2017). In ageing due to a decline in antioxidant activity of the cells coupled with an increase in oxidative injury the CNS is unprotected to free radicals, leading to culmination of several progressive disorders, like Parkinson's and Alzheimer's disease, loss of memory, depression, and anxiety. Apocarotenoids have been noted as a protective and curative agent against neurodegenerative disorders. The intake of apocarotenoids in saffron has been involved in memory impairment as they have been noted to reverse the changes induced by excess of aluminium, in the activity of MAO (monoamine oxidase) along with the levels of RONS stress related biomarkers like MDA and GSH (Linardaki et al., 2013). Alzheimer's and Parkinson's disease are linked with development of amyloid structures bound to hydrophobic proteins. Crocin, owing to its antioxidant properties and amphiphilic nature, has been reported to prevent the formation of amyloid structures due to its inhibitory action on A-beta fibrillogenesis (Alavizadeh and Hosseinzadeh 2014). Crocetin has been noted for its bioactivity in Parkinson's disease by enhancing the antioxidant activity and inhibiting 6-hydroxy dopamine, which plays a vital role in inducing Parkinson's disease, along with depletion of TBARS (Thiobarbituric acid) (Ahmad et al., 2005). As per a recent trial, administration of crocin at a dose of 30 mg/day to volunteers reduced the depression symptomatically along with the complications of the metabolic syndrome (Jam et al., 2017). Administration of Tocotrienol extract from bixa seeds has been reported to promote bone health in rats with testosterone deficiencies (Chin and Ima-Nirwana, 2014).

Role of apocarotenoids in impaired metabolism of macronutrients

Ageing results in impaired metabolism of macromolecules like carbohydrate, proteins and lipids, and results in various metabolic diseases as a consequence of impaired metabolism and increased cellular uptake of these macromolecules, resulting in ailments like cardiovascular diseases, which is predisposed by obesity, diabetes and hypertension; these are metabolic diseases resulting from impaired metabolism of carbohydrates and fats. RONS affects the metabolism of these macromolecules, and with diminishing antioxidant capacity through ageing there is a higher incidence of these diseases. With advancing age, these metabolic diseases act as silent killers and affect vital internal organs such as the heart, kidneys, and liver, and thereby affect their performance and functionality in the body.

Weakened carbohydrate metabolism with ageing

Diabetes is a metabolic disease associated with decreased antioxidant capacity, and involves multiple organ complications (Bashan et al., 2009). Chronic hyperglycemia leads to escalated glucose levels in the cells leading to an exponential increase in RONS beyond the cellular neutralizing capacity. The presence of an increased concentration of RONS leads to activation of multiple pathways such as PKC (protein kinase C) elevated advanced glycation end products (AGE) along with polyol and hexosamine pathway flux (Golbidi et al., 2011). The activation of AGE specifically can alter cellular signalling among cells and proteins like albumin, making them bound to receptors of AGE on macrophages, and induce growth factor production as well as cytokines; inflammatory moieties. Diabetes mellitus is associated with chromic impairment in metabolism of carbohydrates, protein and fats along with a sluggish defence mechanism related to antioxidants. Physiological stress due to prolonged exposure to free radicals leads to diabetic complications like diabetic encephalopathy. Bixin has recently shown to prevent complications involved with diabetes by scavenging the free radical moieties (Victor et al., 2011). Also, as per recent trials, in diabetic rats bixin was found to decline the levels of AOPP; biomarkers of ROS mediated inflammation and escalated in conditions like diabetes and kidney failure. The effectiveness of bixin in prevention of AOPP was reported at par to antihyperglycemic drugs. However, norbixin was reported to increase the levels of AOPP, therefore highlighting its function as a pro-oxidant (Roehrs et al., 2014). In addition, bixin has also been associated with inhibition of advanced glycation end products' development *in vitro* (Gutierrez et al., 2011).

Crocetin has been reported to prevent cardiovascular complications related to diabetes by preventing induction of BEC (bovine endothelial cell) apoptosis by stabilizing the calcium ions and inhibiting the ROS (Xiang et al., 2006). Crocetin has also been noted for its potential benefits in abnormalities related to insulin resistance, hypertension, dyslipidemia, and hyperinsulinemia. Crocetin was reported to be involved in reducing insulin insensitivity induced by free fatty acids and dysregulated expression of adiponectin, leptin and TNF-α (Xi et al., 2007).

VASCULAR REMODELLING IN AGEING

Cardiovascular disease is a result of several multiple chronic complications like diabetes, obesity, hypertension, and atherosclerosis. With the reduction in the concentration of protective antioxidant enzymes the tolerance capacity of the heart to stress due to oxidative mutilation reduces with progressing age. A crucial risk for atherosclerosis is oxidation of LDL leading to escalated arterial stiffness (Brinkley et al., 2009). As per the research outcomes of a recent study, the levels of oxidised LDL increases with advancing age (Paik et al., 2013). In addition, with ageing, RNOS stress adds on to the vascular endothelial dysfunction. Also, vascular remodelling due to enhanced endhotelin-1 (vasoconstrictor) and NT expression has been reported to be linked with progression of cardiovascular disease in ageing (Siti et al., 2015). The major cause for complications related to cardiovascular disease and atherosclerosis with advancement in age is hyperlipidemia, as it is coupled with escalated levels of triglycerides, cholesterol and low density lipoproteins. Hypertension and hypercholesterolemia which are prevalent among the elderly induce the entry of LDL in the blood vessels, where they undergo oxidation and are received by macrophage cells, leading to the development of foam cells. Macrophages in turn promote the release of cytokines (proinflammatory), ultimately leading to extracellular matrix degradation. These deteriorating effects of LDL oxidation are kept in check by nitric oxide (NO) synthesised by nitric oxide synthase secreted by the endothelium (eNOS) (Tabas et al., 2015). Nitric Oxide has been reported to be an inhibitor of inflammation and promoter of vasodilation. However, an isoform of eNOS – iNOS (inducible nitric oxide synthase) – demonstrates proatherosclerotic effects, and leads to escalated production of NO, which inhibits the activity of eNOS and endothelial receptors, and induces the production of peroxy nitrite, leading to oxidation of lipids and LDL moieties. Bixin administration in human subjects has been observed to reduce the biomarkers of RONS mediated stress, the rate of low density lipoprotein oxidation, metabolites of nitric oxide (Conte et al., 2019), and vascular endothelium growth factor (Yong et al., 2013). Apocarotenoids in saffron have been reported to reduce deleterious parameters related to hyperlipidemia such as triglycerides, and enzymes like alkaline phosphatase, alanine aminotransferase and aspartate transaminase along with the markers of oxidative stress like MDA and GSH (Asdaq and Inamdar, 2010). Crocin demonstrates an inhibitory action on pancreatic lipase and enhances the Bcl2/Bax ratio (Xu et al., 2007) thereby exhibiting a preventive role in hyperlipidemia (Sheng et al., 2006) Crocetin has also been associated with reducing the blood cholesterol level, and ultimately prevent the occurrence of atherosclerosis by suppressing NF-kB pathway and declining the expression of vascular cell adhesion molecule (VCAM-1) (Zheng et al., 2005). In addition, apo-10′-lycopenoic acid has been linked to prevention of fatty liver as it increases the activity of the enzyme SIRT1 (Sirtuin1); a regulator of lipid metabolism (Chung et al., 2012).

The bioactivity of β-apocarotenals in differentiation of adipocytes, and thereby in prevention of lipid accumulation, has been examined. β-apo-14′-carotenal was found to suppress adipogenesis independent of the RARα mechanism. Since RXR, RXR-PPAR (peroxisome proliferator-activated receptor), or RXR partnership with other receptors in the nucleus plays a vital part in regulating adipogenesis, initiation of RXRα, PPARα, PPARδ, and PPARγ activity by their synthetic agonist in the presence of β-apo-carotenals was studied. β-apo-14′-carotenal was found to inhibit the activation of RARα, PPARα, PPARδ, and PPARγ, which was induced by agonists respectively (Ziouzenkova et al., 2007). Several other trials indicate the functions of β-apocarotenoids as antagonists of receptors in the nucleus (Eroglu et al., 2010).

Even at nanomolar concentrations β-apocarotenoids such as β-apo-14′-carotenal, β-apo-14′-carotenoic acid, and β-apo-13-carotenone have been reported to antagonize atRA-induced activation of RARs (Eroglu et al., 2012). Studies performed by molecular modelling confirm the interaction of β-apo-13-carotenone and the site of ligand attachment for retinoid receptors. Therefore, β-apocarotenoids can play a vital role by being a natural retinoid antagonist with profound implications in the functional activities of β-carotene as provitamin A and in regulation for risk of degenerative diseases like cancer and cardiovascular disease.

Apocarotenoids Inhibit Proinflammatory Mechanisms

Inflammatory diseases like arthritis and rheumatoid arthritis tend to progress with advancement in age. The anti-inflammatory properties of apocarotenoids are effective against culmination of inflammatory diseases like arthritis and rheumatoid arthritis. The cyclooxygenase pathway activates several inflammatory responses, such as cytokines. Crocin has been found to demonstrate an anti-inflammatory response, by inhibition of prostaglandin E-2 production, along with two vital enzymes of the pathway: COX1 and COX2 (cyclooxygenase 1 and 2) (Hemshekhar et al., 2012). In addition, crocin has been reported to inhibit escalated levels of cathepsin D, and expression of enzymes like MMP-1 and MMP-3 to impart protection to bone joints (Ding et al., 2013). Also, apo-10′-lycopenoic acid has been noted to transactivate RAR, thereby reducing the concentration of pro-inflammation markers in adipose tissue (Gouranton et al., 2011). β-apocarotenoids have also been studied for their role in preventing toxic substances to get accumulated in the liver by activation of

PXR (human pregnane receptor). However, the activation of PXR by β-apocarotenals was found to be lower in comparison to the parent compound β-carotene and rifampicin (standard PXR agonist) (Rühl et al., 2004).

APOCAROTENOIDS PREVENT AGEING-RELATED DETERIORATION OF VISION

Vitamin A is an important fat-soluble vitamin which performs varied functions most importantly in vision, boosting immunity, cellular differentiation and proliferation, epithelial tissue maintenance and promoting the functions of brain. Vitamin A deficiency is a major threat to nutritional and physiological well-being globally. Apocarotenoids play a major role in combating vitamin A deficiency because of their provitamin A activity. The nutritional significance of apocarotenoids is mainly due to their provitamin A activity which is induced by breakdown of carotenoids with β rings in mammals. The provitamin A carotenoids undergo conversions to generate retinol, retinoids, retinoic acid, and 11-cis-retinal which play a vital part in regulation of genes associated with the visual cycle (Grune et al., 2010). Continuous exposure to light leads to retinal stress and deterioration of the photoreceptors. Crocin has been found to activate the PI3K/AKT signalling pathway, thereby promoting proliferation and acts as a protective agents against ischaemic retinopathy (Qi et al., 2013). Oral crocetin administration has been noted to prevent retinal damage induced by NMDA (N-methyl-D-aspartate) by inhibiting the caspase pathway (Ohno et al., 2012).

CONCLUSION

Apocarotenoids play a potent role as an anti-ageing agent by neutralizing the delecterios impact of RONS due to their antioxidant properties. Their cellular signalling, as well as interactions with specific proinflammatory transcription factors, declines the levels of biomarkers of oxidative injury in the body, and thereby they support the weak and diminishing levels of defence mechanisms with advancing age. In addition, apocarotenoids demonstrate well-defined mechanisms in prevention and advancement of oxidative stress-related degnerative disorders. The phenomenon of ageing makes the damage by reactive oxygen species chronic and repeated in nature, which is difficult to reverse by cellular defence mechanisms. However, if the anti-ageing agents (antioxidants) are utilized, they can prevent the damage to skin and internal organs and thereby delay the effects of ageing. Therefore, an increased intake of anti-ageing apocarotenoids like bixin, crocin, crocetin, safranal, apolycopenoid, and β apocarotenals could prevent the chronic nature of the damage by free radicals as these apocarotenoid participate in inhibiting the impact of RONS and support the army of cellular defence mechanisms.

REFERENCES

Ahmad AS, Ansari MA, Ahmad M, Saleem S, Yousuf S, Hoda MN, Islam F. 2005. Neuroprotection by crocetin in a hemi-parkinsonian rat model. *Pharmacology Biochemistry and Behavior* 81: 805–813.

Alavizadeh SH, Hosseinzadeh H. 2014. Bioactivity assessment and toxicity of crocin: a comprehensive review. *Food and Chemical Toxicology* 64: 65–80.

Ansari M, Emami S. 2016. β-Ionone and its analogs as promising anticancer agents. *European Journal of Medicinal Chemistry* 123: 141–154.

Aranda A, Pascual A. 2001. Nuclear hormone receptors and gene expression. *Physiological Reviews* 81: 1269–1304.

Asdaq SMB, Inamdar MN. 2010. Potential of Crocus sativus (saffron) and its constituent, crocin, as hypolipidemic and antioxidant in rats. *Applied Biochemistry and Biotechnology* 162: 358–372.

Baierle M, Nascimento SN, Moro AM, Brucker N, Freitas F, Gauer B, Durgante J, Bordignon S, Zibetti M, Trentini CM. 2015. Relationship between inflammation and oxidative stress and cognitive decline in the institutionalized elderly. *Oxidative Medicine and Cellular Longevity* 2015.

Bashan N, Kovsan J, Kachko I, Ovadia H, Rudich A. 2009. Positive and negative regulation of insulin signaling by reactive oxygen and nitrogen species. *Physiological Reviews* 89: 27–71.

Ben-Dor A, Nahum A, Danilenko M, Giat Y, Stahl W, Martin H D, Emmerich T, Noy N, Levy J, Sharoni Y. 2001. Effects of acyclo-retinoic acid and lycopene on activation of the retinoic acid receptor and proliferation of mammary cancer cells. *Archives of Biochemistry and Biophysics* 391: 295–302.

Birben E, Sahiner UM, Sackesen C, Erzurum S, Kalayci O. 2012. Oxidative stress and antioxidant defense. *World Allergy Organization Journal* 5: 9–19.

Bonet ML, Ribot J, Palou A. 2012. Lipid metabolism in mammalian tissues and its control by retinoic acid. *Biochimica et Biophysica Acta (BBA)-Molecular and Cell Biology of Lipids* 1821: 177–189.

Brinkley TE, Nicklas BJ, Kanaya AM, Satterfield S, Lakatta EG, Simonsick EM, Sutton-Tyrrell K, Kritchevsky SB. 2009. Plasma oxidized low-density lipoprotein levels and arterial stiffness in older adults: the health, aging, and body composition study. *Hypertension* 53: 846–852.

Bukhari SI, Manzoor M, Dhar M. 2018. A comprehensive review of the pharmacological potential of Crocus sativus and its bioactive apocarotenoids. *Biomedicine & Pharmacotherapy* 98: 733–745.

Bushue N, Wan Y-JY. 2010. Retinoid pathway and cancer therapeutics. *Advanced Drug Delivery Reviews* 62: 1285–1298.

Chen L, Liu B. 2017. Relationships between stress granules, oxidative stress, and neurodegenerative diseases. *Oxidative Medicine and Cellular Longevity* 2017.

Chin K-Y, Ima-Nirwana S. 2014. Effects of annatto-derived tocotrienol supplementation on osteoporosis induced by testosterone deficiency in rats. *Clinical Interventions in Aging* 9: 1247.

Chung J, Koo K, Lian F, Hu KQ, Ernst H, Wang X-D. 2012. Apo-10'-lycopenoic acid, a lycopene metabolite, increases sirtuin 1 mRNA and protein levels and decreases hepatic fat accumulation in ob/ob mice. *The Journal of Nutrition* 142: 405–410.

Conte L, Somacal S, Nichelle SM, Rampelotto C, Robalo SS, Roehrs M, Emanuelli T. 2019. Short-term bixin supplementation of healthy subjects decreases the susceptibility of LDL to Cu2+-induced oxidation ex vivo. *Journal of Nutrition and Metabolism* 2019.

Ding Q, Zhong H, Qi Y, Cheng Y, Li W, Yan S, Wang X. 2013. Anti-arthritic effects of crocin in interleukin-1β-treated articular chondrocytes and cartilage in a rabbit osteoarthritic model. *Inflammation Research* 62: 17–25.

Erdman Jr JW, Ford NA, Lindshield BL. 2009. Are the health attributes of lycopene related to its antioxidant function? *Archives of Biochemistry and Biophysics* 483: 229–235.

Eroglu A, Harrison EH. 2013. Carotenoid metabolism in mammals, including man: formation, occurrence, and function of apocarotenoids thematic review series: fat-soluble vitamins: vitamin A. *Journal of Lipid Research* 54: 1719–1730.

Eroglu A, Hruszkewycz DP, Curley Jr RW, Harrison EH. 2010. The eccentric cleavage product of β-carotene, β-apo-13-carotenone, functions as an antagonist of RXRα. *Archives of Biochemistry and Biophysics* 504: 11–16.

Eroglu A, Hruszkewycz DP, dela Sena C, Narayanasamy S, Riedl KM, Kopec RE, Schwartz SJ, Curley RW, Harrison EH. 2012. Naturally occurring eccentric cleavage products of provitamin A β-carotene function as antagonists of retinoic acid receptors. *Journal of Biological Chemistry* 287: 15886–15895.

Frijhoff J, Winyard PG, Zarkovic N, Davies SS, Stocker R, Cheng D, Knight AR, Taylor EL, Oettrich J, Ruskovska T. 2015. Clinical relevance of biomarkers of oxidative stress. *Antioxidants & Redox Signaling* 23: 1144–1170.

Golbidi S, Alireza Ebadi S, Laher I. 2011. Antioxidants in the treatment of diabetes. *Current Diabetes Reviews* 7: 106–125.

Gouranton E, Aydemir G, Reynaud E, Marcotorchino J, Malezet C, Caris-Veyrat C, Blomhoff R, Landrier J-F, Rühl R. 2011. Apo-10'-lycopenoic acid impacts adipose tissue biology via the retinoic acid receptors. *Biochimica et Biophysica Acta (BBA)-Molecular and Cell Biology of Lipids* 1811: 1105–1114.

Grune T, Lietz G, Palou A, Ross AC, Stahl W, Tang G, Thurnham D, Yin S-a, Biesalski HK. 2010. β-Carotene is an important vitamin A source for humans. *The Journal of Nutrition* 140: 2268S–2285S.

Gutierrez RMP, Baez EG, Cortez MdSL, Cárdenas SA. 2011. Extracts of bixa inhibit glycation and AGEs formation *in vitro*. *Journal of Medicinal Plants Research* 5: 942–948.

Harrison EH, Quadro L. 2018. Apocarotenoids: emerging roles in mammals. *Annual Review of Nutrition* 38: 153–172.

Hemshekhar M, Santhosh MS, Sunitha K, Thushara R, Kemparaju K, Rangappa K, Girish K. 2012. A dietary colorant crocin mitigates arthritis and associated secondary complications by modulating cartilage deteriorating enzymes, inflammatory mediators and antioxidant status. *Biochimie* 94: 2723–2733.

Jabini R, Ehtesham-Gharaee M, Dalirsani Z, Mosaffa F, Delavarian Z, Behravan J. 2017. Evaluation of the cytotoxic activity of crocin and safranal, constituents of saffron, in oral squamous cell carcinoma (KB cell line). *Nutrition and Cancer* 69: 911–919.

Jacob KD, Hooten NN, Trzeciak AR, Evans MK. 2013. Markers of oxidant stress that are clinically relevant in aging and age-related disease. *Mechanisms of Ageing and Development* 134: 139–157.

Jam IN, Sahebkar AH, Eslami S, Mokhber N, Nosrati M, Khademi M, Foroutan-Tanha M, Ghayour-Mobarhan M, Hadizadeh F, Ferns G. 2017. The effects of crocin on the symptoms of depression in subjects with metabolic syndrome. *Advances in Clinical and Experimental Medicine* 26: 925–930.

Kanakis CD, Tarantilis PA, Tajmir-Riahi HA, Polissiou MG. 2007. Crocetin, dimethylcrocetin, and safranal bind human serum albumin: stability and antioxidative properties. *Journal of Agricultural and Food Chemistry* 55: 970–977.

Kaulmann A, Bohn T. 2014. Carotenoids, inflammation, and oxidative stress—implications of cellular signaling pathways and relation to chronic disease prevention. *Nutrition Research* 34: 907–929.

Khansari N, Shakiba Y, Mahmoudi M. 2009. Chronic inflammation and oxidative stress as a major cause of age-related diseases and cancer. *Recent Patents on Inflammation & Allergy Drug Discovery* 3: 73–80.

Kotake-Nara E, Kim S-J, Kobori M, Miyashita K, Nagao A. 2002. Acyclo-retinoic acid induces apoptosis in human prostate cancer cells. *Anticancer Research* 22: 689–695.

Lian F, Smith DE, Ernst H, Russell RM, Wang X-D. 2007. Apo-10′-lycopenoic acid inhibits lung cancer cell growth *in vitro*, and suppresses lung tumorigenesis in the A/J mouse model *in vivo*. *Carcinogenesis* 28: 1567–1574.

Lian F, Wang XD. 2008. Enzymatic metabolites of lycopene induce Nrf2-mediated expression of phase II detoxifying/antioxidant enzymes in human bronchial epithelial cells. *International Journal of Cancer* 123: 1262–1268.

Liguori I, et al. 2018. Oxidative stress, aging, and diseases. *Clinical Interventions in Aging* 13: 757–772.

Linardaki ZI, Orkoula MG, Kokkosis AG, Lamari FN, Margarity M. 2013. Investigation of the neuroprotective action of saffron (Crocus sativus L.) in aluminum-exposed adult mice through behavioral and neurobiochemical assessment. *Food and Chemical Toxicology* 52: 163–170.

Lindshield BL, Canene-Adams K, Erdman Jr JW. 2007. Lycopenoids: are lycopene metabolites bioactive? *Archives of Biochemistry and Biophysics* 458: 136–140.

Lobo GP, Isken A, Hoff S, Babino D, von Lintig J. 2012. BCDO2 acts as a carotenoid scavenger and gatekeeper for the mitochondrial apoptotic pathway. *Development* 139: 2966–2977.

Madan K, Nanda S. 2018. In-vitro evaluation of antioxidant, anti-elastase, anti-collagenase, anti-hyaluronidase activities of safranal and determination of its sun protection factor in skin photoaging. *Bioorganic Chemistry* 77: 159–167.

Marseglia L, Manti S, D'Angelo G, Nicotera A, Parisi E, Di Rosa G, Gitto E, Arrigo T. 2015. Oxidative stress in obesity: a critical component in human diseases. *International Journal of Molecular Sciences* 16: 378–400.

Meléndez-Martínez AJ. 2019. An overview of carotenoids, apocarotenoids and vitamin A in agro-food, nutrition, health and disease. *Molecular Nutrition & Food Research.* 1801045.

Miquel J. 2009. An update of the oxidation-inflammation theory of aging: the involvement of the immune system in oxi-inflamm-aging. *Current Pharmaceutical Design* 15: 3003–3026.

Mousavi SH, Moallem SA, Mehri S, Shahsavand S, Nassirli H, Malaekeh-Nikouei B. 2011. Improvement of cytotoxic and apoptogenic properties of crocin in cancer cell lines by its nanoliposomal form. *Pharmaceutical Biology* 49: 1039–1045.

Ochiai T, Ohno S, Soeda S, Tanaka H, Shoyama Y, Shimeno H. 2004. Crocin prevents the death of rat pheochromyctoma (PC-12) cells by its antioxidant effects stronger than those of α-tocopherol. *Neuroscience Letters* 362: 61–64.

Ohno Y, Nakanishi T, Umigai N, Tsuruma K, Shimazawa M, Hara H. 2012. Oral administration of crocetin prevents inner retinal damage induced by N-methyl-D-aspartate in mice. *European Journal of Pharmacology* 690: 84–89.

Paik J, Chae J, Kang R, Kwon N, Lee S-H, Lee JH. 2013. Effect of age on atherogenicity of LDL and inflammatory markers in healthy women. *Nutrition, Metabolism and Cardiovascular Diseases* 23: 967–972.

Pole A, Dimri M, Dimri GP. 2016. Oxidative stress, cellular senescence and ageing. *AIMS Molecular Science* 3: 300–324.

Prakash P, Liu C, Hu K-Q, Krinsky NI, Russell RM, Wang X-D. 2004. β-Carotene and β-apo-14′-carotenoic acid prevent the reduction of retinoic acid receptor β in benzo [a] pyrene-treated normal human bronchial epithelial cells. *The Journal of Nutrition* 134: 667–673.

Qi Y, Chen L, Zhang L, Liu W-B, Chen X-Y, Yang X-G. 2013. Crocin prevents retinal ischaemia/reperfusion injury-induced apoptosis in retinal ganglion cells through the PI3K/AKT signalling pathway. *Experimental Eye Research* 107: 44–51.

Ramaswami M, Taylor JP, Parker R. 2013. Altered ribostasis: RNA-protein granules in degenerative disorders. *Cell* 154: 727–736.

Reynaert NL, Gopal P, Rutten EP, Wouters EF, Schalkwijk CG. 2016. Advanced glycation end products and their receptor in age-related, non-communicable chronic inflammatory diseases; Overview of clinical evidence and potential contributions to disease. *The International Journal of Biochemistry & Cell Biology* 81: 403–418.

Rios AdO, Antunes LM, de LP Bianchi M. 2009. Bixin and lycopene modulation of free radical generation induced by cisplatin–DNA interaction. *Food Chemistry* 113: 1113–1118.

Rivera-Madrid R, Aguilar-Espinosa M, Cárdenas-Conejo Y, Garza-Caligaris LE. 2016. Carotenoid derivates in achiote (*Bixa orellana*) seeds: synthesis and health promoting properties. *Frontiers in Plant Science* 7: 1406.

Rodriguez-Concepcion M, Avalos J, Bonet ML, Boronat A, Gomez-Gomez L, Hornero-Mendez D, Limon MC, Meléndez-Martínez AJ, Olmedilla-Alonso B, Palou A. 2018. A global perspective on carotenoids: metabolism, biotechnology, and benefits for nutrition and health. *Progress in Lipid Research* 70: 62–93.

Roehrs M, Figueiredo CG, Zanchi MM, Bochi GV, Moresco RN, Quatrin A, Somacal S, Conte L, Emanuelli T. 2014. Bixin and norbixin have opposite effects on glycemia, lipidemia, and oxidative stress in streptozotocin-induced diabetic rats. *International Journal of Endocrinology* 2014.

Rojo de la Vega M, Zhang DD, Wondrak GT. 2018. Topical bixin confers NRF2-dependent protection against photodamage and hair graying in mouse skin. *Frontiers in Pharmacology* 9: 287.

Rühl R, Sczech R, Landes N, Pfluger P, Kluth D, Schweigert FJ. 2004. Carotenoids and their metabolites are naturally occurring activators of gene expression via the pregnane X receptor. *European Journal of Nutrition* 43: 336–343.

Salisbury D, Bronas U. 2015. Reactive oxygen and nitrogen species: impact on endothelial dysfunction. *Nursing Research* 64: 53–66.

Samarghandian S, Shoshtari ME, Sargolzaei J, Hossinimoghadam H, Farahzad JA. 2014. Anti-tumor activity of safranal against neuroblastoma cells. *Pharmacognosy Magazine* 10: S419.

Sharoni Y, Linnewiel-Hermoni K, Khanin M, Salman H, Veprik A, Danilenko M, Levy J. 2012. Carotenoids and apocarotenoids in cellular signaling related to cancer: a review. *Molecular Nutrition & Food Research* 56: 259–269.

Sheng L, Qian Z, Zheng S, Xi L. 2006. Mechanism of hypolipidemic effect of crocin in rats: crocin inhibits pancreatic lipase. *European Journal of Pharmacology* 543: 116–122.

Siti HN, Kamisah Y, Kamsiah J. 2015. The role of oxidative stress, antioxidants and vascular inflammation in cardiovascular disease (a review). *Vascular Pharmacology* 71: 40–56.

Sosa V, Moliné T, Somoza R, Paciucci R, Kondoh H, LLeonart ME. 2013. Oxidative stress and cancer: an overview. *Ageing Research Reviews* 12: 376–390.

Tabas I, García-Cardeña G, Owens GK. 2015. Recent insights into the cellular biology of atherosclerosis. *J Cell Biol* 209: 13–22.

Tibaduiza EC, Fleet JC, Russell RM, Krinsky NI. 2002. Excentric cleavage products of β-carotene inhibit estrogen receptor positive and negative breast tumor cell growth *in vitro* and inhibit activator protein-1-mediated transcriptional activation. *The Journal of Nutrition* 132: 1368–1375.

Tibodeau JD, Isham CR, Bible KC. 2010. Annatto constituent cis-bixin has selective antimyeloma effects mediated by oxidative stress and associated with inhibition of thioredoxin and thioredoxin reductase. *Antioxidants & Redox Signaling* 13: 987–997.

Tönnies E, Trushina E. 2017. Oxidative stress, synaptic dysfunction, and Alzheimer's disease. *Journal of Alzheimer's Disease* 57: 1105–1121.

Tougu V, Karafin A, Palumaa P. 2008. Binding of zinc (II) and copper (II) to the full-length Alzheimer's amyloid-β peptide. *Journal of Neurochemistry* 104: 1249–1259.

Trpkovic A, Resanovic I, Stanimirovic J, Radak D, Mousa SA, Cenic-Milosevic D, Jevremovic D, Isenovic ER. 2015. Oxidized low-density lipoprotein as a biomarker of cardiovascular diseases. *Critical Reviews in Clinical Laboratory Sciences* 52: 70–85.

Umigai N, Tanaka J, Tsuruma K, Shimazawa M, Hara H. 2012. Crocetin, a carotenoid derivative, inhibits VEGF-induced angiogenesis via suppression of p38 phosphorylation. *Current Neurovascular Research* 9: 102–109.

Van Helden YG, Keijer J, Heil SG, Picó C, Palou A, Oliver P, Munnia A, Briedé JJ, Peluso M, Franssen-van Hal NL. 2009. Beta-carotene affects oxidative stress-related DNA damage in lung epithelial cells and in ferret lung. *Carcinogenesis* 30: 2070–2076.

Victor R-JJ, Araújo GR, da Cruz Pádua B, Chaves MM, Pedrosa ML, Silva ME, Costa DC. 2011. Annato extract and β-carotene modulate the production of reactive oxygen species/nitric oxide in neutrophils from diabetic rats. *Journal of Clinical Biochemistry and Nutrition*. 1112050127.

Walter MH, Strack D. 2011. Carotenoids and their cleavage products: biosynthesis and functions. *Natural Product Reports* 28: 663–692.

Watson J. 2013. Oxidants, antioxidants and the current incurability of metastatic cancers. *Open Biology* 3: 120144.

Wu JQ, Kosten TR, Zhang XY. 2013. Free radicals, antioxidant defense systems, and schizophrenia. *Progress in Neuro-Psychopharmacology and Biological Psychiatry* 46: 200–206.

Xi L, Qian Z, Xu G, Zheng S, Sun S, Wen N, Sheng L, Shi Y, Zhang Y. 2007. Beneficial impact of crocetin, a carotenoid from saffron, on insulin sensitivity in fructose-fed rats. *The Journal of Nutritional Biochemistry* 18: 64–72.

Xia Y, Shen S, Verma IM. 2014. NF-κB, an active player in human cancers. *Cancer Immunology Research* 2: 823–830.

Xiang M, Yang M, Zhou C, Liu J, Li W, Qian Z. 2006. Crocetin prevents AGEs-induced vascular endothelial cell apoptosis. *Pharmacological Research* 54: 268–274.

Xu G, Gong Z, Yu W, Gao L, He S, Qian Z. 2007. Increased expression ratio of Bcl-2/Bax is associated with crocin-mediated apoptosis in bovine aortic endothelial cells. *Basic & Clinical Pharmacology & Toxicology* 100: 31–35.

Yong YK, Zakaria ZA, Kadir AA, Somchit MN, Lian GEC, Ahmad Z. 2013. Chemical constituents and antihistamine activity of Bixa orellana leaf extract. *BMC Complementary and Alternative Medicine* 13: 32.

Zheng S, Qian Z, Tang F, Sheng L. 2005. Suppression of vascular cell adhesion molecule-1 expression by crocetin contributes to attenuation of atherosclerosis in hypercholesterolemic rabbits. *Biochemical Pharmacology* 70: 1192–1199.

Ziouzenkova O, Orasanu G, Sukhova G, Lau E, Berger JP, Tang G, Krinsky NI, Dolnikowski GG, Plutzky J. 2007. Asymmetric cleavage of β-carotene yields a transcriptional repressor of retinoid X receptor and peroxisome proliferator-activated receptor responses. *Molecular Endocrinology* 21: 77–88.

6 An Approach to Vitamin A and Retinoids and Their Importance in Human Health

Víctor Manuel Carballo-Uicab, Rosa Us-Camas and Renata Rivera-Madrid

CONTENTS

PROVITAMIN A AND RETINOIDS

Carotenoids are part of group of pigments naturally located in plants, algae, fungi, and bacteria, which are synthesized *de novo*, and play essential functions in development and photosynthesis. In plants, carotenoids form part of the photosynthetic apparatus to act as accessory pigment for photosynthesis, photoprotection, and they are precursors for signalling molecules such as strigolactone (SL) and abscisic acid (ABA) which play an essential function in development and response at biotic/abiotic stress (Park et al., 2002; Cazzonelli and Pogson, 2010).

In animals, the carotenoids are an integral part of a healthy nutrition, functioning as nutritional antioxidants and vitamin A (VA) precursors (Giuliano, 2017; Harrison and Quadro, 2018; Sun et al., 2018). Nevertheless, animals (including humans) are not capable of synthesizing carotenoids *de novo*; consequently, they have to be acquired in food. For example, bird feathers' pigment is caused by a diet rich in carotenoid (Farré et al., 2010) (Figure 6.1).

There are more than 700 carotenoids identified in nature; however, only some of these carotenoids are included in our foods to be absorbed, metalized, and/or used by our body. Some of the carotenoids commonly present in green vegetables are the α-carotene, β-carotene, lycopene, β-cryptoxanthin, zeaxanthin, and lutein (Rao and Rao, 2007; Fernández-García et al., 2012). The carotenoid concentration largely depends on the type of food, tissue absorption, carotenoid-release from tissues to blood, and catabolism rate (Perera and Yen, 2007). About 90% of carotenoids are found in body tissues, and 10% in blood plasma (Perera and Yen, 2007). Lutein, lycopene, zeaxanthin, β-cryptoxanthin, β-carotene, and α-carotene cover 60%–70% of the carotenoid content in the blood plasma (Khachik et al., 1992). These carotenoids are found in all tissues, especially in adult's adipose tissue, liver, and blood plasma. β-carotene, lutein, lycopene, and canthaxanthin are found in the skin.

The carotenoids play diverse important roles in people's health. The VA or retinol (ROL) is of vital importance in people's health, for example for the eye health (During and Harrison, 2007). In our

FIGURE 6.1 Some foods that we commonly consume are essential for production of provitamin A (PVA) and retinoids.

diet of animal source, it is found as retinol esters (REs) and in plant food sources such as provitamin A (PVA) (Penniston and Tanumihardjo, 2006). The carotenoids that have vitamin A activity are known as PVA (Díaz-Gómez et al., 2017). The principal requisite PVA function is that carotenoids need to have a β-ionone ring residue, without oxygenated functional groups. The β-carotene is formed by two rings in its structure and is found in most diets (Gerster, 1997; Fernández-García et al., 2012; De Moura et al., 2015).

The most essential PVA carotenoid is the β-carotene, because it is the best potent dietary precursor of VA and is the only carotenoid capable of yielding two molecules of VA (Krinsky and Johnson, 2005), also functioning as an antioxidant, and is an efficient singlet oxygen quencher, a major present carotenoid in skin and scavenges reactive oxygen species (Grune et al., 2010). Although carotenoids primarily implement antioxidant effects, some carotenoids possess PVA activity; for example: α-carotene, β-carotene, and β-cryptoxanthin (Díaz-Gómez et al., 2017). Others, for example, zeaxanthin and lutein (constitute macular pigment in the eye) have antioxidant functions, but not the ability to convert into VA (Farré et al., 2010; Eggersdorfer and Wyss, 2018; Toti et al., 2018). Both VA and PVA are processed to retinal and retinoic acid (RA) (the functional forms of VA) to provide the vitamin necessary for vital functions (Ghosh et al., 2019).

VA is a term used to describe any compound with the biological function of VA or having structural similarity, and/or ability to act as VA. The retinol, retinal, and RA are collectively known as VA; in other words, the expression 'retinoid' denotes analogues of VA (Blomhoff and Blomhoff, 2006; Poulain et al., 2009; Gonçalves et al., 2016; Gysel, 2016; Zasada and Budzisz, 2019). In accordance with IUPAC (International Union of Pure and Applied Chemistry) and IUBMB (International Union of Biochemistry and Molecular Biology), the retinoids are structurally composed of a cyclic end group, a polyene side chain, and a polar end group (Dixon, 1983; Meléndez-Martínez, 2019).

BIOSYNTHESIS OF VITAMIN A

The oxidative breakdown of carotenoids by the carotenoid cleavage dioxygenases (CCDs) family is the main pathway for the formation of VA to form PVA (von Lintig and Wyss, 2001). The characterization of CCDs was originally studied in plants (Schwartz et al., 1997); consequently, the identification of these enzymes led to the discovery of CCDs in other organisms, for example, bacteria, fungi and animals (Lintig and Vogt, 2000, 2004; Ruch et al., 2005; Prado-Cabrero et al., 2007; Bruno et al., 2016). The CCDs belong to a class of enzymes implicated in the biosynthesis of secondary metabolites well-known as apocarotenoids. In plants, some CCDs are able to cleave precise double bonds of the carotenoid structure (Rodríguez-Ávila et al., 2011), and are also implicated in the construction of numerous and vital metabolites like VA and precursors for signalling molecules, including ABA, SL, and trisporic acid in fungi, which controls morphogenesis and retinoids (VA) in animals (Nagao, 2004; Blomhoff and Blomhoff, 2006; Nisar et al., 2015; Giuliano, 2017; Harrison and Quadro, 2018).

The mammalian retinal pigment epithelium (RPE65) protein was the first animal 9-cis-epoxy-carotenoid dioxygenases (NCED) homologue to be described in human (Giuliano et al., 2003). Because of the importance of CCDs in the biosynthesis of VA in animals, other homologue CCDs known as β-carotene oxygenases (**BCOs**) have been analysed and identified from several animal kinds based on the similarity of plant CCDs (von Lintig and Wyss, 2001; Giuliano et al., 2003; DellaPenna and Pogson, 2006). In humans, there exist two CCD enzymes known as β-carotene 15,15′ oxygenase-1 (**BCO1**) and β-carotene 9′,10′ oxygenase-2 (**BCO2**), that use β-carotene as the substrate to produce the PVA (Raghuvanshi et al., 2015).

Because the β-carotene is the principal precursor of VA, two ways to converting β-carotene to retinoids (VA) have been described: the first route involves a central or symmetrical section, and the second route includes an asymmetric (eccentric) section (Karrer et al., 1930; Glover, 1961; Eroglu and Harrison, 2013; Pérez-Gálvez et al., 2018):

1. A symmetrical section by a cytosolic enzyme BCO1 on the β-carotene produces two VA molecules (in the liver) to generate retinal precursors of retinol and retinoic acid (Nagao, 2004; Harrison, 2012); however, carotenoids (for example the α-carotene and β-cryptoxanthin) can generate a single VA molecule. These carotenoids have been found mainly in vegetable foods (De Moura et al., 2015) (Figure 6.2).
2. The second route is the asymmetric section of β-carotene by BCO2, the aleatory section of its conjugated double bond producing different β-apo-carotenals fragments (Nagao, 2004; Harrison, 2012). Nonetheless, BCO2 is able to section lineal carotenoids such as lycopene (Wang et al., 1991; Schwartz et al., 2003; Fraser and Bramley, 2004; Nagao, 2004).

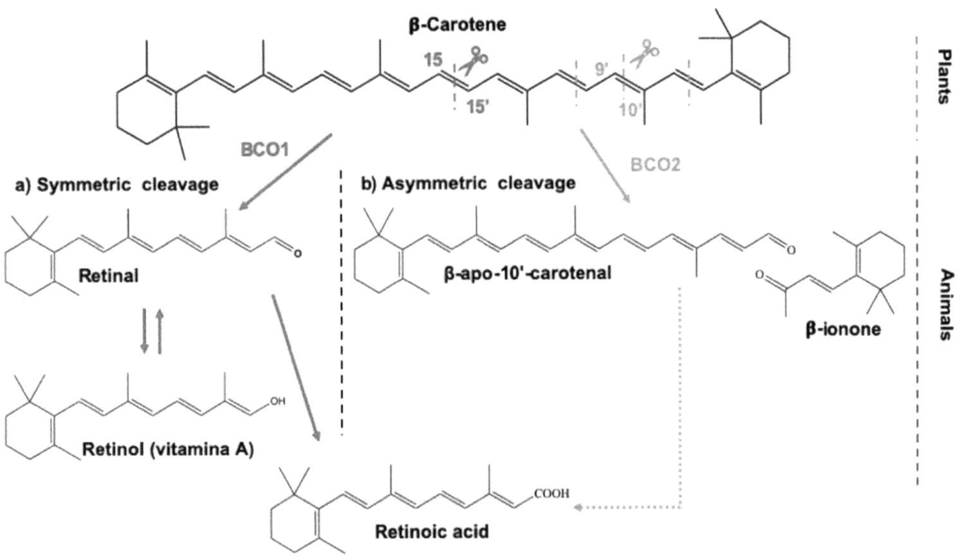

FIGURE 6.2 Pathway of retinoids and VA form β-carotene. (a) Two molecules of retinal are generated by symmetric scission of β-carotene at the 15,15′ position by the enzyme BCO1. (b) β-apo-10′-carotenal and β-ionone are generated by asymmetric scission of β-carotene at the 9′,10′ positions by BCO2.

BCO1 and BCO2 enzymes present 40% sequence homology and are also located in internal organs, for example, small intestine, liver, kidneys, and lungs. BCO1 and BCO2 are principally located in the liver and intestine. These two organs are important for the absorption of VA, but the liver represents the majority of retinoid reserves (Harrison, 2012; Toti et al., 2018). BCO1 localizes in the cytoplasm and transforms some carotenoids, like as β-carotene to retinoids, while BCO2 is located in the mitochondria, and has a broad substrate specificity (Lindqvist and Andersson, 2002; Amengual et al., 2011).

VITAMIN A DEFICIENCY

Vitamin deficiencies cause severe mortal diseases in humans (Asensi-Fabado and Munné-Bosch, 2010). Vitamin A deficiency (VAD) is recognized as a social problem in emerging countries. The importance of VA goes beyond vision health (Gonçalves et al., 2016). It generally has an impact on infants and pregnant women in developing nations. Consequently, VAD causes loss of vision, intrauterine growth restriction (IUGR), xerophthalmia (dry eyes), and increases the risk of other epidemic diseases (Gonçalves et al., 2016; Maqbool et al., 2018). VA is important to maintain normal reproduction, development, epithelial cell integrity and immunity, and prevent abnormalities, spermatogenesis, and reproductive failures that compromise implantation and fetal development (aberration of tissues and organs) (White et al., 2000). It also alters cellular differentiation (Sommer, 2008), retards growth and development, and impairs the immune system, causing anemia and infections (Sommer and Davidson, 2002).

Approximately 250 million kindergarten-age children experience VAD worldwide. In Africa, about 40%–60% of children are affected by VAD (Wiseman et al., 2017). In emerging countries (especially in Africa and South East Asia), 10%–20% of women in pregnancy experience VAD (Wiseman et al., 2017). In South East Asia, the incidence of VAD appears to be resolute by economic limitations, poverty, race, a diet poor in VA is limited due to the quality of food and its assimilation in the body (Akhtar et al., 2013). VAD can occur in any age, but children under 6 years are the most vulnerable (FAO and WHO, 1998). In Bangladesh and India, the highest mortality and VAD index is very common among babies and children around 5 years of age, which is reflected as one third of the worldwide mortality (Akhtar et al., 2013). A Russian boy (8 years) with discomfort in the eyes due to VAD (<2 mg/dL), received a VA supplement over 3 months, and he showed improvement in his health status, followed by prompt recovery (Hos and Cursiefen, 2014).

In Sri Lanka, a study on pregnant women showed that 1.2% suffer night blindness because of poor VA levels in the blood (<20 μg/dL) (Jayasekera et al., 1991). This coincides with the result in China, where serum retinol concentrations were measured in 1,209 pregnant women in 31 provinces, considering age, gestational age, parity and education levels. The results showed that the average VA was 1.63 μmol/L, and only 5.3% had VAD. According to WHO, the lowest serum retinol, <0.70 μmol/L is to be considered VAD in pregnant women (WHO, 2009). The lower frequency of VAD is due to the improvement of the socioeconomic level; age and vitamin A status which did not correlate with the level of vitamin A declined with age (Yang et al., 2016). Recently in Brazil, 442 pregnant women showed retinol concentrations of 1.05 μmol/L, as a result of the change in plasma volume during pregnancy, absorption rate, and the bioconversion factor to retinol (Neves et al., 2019).

In Africa and Asia, 85% of plant-based foods contain PVA carotenoids. But not all of these foods contain the appropriate requirements for a balanced diet, nor are they rich in VA, nor they are accessible for consumption (Schweiggert and Carle, 2017). The assimilation of β-carotene in humans from plant-based foods is 5%–65% (Haskell, 2012). In contrast, the majority of people in first-world countries need to consume >70% PVA in their diet (FAO and WHO, 1998). However, people that manage highly limiting diets or eliminate some foods are vulnerable VAD (Weber and Grune, 2012).

On the other hand, the diet in first-world countries represents 66%–80% of VA consumption, while 20%–34% represents PVA (Van Loo-Bouwman et al., 2014). A diet based on β-carotene and VA supplementation has shown an increase in VA concentration in men (68.4%) and women

(63.8%), and PVA concentration in men (31.6%) and women (36.2%) (11 case studies from 8 countries) although the VA intake varies according to country, and for the dietary habits and food processing therein (Weber and Grune, 2012).

The bioavailability is the minimum required amount of a consumed carotenoid for vital functions or storage, and the bioaccessibility is the intake of carotenoids that can be absorbed by the body (La Frano et al., 2014). A study comparing the bioavailability of carotenoids in principal food sources of β-carotene such as carrot, tomato, and papaya in 16 participants showed that papaya represents a very important source of PVA (Schweiggert et al., 2014). So, 1 μg of VA in the human body from a minimal dose of β-carotene is defined as the equivalent of β-carotene (VEB) to be transformed into VA (Van Loo-Bouwman et al., 2014); however, this depends on diet and other physiological factors (Haskell, 2012). There are reports that children can transform β-carotene more efficiently than adults (Van Loo-Bouwman et al., 2014).

Recently, a study in France evaluated the key nutrients and VA bioavailability factors diet of adults (age >18) and their association with the consumption of food of plant and animal origin, showing that there is an association with diet type and VEB. Thus, the VEB estimated was 17:1 of β-carotene and 26:1 for other types of PVA carotenoids (Perignon et al., 2018). A similar case was reported in Bangladeshi women with a diet of biofortified orange sweet potato, showing an increase in VA production in the body, but this increase was not significant when the diet was white-fleshed sweet potatoes (Jamil et al., 2012).

In other analyses, using a mixture of plant-based foods as a diet in Indonesian breastfeeding women for 5 days/week, 12 weeks later, the VA and β-carotene content in the blood was unchanged. In contrast, β-carotene and VA concentrations in a group with a supplemental diet showed an increase in β-carotene content (de Pee et al., 1995). The VEB is more efficient in biofortified crops than non-biofortified crops (La Frano et al., 2014; Giuliano, 2017).

VITAMIN A REQUIREMENTS

The vitamin A equivalency ratio (ER) is the β-carotene dose that corresponds to a unit of retinol (VA) (Giuliano, 2017). This relation varies from 3.8:1 to 28:1 in humans (Haskell, 2012). In babies and children, the recommended safe intake (0–6 months) is 375 μg ER/day, in adolescents (10–18 years) 600 μg ER/day, in adults: females (19–65 years) 500 μg ER/day, males (19–65 years) 600 μg ER/day, in pregnant women 800 μg ER/day, and in lactating women 850 μg ER/day (FAO and WHO, 1998). According to the Institute of Medicine of the National Academies, a dose of 300 μg/day of retinol is recommended for children 1–3 years of age, and 800 μg for adults (Giuliano et al., 2008).

BIOFORTIFIED CROPS

Biofortification is the process of crop improvement through the addition of vitamins, increasing the nutritional quality (Asensi-Fabado and Munné-Bosch, 2010; Bouis and Welch, 2010). According to the World Health Organization (WHO), the process of biofortification of crops can be achieved by using biological techniques like transgenic techniques (WHO, 2019). Intake of carotenoid-rich foods has been shown to have health benefits (Schweiggert and Carle, 2017). The generation of improved crops to increase their nutritional value as PVA (mainly β-carotene) through conventional techniques or the generation of transgenic crops has led to the appearance of 'golden' crops (Asensi-Fabado and Munné-Bosch, 2010; Beyer, 2010; Bouis and Welch, 2010). Golden rice (Ye et al., 2000) and sorghum (Lipkie et al., 2013) are the main products representing the generation of genetically improved crops (golden crops), with high PVA nutritional values (Kumar et al., 2019).

With the 'golden rice' achievement, new crops have been generated based on metabolic engineering of carotenoids, as well as conventional breeding in potato, canola, tomato, carrot, and cauliflower (Giuliano et al., 2008; Giuliano, 2014, 2017; Kumar et al., 2019). Because the VA is deficient

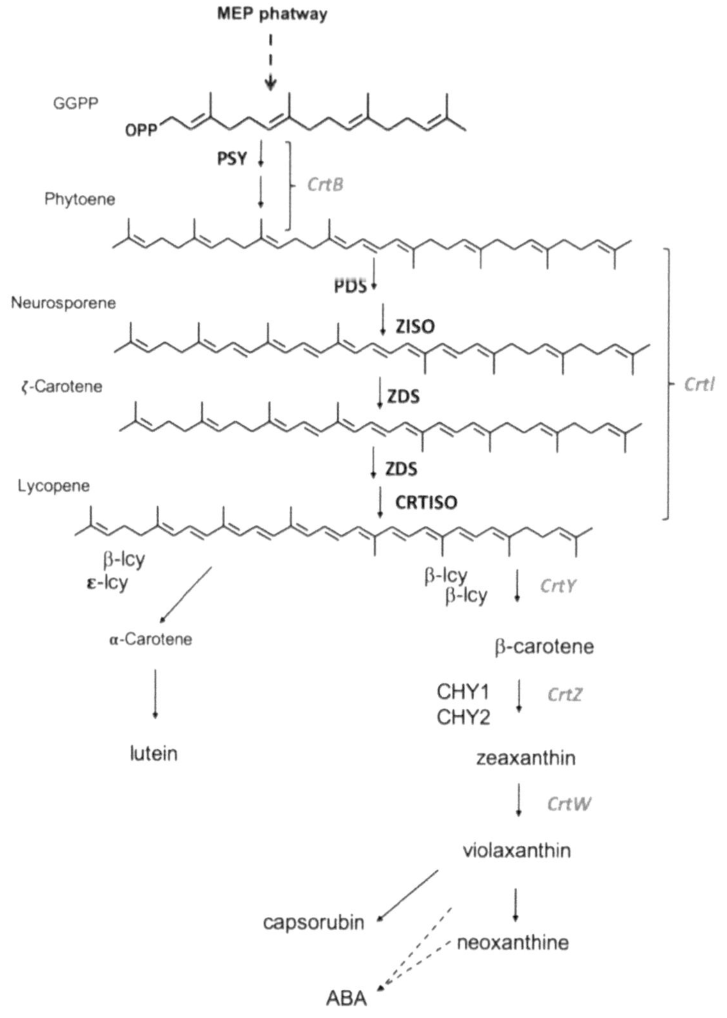

FIGURE 6.3 Pathway of carotenoid biosynthesis in plants and bacteria. Enzyme synthesis is shown in bold. PSY/CrtB, phytoene synthase; PDS/Crtl phytoene desaturase phytoene; ZDS, ζ carotene desaturase; CRTISO, carotenoid isomerases; β-lyc/CrtY, lycopene β-cyclase; β-lyc, lycopene ε-cyclase; CrtZ: β-carotene hydroxylase, CrtW, cetolase. Enzyme names are shown in black (plant enzymes) and red (bacterial enzymes).

in some cultivars, metabolic engineering, such as the overexpression of carotenoid pathway gene bacterial in plants, has been a successful strategy used (Figure 6.3). The golden rice (*Oryza sativa*) was created from the simultaneous expression of two different *PSY* (bacteria and plants), resulting in increased β-carotenoids (1.6 µg/g) in the endosperm (Ye et al., 2000) (Table 6.1).

In a new study with a different *PSY* expression from maize (*Zea mays*) increased total carotenoid levels 23-fold (37 µg/g), accumulating mainly β-carotene. This new modification plant was named 'Golden Rice 2' (Paine et al., 2005). In the case of sorghum (*Sorghum bicolor* L.) using the similar strategy as the Golden Rice 2 with certain modifications, the content of β-carotene was 14 µg/g DW (Lipkie et al., 2013).

In wheat endosperm (*Triticum aestivum* L. cv. Bobwhite) the repress of carotenoid hydroxylase (*TaHYD*) generated increased the β-carotene by 10.5-fold to 1.76 µg/g, while overexpression of *CrtB* increased β-carotene by 14.6-fold to 2.45 µg/g, and the combination of this strategy increased

TABLE 6.1
Carotenoid biofortification in crops increases β-carotene

Crops	Gene	β-carotene accumulation (μg/g)	Tissue accumulation	Source
Rice (*Oryza sativa*)	Ov. of *PSY* + *Crtl*	1.6 DW	endosperm	Ye et al. (2000)
Rice (2 *Oryza sativa*) (endosperm)	Ov. of *PSY* + *Crtl*	37 DW	endosperm	Paine et al. (2005)
Sorghum (Sorghum bicolor)	Ov. *PSY* + *Crtl*	14 DW	grain	Lipkie et al. (2013)
Potato (*Solanum tuberosum*)	Ov. *CrtB* + *Crtl* + *CrtY*	47 DW	tuber	Diretto et al. (2007b)
Canola (*Brassica napus*)	Ov. *CrtB* + *CrtI*	857 FW	seed	Ravanello et al. (2003)
Banana (Transgenic *Musa acuminata*)	Ov. *MtPsy2a or PSY*	55 DW	fruit	Paul et al. (2017)
Wheat (*Triticum aestivum* L.)	Ov. *CrtB* +/or silencing of *TaHYD*	5.06 DW	endosperm	Zeng et al. (2015b)
Maize (*P. ananatis*)	*CrtB* + *CrtI*	9.8 DW	endosperm	Aluru et al. (2008)

Ov: overexpression, DW: dry weight, FW: fresh weight.

β-carotene 31-fold to 5.06 μg/g (Zeng et al., 2015b). The repress of carotenoid hydroxylase (*CHY1* and *CHY2*) in potato tuber increased β-carotene 38-fold, and the total of carotenoids increased 4.5-fold (Diretto et al., 2007b). In the case of wheat kernels with poor carotenoid values and PVA activity, the silencing of lycopene β-cyclase gene (*TaLCYB*) confirmed a reduction of β-carotene content (Zeng et al., 2015a). An analysis in carrot (*Daucus carota*) showed that the total carotenoid and β-carotene levels increased in leaves and roots when the endogenous *LCYB1* gen was overexpressed (Moreno et al., 2013).

The overexpression of bacterial phytoene synthase gene (*CrtB*) in canola seed showed a 50-fold increase in carotenoids, especially β-carotene (Shewmaker et al., 1999). This has been improved using the pooled overexpression of *CrtB* and *CrtI* increasing β-carotene to 857 μg/g FW (Ravanello et al., 2003). Similar results were reported with the overexpression of *CrtB,* of *CrtI* and *CrtY* in potato (*Solanum tuberosum*) to generate the 'Golden potato'. The overexpression of these three genes increased the concentration of β-carotene to 47 μg/g DW. This would be enough to supply 50% of the recommended daily dose (Diretto et al., 2007a). Others had the 'golden banana' which had generation of lines with PVA levels of 55 μg/g (DW) β-carotene equivalent (Paul et al., 2017). In the maize endosperm the overexpression of *CrtB* and *CrtI* genes and γ-zein promoter, increasing the total carotenoid content to 33.6 g/g DW and increased the β-carotene 34-fold to 9.8 μg/g DW (Aluru et al., 2008).

Bioavailability reports have confirmed the efficiency of golden crops to improve and increase VA content (Giuliano, 2017). Although biofortification can increase the PVA contained in a basic food, it is still a challenge to determine the necessary amount absorbed by the human body to improve micronutrient levels (Bechoff and Dhuique-Mayer, 2017).

Currently, the VA values acquired in a diet of plant-based foods, as well as the specific amount or benefits of any carotenoid absorbed in the body during the intake, have not been specifically determined (Perera and Yen, 2007). The carotenoid bioavailability can be disturbed in several circumstances; this has been widely discussed (Desmarchelier and Borel, 2017; Grune et al., 2010), and can be summarized by the proposal of 'SLAMENGHI' (Castenmiller and West, 1998).

CONCLUSION

Vitamins are important for the proper functioning of our body. Micronutrients such as PVA and VA have become the main target in the biofortification of staple crops, to combat malnutrition in vulnerable populations, such as children and women in undeveloped countries. However, certain aspects, such as nutrient absorption in biofortified crops, and their relationship with other factors, are still required for metabolic engineering of carotenoids to become a viable technology. The new genome editing tools, such as CRISPR-Cas, offer great potential in crop improvement to increase VA production.

ACKNOWLEDGEMENTS

The authors thank Consejo Nacional de Ciencia y Tecnología (CONACYT) for their support with the project: Fronteras de la Ciencia No. 2016-01-1716.

REFERENCES

Akhtar, S., Ahmed, A., Randhawa, M. A., Atukorala, S., Arlappa, N., Ismail, T. et al. (2013). Prevalence of vitamin A deficiency in South Asia: Causes, outcomes, and possible remedies. *J. Health Popul. Nutr.* 31, 413–423.

Aluru, M., Xu, Y., Guo, R., Wang, Z., Li, S., White, W. et al. (2008). Generation of transgenic maize with enhanced provitamin A content. *J. Exp. Bot.* 59, 3551–3562. doi:10.1093/jxb/ern212.

Amengual, J., Lobo, G. P., Golczak, M., Li, H. N. M., Klimova, T., Hoppel, C. L. et al. (2011). A mitochondrial enzyme degrades carotenoids and protects against oxidative stress. *FASEB J.* 25, 948–959. doi:10.1096/fj.10-173906.

Asensi-Fabado, M. A., and Munné-Bosch, S. (2010). Vitamins in plants: Occurrence, biosynthesis and antioxidant function. *Trends Plant Sci.* 15, 582–592. doi:10.1016/j.tplants.2010.07.003.

Bechoff, A., and Dhuique-Mayer, C. (2017). Factors influencing micronutrient bioavailability in biofortified crops. *Ann. N. Y. Acad. Sci.* 1390, 74–87. doi:10.1111/nyas.13301.

Beyer, P. (2010). Golden Rice and 'Golden' crops for human nutrition. *N. Biotechnol.* 27, 478–481. doi:10.1016/j.nbt.2010.05.010.

Blomhoff, R., and Blomhoff, H. K. (2006). Overview of retinoid metabolism and function Rune. *J. Neurobiol.* 14, 1564–1583. doi:10.1002/neu.

Bouis, H. E., and Welch, R. M. (2010). Biofortification—A sustainable agricultural strategy for reducing micronutrient malnutrition in the global south. *Crop Sci.* 50, S-20–S-32. doi:10.2135/cropsci2009.09.0531.

Bruno, M., Koschmieder, J., Wuest, F., Schaub, P., Fehling-Kaschek, M., Timmer, J. et al. (2016). Enzymatic study on AtCCD4 and AtCCD7 and their potential to form acyclic regulatory metabolites. *J. Exp. Bot.* 67, 5993–6005. doi:10.1093/jxb/erw356.

Castenmiller, J. J. M., and West, C. E. (1998). Bioavailability and bioconversion of carotenoids. *Annu. Rev. Nutr.* 18, 19–38. doi:10.1146/annurev.nutr.18.1.19.

Cazzonelli, C. I., and Pogson, B. J. (2010). Source to sink: Regulation of carotenoid biosynthesis in plants. *Trends Plant Sci.* 15, 266–274. doi:10.1016/j.tplants.2010.02.003.

De Moura, F. F., Miloff, A., and Boy, E. (2015). Retention of provitamin A carotenoids in staple crops targeted for biofortification in Africa: Cassava, maize and sweet potato. *Crit. Rev. Food Sci. Nutr.* 55, 1246–1269. doi:10.1080/10408398.2012.724477.

de Pee, S., West, C. E., Hautvast, J. G. A. J., Muhilal, K. D., and West, C. E. (1995). Lack of improvement in vitamin A status with increased consumption of dark-green leafy vegetables. *Lancet* 346, 75–81. doi:10.1016/S0140-6736(95)92111-7.

DellaPenna, D., and Pogson, B. J. (2006). Vitamin synthesis in plants: Tocopherols and carotenoids. *Annu. Rev. Plant Biol.* 57, 711–738. doi:10.1146/annurev.arplant.56.032604.144301.

Desmarchelier, C., and Borel, P. (2017). Overview of carotenoid bioavailability determinants: From dietary factors to host genetic variations. *Trends Food Sci. Technol.* 69, 270–280. doi:10.1016/j.tifs.2017.03.002.

Díaz-Gómez, J., Moreno, J. A., Angulo, E., Sandmann, G., Zhu, C., Capell, T. et al. (2017). Provitamin A carotenoids from an engineered high-carotenoid maize are bioavailable and zeaxanthin does not compromise β-carotene absorption in poultry. *Transgenic Res.* 26, 591–601. doi:10.1007/s11248-017-0029-y.

Diretto, G., Al-Babili, S., Tavazza, R., Papacchioli, V., Beyer, P., and Giuliano, G. (2007a). Metabolic engineering of potato carotenoid content through tuber-specific overexpression of a bacterial mini-pathway. *PLoS One* 2, 1–8. doi:10.1371/journal.pone.0000350.

Diretto, G., Welsch, R., Tavazza, R., Mourgues, F., Pizzichini, D., Beyer, P. et al. (2007b). Silencing of beta-carotene hydroxylase increases total carotenoid and beta-carotene levels in potato tubers. *BMC Plant Biol.* 7, 11. doi:10.1186/1471-2229-7-11.

Dixon, H. B. F. (1983). Nomenclature of retinoids. *Pure Appl. Chem.* 55, 721–726. doi:10.1351/pac198855040721.

During, A., and Harrison, E. H. (2007). Mechanisms of provitamin A (carotenoid) and vitamin A (retinol) transport into and out of intestinal Caco-2 cells. *J. Lipid Res.* 48, 2283–2294. doi:10.1194/jlr.m700263-jlr200.

Eggersdorfer, M., and Wyss, A. (2018). Carotenoids in human nutrition and health. *Arch. Biochem. Biophys.* 652, 18–26. doi:10.1016/j.abb.2018.06.001.

Eroglu, A., and Harrison, E. H. (2013). Carotenoid metabolism in mammals, including man: Formation, occurrence, and function of apocarotenoids. *J. Lipid Res.* 54, 1719–1730. doi:10.1194/jlr.R039537.

FAO, and WHO. (1998). *Vitamin and Mineral Requirements in Human Nutrition*, Second edition. World Health Organization, Bangkok, Thailand: 1–20. doi:92 4 154612 3.

Farré, G., Sanahuja, G., Naqvi, S., Bai, C., Capell, T., Zhu, C. et al. (2010). Travel advice on the road to carotenoids in plants. *Plant Sci.* 179, 28–48. doi:10.1016/j.plantsci.2010.03.009.

Fernández-García, E., Carvajal-Lérida, I., Jarén-Galán, M., Garrido-Fernández, J., Pérez-Gálvez, A., and Hornero-Méndez, D. (2012). Carotenoids bioavailability from foods: From plant pigments to efficient biological activities. *Food Res. Int.* 46, 438–450. doi:10.1016/j.foodres.2011.06.007.

Fraser, P. D., and Bramley, P. M. (2004). The biosynthesis and nutritional uses of carotenoids. *Prog. Lipid Res.* 43, 228–265. doi:10.1016/j.plipres.2003.10.002.

Gerster, H. (1997). Vitamin A—Functions, dietary requirements and safety in humans. *Int. J. Vitam. Nutr. Res.* 67, 71–90. Available at: http://ovidsp.ovid.com/ovidweb.cgi?T=JS&PAGE=reference&D=emed4&NEWS=N&AN=9129249.

Ghosh, S., Adak, K., Saha, P., Upadhyay, S., Ghosh, D. P. et al. (2019). Beta-carotene retention as retinol activity equivalent at different cooking and storage variants. *Int. Food Res. J.* 26, 355–361. Available at: http://www.ifrj.upm.edu.my/26(01)2019/(40).pdf.

Giuliano, G. (2014). Plant carotenoids: Genomics meets multi-gene engineering. *Curr. Opin. Plant Biol.* 19, 111–117. doi:10.1016/j.pbi.2014.05.006.

Giuliano, G. (2017). Provitamin A biofortification of crop plants: a gold rush with many miners. *Curr. Opin. Biotechnol.* 44, 169–180. doi:10.1016/j.copbio.2017.02.001.

Giuliano, G., Al-Babili, S., and von Lintig, J. (2003). Carotenoid oxygenases: Cleave it or leave it. *Trends Plant Sci.* 8, 143–145. doi:10.1016/S1360-1385(03)00012-8.

Giuliano, G., Tavazza, R., Diretto, G., Beyer, P., and Taylor, M. A. (2008). Metabolic engineering of carotenoid biosynthesis in plants. *Trends Biotechnol.* 26, 139–145. doi:10.1016/j.tibtech.2007.12.003.

Glover, J. (1961). The Conversion of β-Carotene into Vitamin A. In *Vitamins & Hormones* (Vol. 18, pp. 371–386). Academic Press. doi:10.1016/S0083-6729(08)60869-9.

Gonçalves, A., Estevinho, B. N., and Rocha, F. (2016). Microencapsulation of vitamin A: A review. *Trends Food Sci. Technol.* 51, 76–87. doi:10.1016/j.tifs.2016.03.001.

Grune, T., Lietz, G., Palou, A., Ross, A. C., Stahl, W., Tang, G. et al. (2010). b-carotene is an important vitamin A source. *J. Nutr.* 140, 2268S–2285S. doi:10.3945/jn.109.119024.ants.

Gysel, D. Van (2016). *Practical Pediatric Dermatology: Controversies in Diagnosis and Treatment.* doi:10.2307/3461527.

Harrison, E. H. (2012). Mechanisms involved in the intestinal absorption of dietary vitamin A and provitamin A carotenoids. *Biochim. Biophys. Acta - Mol. Cell Biol. Lipids* 1821, 70–77. doi:10.1016/j.bbalip.2011.06.002.

Harrison, E. H., and Quadro, L. (2018). Apocarotenoids: Emerging roles in mammals. *Annu. Rev. Nutr.* 38, 153–172. doi:10.1146/annurev-nutr-082117-051841.

Haskell, M. J. (2012). The challenge to reach nutritional adequacy for vitamin A: b-carotene bioavailability and conversion—evidence in humans. *Am. Soc. Nutr.* 96, 1193S–1203S. doi:10.3945/ajcn.112.034850.1.

Hos, D., and Cursiefen, C. (2014). Severe vitamin A deficiency in a child presenting as Xerophthalmia. *J. Pediatr.* 165, 875. doi:10.1016/j.jpeds.2014.06.054.

Jamil, K. M., Brown, K. H., Jamil, M., Peerson, J. M., Keenan, A. H., Newman, J. W. et al. (2012). Daily consumption of orange-fleshed sweet potato for 60 days increased plasma β-carotene concentration but did not increase total body vitamin A pool size in Bangladeshi women. *J. Nutr.* 142, 1896–1902. doi:10.3945/jn.112.164830.

Jayasekera, J. P. D. J. S., Atukorala, T. M. S., and Seneviratne, H. R. (1991). Vitamin A status of pregnant women in five districts of Sri Lanka. *Asia-Oceania J. Obstet. Gynaecol.* 17, 217–224. doi:10.1111/j.1447-0756.1991.tb00263.x.

Karrer, P., Helfenstein, A., Wehrli, H., and Wettstein, A. (1930). Pflanzenfarbstoffe XXV. Über die Konstitution des Lycopins und Carotins. *Helv. Chim. Acta* 13, 1084–1099. doi:10.1002/hlca.19300130532.

Khachik, F., Beecher, G. R., Goli, M. B., Lusby, W. R., Smith, J. C., Beecher, G. R. et al. (1992). Separation and identification of carotenoids and their oxidation products in the extracts of human plasma. *Anal. Chem.* 64, 2111–2122. doi:10.1021/ac00042a016.

Krinsky, N. I., and Johnson, E. J. (2005). Carotenoid actions and their relation to health and disease. *Mol. Aspects Med.* 26, 459–516. doi:10.1016/j.mam.2005.10.001.

Kumar, S., Palve, A., Joshi, C., Srivastava, R. K., and Rukhsar, R. K. (2019). Crop biofortification for iron (Fe), zinc (Zn) and vitamin A with transgenic approaches. *Heliyon* 5, e01914. doi:10.1016/j.heliyon.2019.e01914.

La Frano, M. R., de Moura, F. F., Boy, E., Lönnerdal, B., and Burri, B. J. (2014). Bioavailability of iron, zinc, and provitamin A carotenoids in biofortified staple crops. *Nutr. Rev.* 72, 289–307. doi:10.1111/nure.12108.

Lindqvist, A., and Andersson, S. (2002). Biochemical properties of purified recombinant human β-carotene 15,15′-monooxygenase. *J. Biol. Chem.* 277, 23942–23948. doi:10.1074/jbc.M202756200.

Lintig, J. Von, and Vogt, K. (2000). Filling the gap in vitamin A research. *J. Biol. Chem.* 275, 11915–11920. doi:10.1074/jbc.275.16.11915.

Lintig, J. von, and Vogt, K. (2004). Vitamin A formation in animals: Molecular identification and functional characterization of carotene cleaving enzymes. *Am. Soc. Nutr. Sci.*, 134, 246–250.

Lipkie, T. E., De Moura, F. F., Zhao, Z. Y., Albertsen, M. C., Che, P., Glassman, K. et al. (2013). Bioaccessibility of carotenoids from transgenic provitamin A biofortified sorghum. *J. Agric. Food Chem.* 61, 5764–5771. doi:10.1021/jf305361s.

Maqbool, M. A., Aslam, M., Beshir, A., and Khan, M. S. (2018). Breeding for provitamin A biofortification of maize (Zea mays L.). *Plant Breed.* 137, 451–469. doi:10.1111/pbr.12618.

Meléndez-Martínez, A. J. (2019). An overview of carotenoids, apocarotenoids and vitamin A in agro-food, nutrition, health and disease. *Mol. Nutr. Food Res.* 1801045. doi:10.1002/mnfr.201801045.

Moreno, J. C., Pizarro, L., Fuentes, P., Handford, M., Cifuentes, V., and Stange, C. (2013). Levels of lycopene β-cyclase 1 modulate carotenoid gene expression and accumulation in *Daucus carota*. *PLoS One* 8, 1–16. doi:10.1371/journal.pone.0058144.

Nagao, A. (2004). Oxidative conversion of carotenoids to retinoids and other products. *J. Nutr.* 134, 237–240.

Neves, P. A. R., Campos, C. A. S., Malta, M. B., Lourenço, B. H., Castro, M. C., and Cardoso, M. A. (2019). Predictors of vitamin A status among pregnant women in Western Brazilian Amazon. *Br. J. Nutr.* 121, 202–211. doi:10.1017/S0007114518003112.

Nisar, N., Li, L., Lu, S., Khin, N. C., and Pogson, B. J. (2015) . Carotenoid metabolism in plants. *Mol. Plant.* 8, 68–82. doi:10.1016/j.molp.2014.12.007.

Paine, J. A., Shipton, C. A., Chaggar, S., Howells, R. M., Kennedy, M. J., Vernon, G. et al. (2005). Improving the nutritional value of Golden Rice through increased pro-vitamin A content. *Nat. Biotechnol.* 23, 482–487. doi:10.1038/nbt1082.

Park, H., Kreunen, S. S., Cuttriss, A. J., DellaPenna, D., and Pogson, B. J. (2002). Identification of the carotenoid isomerase provides insight into carotenoid biosynthesis, prolamellar body formation, and photomorphogenesis. *Plant Cell* 14, 321–332. doi:10.1105/tpc.010302.

Paul, J. Y., Khanna, H., Kleidon, J., Hoang, P., Geijskes, J., Daniells, J. et al. (2017). Golden bananas in the field: elevated fruit pro-vitamin A from the expression of a single banana transgene. *Plant Biotechnol. J.* 15, 520–532. doi:10.1111/pbi.12650.

Penniston, K. L., and Tanumihardjo, S. A. (2006). The acute and chronic toxic effects of vitamin A. *Am. J. Clin. Nutr.* 83, 191–201. Available at: https://academic.oup.com/ajcn/article-abstract/83/2/191/4649798.

Perera, C. O., and Yen, G. M. (2007). Functional properties of carotenoids in human health. *Int. J. Food Prop.* 10, 201–230. doi:10.1080/10942910601045271.

Pérez-Gálvez, A., Sánchez-García, A., Garrido-Fernández, J., and Ríos, J. J. (2018). MS tools for a systematic approach in survey for carotenoids and their common metabolites. *Arch. Biochem. Biophys.* 650, 85–92. doi:10.1016/j.abb.2018.05.010.

Perignon, M., Barré, T., Gazan, R., Amiot, M. J., and Darmon, N. (2018). The bioavailability of iron, zinc, protein and vitamin A is highly variable in French individual diets: Impact on nutrient inadequacy assessment and relation with the animal-to-plant ratio of diets. *Food Chem.* 238, 73–81. doi:10.1016/j.foodchem.2016.12.070.

Poulain, S., Evenou, F., Carré, M. C., Corbel, S., Vignaud, J. M., and Martinet, N. (2009). Vitamin A/retinoids signalling in the human lung. *Lung Cancer* 66, 1–7. doi:10.1016/j.lungcan.2009.03.008.

Prado-Cabrero, A., Scherzinger, D., Avalos, J., and Al-Babili, S. (2007). Retinal biosynthesis in fungi: Characterization of the carotenoid oxygenase CarX from Fusarium fujikuroi. *Eukaryot. Cell* 6, 650–657. doi:10.1128/EC.00392-06.

Raghuvanshi, S., Reed, V., Blaner, W. S., and Harrison, E. H. (2015). Cellular localization of β-carotene 15,15′ oxygenase-1 (BCO1) and β-carotene 9′,10′ oxygenase-2 (BCO2) in rat liver and intestine. *Arch. Biochem. Biophys.* 572, 19–27. doi:10.1016/j.abb.2014.12.024.

Rao, A. V., and Rao, L. G. (2007). Carotenoids and human health. *Pharmacol. Res.* 55, 207–216. doi:10.1016/j.phrs.2007.01.012.

Ravanello, M. P., Ke, D., Alvarez, J., Huang, B., and Shewmaker, C. K. (2003). Coordinate expression of multiple bacterial carotenoid genes in canola leading to altered carotenoid production. *Metab. Eng.* 5, 255–263. doi:10.1016/j.ymben.2003.08.001.

Rodríguez-Ávila, N. L., Narváez-Zapata, J. a., Ramírez-Benítez, J. E., Aguilar-Espinosa, M. L., and Rivera-Madrid, R. (2011). Identification and expression pattern of a new carotenoid cleavage dioxygenase gene member from *Bixa orellana. J. Exp. Bot.* 62, 5385–5395. doi:10.1093/jxb/err201.

Ruch, S., Beyer, P., Ernst, H., and Al-Babili, S. (2005). Retinal biosynthesis in Eubacteria: *In vitro* characterization of a novel carotenoid oxygenase from Synechocystis sp. PCC 6803. *Mol. Microbiol.* 55, 1015–1024. doi:10.1111/j.1365-2958.2004.04460.x.

Schwartz, S. H., Tan, B. C., Gage, D. A., Zeevaart, J. A., and McCarty, D. R. (1997). Specific oxidative cleavage of carotenoids by VP14 of maize. *Science (80-.).* 276, 1872–1874. doi:10.1126/science.276.5320.1872.

Schwartz, S. H., Tan, B. C., McCarty, D. R., Welch, W., and Zeevaart, J. A. D. (2003). Substrate specificity and kinetics for VP14, a carotenoid cleavage dioxygenase in the ABA biosynthetic pathway. *Biochim. Biophys. Acta - Gen. Subj.* 1619, 9–14. doi:10.1016/S0304-4165(02)00422-1.

Schweiggert, R. M., and Carle, R. (2017). Carotenoid deposition in plant and animal foods and its impact on bioavailability. *Crit. Rev. Food Sci. Nutr.* 57, 1807–1830. doi:10.1080/10408398.2015.1012756.

Schweiggert, R. M., Kopec, R. E., Villalobos-Gutierrez, M. G., Högel, J., Quesada, S., Esquivel, P. et al. (2014). Carotenoids are more bioavailable from papaya than from tomato and carrot in humans: A randomised cross-over study. *Br. J. Nutr.* 111, 490–498. doi:10.1017/S0007114513002596.

Shewmaker, C. K., Sheehy, J. A., Daley, M., Colburn, S., and Ke, D. Y. (1999). Seed-specific overexpression of phytoene synthase: Increase in carotenoids and other metabolic effects. *Plant J.* 20, 401–412. doi:10.1046/j.1365-313X.1999.00611.x.

Sommer, A. (2008). Vitamin A deficiency and clinical disease: An historical overview. *J. Nutr.* 138, 1835–1839. doi:10.1093/jn/138.10.1835.

Sommer, A., and Davidson, F. R. (2002). Assessment and control of vitamin A deficiency: The Annecy Accords. *J. Nutr.* 132, 2845S–2850S. doi:10.1093/jn/132.9.2845s.

Sun, T., Yuan, H., Cao, H., Yazdani, M., Tadmor, Y., and Li, L. (2018). Carotenoid metabolism in plants: The role of plastids. *Mol. Plant* 11, 58–74. doi:10.1016/j.molp.2017.09.010.

Toti, E., Chen, C.-Y. O., Palmery, M., Villaño Valencia, D., and Peluso, I. (2018). Non-provitamin A and provitamin A carotenoids as immunomodulators: Recommended dietary allowance, therapeutic index, or personalized nutrition? *Oxid. Med. Cell. Longev.* 2018, 1–20. doi:10.1155/2018/4637861.

Van Loo-Bouwman, C. A., Naber, T. H. J., and Schaafsma, G. (2014). A review of vitamin A equivalency of β-carotene in various food matrices for human consumption. *Br. J. Nutr.* 111, 2153–2166. doi:10.1017/S0007114514000166.

von Lintig, J., and Wyss, A. (2001). Molecular analysis of vitamin A formation: Cloning and characterization of β-carotene 15,15′-dioxygenases. *Arch. Biochem. Biophys.* 385, 47–52. doi:10.1006/abbi.2000.2096.

Wang, X. D., Tang, G. W., Fox, J. G., Krinsky, N. I., and Russell, R. M. (1991). Enzymatic conversion of β-carotene into β-apo-carotenals and retinoids by human, monkey, ferret, and rat tissues. *Arch. Biochem. Biophys.* 285, 8–16. doi:10.1016/0003-9861(91)90322-A.

Weber, D., and Grune, T. (2012). The contribution of β-carotene to vitamin A supply of humans. *Mol. Nutr. Food Res.* 56, 251–258. doi:10.1002/mnfr.201100230.

White, J. C., Highland, M., Kaiser, M., and Clagett-Dame, M. (2000). Vitamin A deficiency results in the dose-dependent acquisition of anterior character and shortening of the caudal hindbrain of the rat embryo. *Dev. Biol.* 220, 263–284. doi:10.1006/dbio.2000.9635.

WHO. (2009). Global prevalence of vitamin A deficiency in populations at risk 1995-2005. *WHO Glob. Database Vitam. A Deficiency. Newsletter*, 4(2), 5–6. doi:978 92 4 159801 9.

WHO. (2019). WHO Biofortification of staple crops. World Health Organization. Available at: https://www.who.int/elena/titles/biofortification/en/ [Accessed October 3, 2019].

Wiseman, E. M., Bar-El Dadon, S., and Reifen, R. (2017). The vicious cycle of vitamin A deficiency: A review. *Crit. Rev. Food Sci. Nutr.* 57, 3703–3714. doi:10.1080/10408398.2016.1160362.

Yang, C., Chen, J., Liu, Z., Yun, C., Piao, J., and Yang, X. (2016). Prevalence and influence factors of vitamin A deficiency of Chinese pregnant women. *Nutr. J.* 15, 1–7. doi:10.1186/s12937-016-0131-7.

Ye, X., Al-Babili, S., Klöti, A., Zhang, J., Paola Lucca, P., Peter B. et al. (2000). Engineering the provitamin A (-Carotene) biosynthetic pathway into (Carotenoid-Free) rice endosperm. *Science (80-.).* 287, 303–305. doi:10.1126/science.287.5451.303.

Zasada, M., and Budzisz, E. (2019). Retinoids: Active molecules influencing skin structure formation in cosmetic and dermatological treatments. *Postepy. Dermatol. Alergol.* 36, 392–398.

Zeng, J., Wang, C., Chen, X., Zang, M., Yuan, C., Wang, X. et al. (2015a). The lycopene β-cyclase plays a significant role in provitamin A biosynthesis in wheat endosperm. *BMC Plant Biol.* 15. doi:10.1186/s12870-015-0514-5.

Zeng, J., Wang, X., Miao, Y., Wang, C., Zang, M., Chen, X. et al. (2015b). Metabolic engineering of wheat provitamin A by simultaneously overexpressing CrtB and silencing carotenoid hydroxylase (TaHYD). *J. Agric. Food Chem.* 63, 9083–9092. doi:10.1021/acs.jafc.5b04279.

7 Epoxy Carotenoid and Its Importance: A Review

Hebron Kirubai Raj and Siva Ramamoorthy

CONTENTS

INTRODUCTION

Carotenoids are a group of naturally occurring tetraterpenoid pigments, which are predominantly present in flowers, fruits, roots, and other parts of the plants, followed by certain other organisms like algae, bacteria, and fungi. They range from bright orange, to yellow, to red in colour. Even though more than 700 of these diversified pigments have been discovered, many have yet not been explored (Agócs et al., 2018). Carotenoids play a key role as precursors in the synthesis of phytohormones and strigolactones; along with their two main roles in organisms: one is photosynthesis and the other is photoprotection that helps the organism for its food and survival.

Carotenoids are industrially important pigments which have high demand in the global markets, as they play a major role in human nutrition and health. Most of the carotenoids were considered to be important for human vision alone, as stated some decades before, with β-carotene being an important carotenoid as the precursor for retinol (Vitamin A) (Kiokias et al., 2016). However, they have many other important roles as natural food colourants, omitting the chemically synthesized food colourants with hazardous side effects (Anantharaman et al., 2014); as anti-tyrosinase inhibitors which help in inhibiting the tyrosinase an enzyme responsible for hyper pigmentation (Anantharaman et al., 2016b); as anti-cancer agents (avoiding synthesized medicines for cancer treatments), they are very effective in many cancer cell lines (Anantharaman et al., 2016a); as chelators in the DNA interaction studies to avoid toxic EtBr-like chemicals which are carcinogenic (Anantharaman et al., 2015) and in interaction studies with various human proteins (Mohan et al., 2018). Apart from the importance of carotenoids in human health, they have usability in the field of cosmetics in lip balms, face creams, etc. which are natural (Torregrosa-Crespo et al., 2018), and in the textile industries for their vibrant colour range, which can be applied with different shades.

As the demand for carotenoids is increasing day by day, the exploration of known carotenoids from other sources and the unexplored carotenoids has to be given importance in order to fulfil the current needs. Many researchers have been targeting the known carotenoids repeatedly; however, not the unknown ones and unconsidered carotenoids were given much priority and studied in depth. This review spotlights: carotenoids and their categories; epoxy carotenoids, their regulation in plants, genes involved; auroxanthin investigation, its sources and biosynthesis; and roles in human health and industrial demand.

CAROTENOID CATEGORIES

Carotenoids mostly have 40 carbons with conjugated double bonds, and their structure decides their biological role, classified on the basis of their chemical structure. Thereby, structurally, they are classified as acyclic, cyclic, hydroxy, epoxy, and site-specific carotenoids.

Acyclic Carotenoids

They are carotenoids which have a linear carbon chain without any benzene rings. Acyclic carotenes include phytoene (colourless compound), lycopene (red coloured pigment), neurosporene and phytofluene (colourless compound). Lycopene, which is predominantly present in tomatoes, is approved as natural food colourant E160d (Rao and Rao, 2018).

Cyclic Carotenoids

Carotenoids that have cyclic compounds like benzene rings in their structures. α-carotene (orange colour), β-carotene (orange colour), γ-carotene (orange colour), δ-carotene, α-zeacarotene and β-zeacarotene fall into this category. β-carotene is among one of the important carotenoids explored so far and is also an approved natural colourant E160a (Rodriguez-Amaya, 2018).

Hydroxy Carotenoids

These carotenoids are known by at least one hydroxyl group in their structure. Hydroxy carotenoids include α-cryptoxanthin (yellow colour), β-cryptoxanthin (orange colour), lutein (yellow colour), lycoxanthin, rubixanthin, zeaxanthin, and zeinoxanthin. Lutein was reported for its presence in our retina along with zeaxanthin, and is also approved for food, feed colourant E161b (Sarialtin and Coban, 2018).

Epoxy Carotenoids

In this category most of them carry an epoxy group or a furanoid ring in their structure (Figure 7.1). Common forms are 5,6-epoxides, 5,8-epoxides, 3,6-epoxides, 4,5-epoxides and 1,2-epoxides;

Structure of epoxy carotenoids:

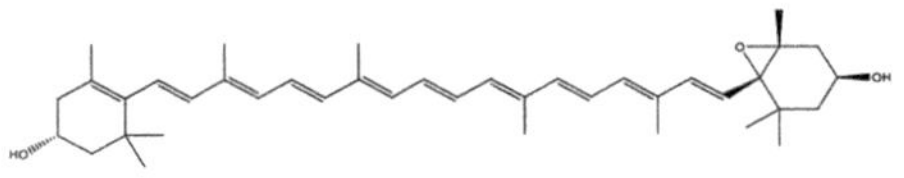

Antheraxanthin

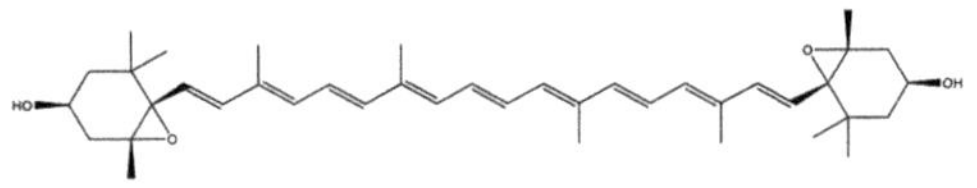

Violaxanthin

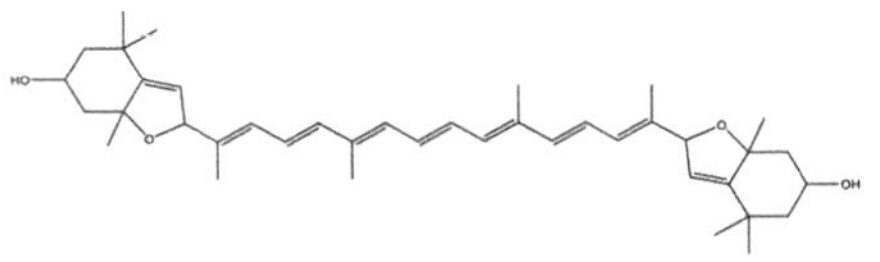

Auroxanthin

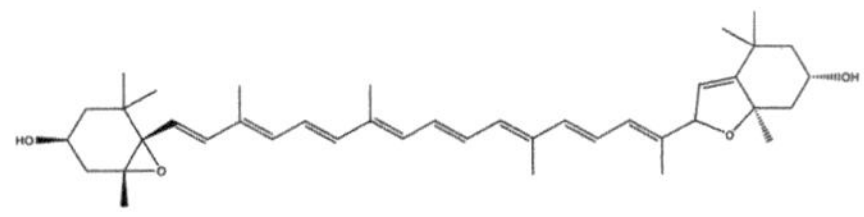

Luteoxanthin

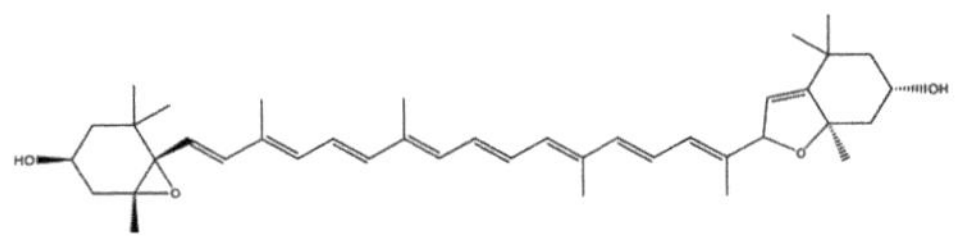

Mutatoxanthin

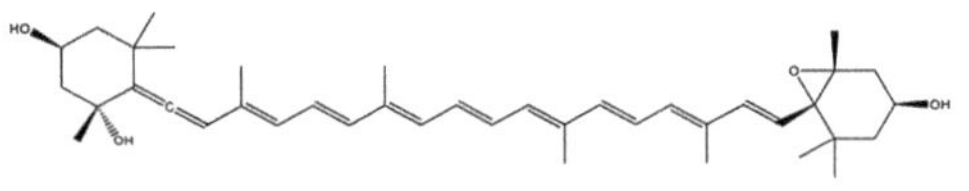

Neoxanthin

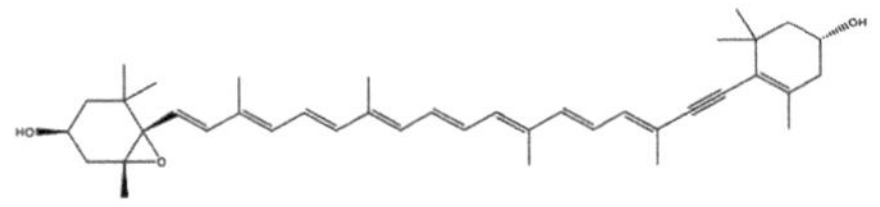

Diadinoxanthin

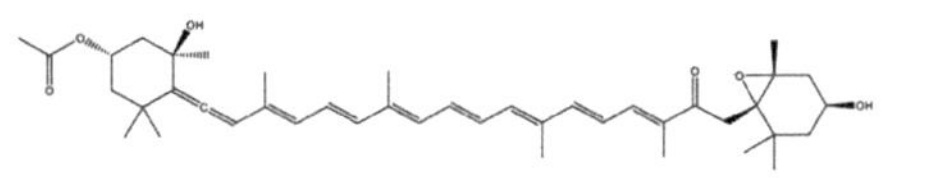

Fucoxanthin

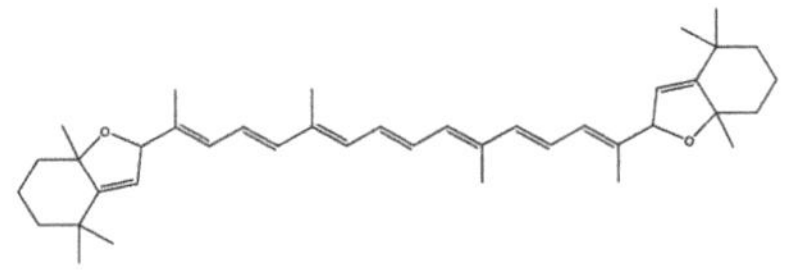

Aurochrome

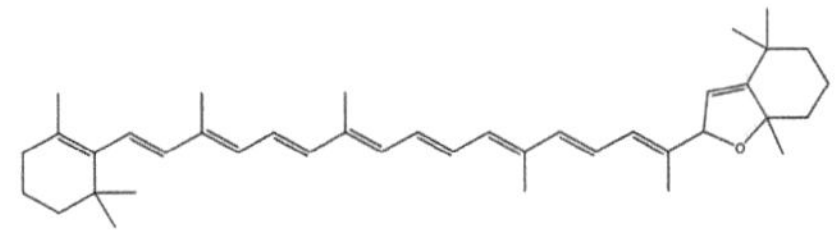

Mutatochrome

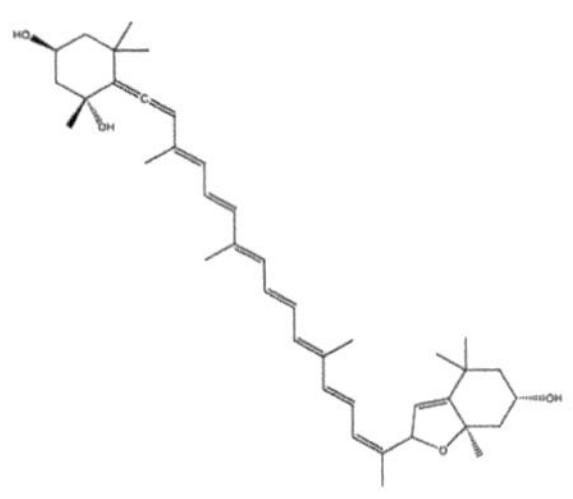

Neochrome

FIGURE 7.1 Structures of epoxy carotenoids. Epoxy carotenoids vary in the arrangement of their epoxy positions. Mainly from 5,6-epoxides to 5,8-epoxides, which occur in their photosynthetic organelles.

TABLE 7.1
List of epoxy carotenoids

S. no.	Name	Molecular formulae	Chemical name	Reference
1.	Antheraxanthin	$C_{40}H_{56}O_3$	5′,6′-dihydro-5′,6′-epoxy-β, β-carotene-3,3′-diol	Shimode et al. (2018)
2.	Violaxanthin	$C_{40}H_{56}O_4$	5,6:5′,6′-diepoxy-5,5′,6,6′-tetrahydro-β, β-carotene-3,3′-diol	Shimode et al. (2018)
3.	Luteoxanthin	$C_{40}H_{56}O_4$	5,6:5′,8′-diepoxy-5,6,5′,8′-tetrahydro-β, β-carotene-3,3′-diol	Chacón-Ordóñez et al. (2017)
4.	Auroxanthin	$C_{40}H_{56}O_4$	5,8:5′,8′-diepoxy-5,8,5′,8′-tetrahydro-β, β-carotene-3,3′-diol	Shimode et al. (2018)
5.	Mutatoxanthin	$C_{40}H_{56}O_3$	5,8-epoxy-5,8-dihydro-β, β-carotene-3,3′-diol	Shimode et al. (2018)
6.	Neoxanthin	$C_{40}H_{56}O_4$	6,7-dihydro-5,5′,6,6′-tetrahydro-5′,6′-epoxy-β, β-carotene-3,3′-triol	Wang et al. (2017)
7.	Diadinoxanthin	$C_{40}H_{54}O_3$	5,6-epoxy-7′,8′-didehyro-5,6-dihydro-β, β-carotene-3,3′-diol	Wang et al. (2014)
8.	Fucoxanthin	$C_{42}H_{58}O_6$	3′,5′-didehydro-8-oxo6′,7′-didehyro-5,5′,6,6′,7,8-hexahydro-5,6-epoxy-β, β-carotene-3′-yl acetate	Miyashita et al. (2011)
9.	Neochrome	$C_{40}H_{55}O_4$	5′,8′-epoxy-6,7-dihydro-5,6,5′,6′-tetrahydro-β, β-carotene-3,3′-triol	Hadjipieri et al. (2017)
10.	Citroxanthin/ Mutatochrome	$C_{40}H_{56}O$	5,8-dihydro-5,8-epoxy-β, β-carotene	Colodel and de Oliveira Petkowicz (2019)
11.	Aurochrome	$C_{40}H_{56}O_2$	5,5′,8,8′-tetrahydro-5,8,5′,8′-diepoxy-β, β-carotene	Price et al. (2018)

predominantly 5,6-epoxides and 5,8-epoxides have been reported (Table 7.1) (Deli and Osz, 2004). In some cases, precursors will have epoxy furanoid rearrangement which will avail as a new compound.

Site-Specific Carotenoids

There are some carotenoids that occur in specific plants like bixin, crocetin, capsanthin, and capsorubin. Bixin is the main apocarotenoid which is found specifically in *Bixa orenella*; it gives bright red colour and is an approved food colourant E160b. Crocetin is present in *Crocus sativus*, which yields saffron colour and is an approved colourant E160c (Priya et al., 2017). Capsanthin and capsorubin are present in *Capsicum annum* sp. and provide a red colour (Jeknić et al., 2012).

SYNTHESIS OF EPOXY CAROTENOIDS

The 5,8 epoxide carotenoids (mutatoxanthin, luteoxanthin, auroxanthin and neochrome) can be generated from their respective 5,6 epoxide precursor pigments (antheraxanthin, violaxanthin and neoxanthin) as reported by Fernandez team (Fernandez-Orozco et al., 2013) (Figure 7.2). Their separation from the crude extract can be attained by using HPLC, C18 column alone as stationary phase based on their polarity and respective retention time (Turcsi et al., 2016). Through column chromatography compounds can be separated using different concentrations of solvents like hexane, ethyl acetate and methanol.

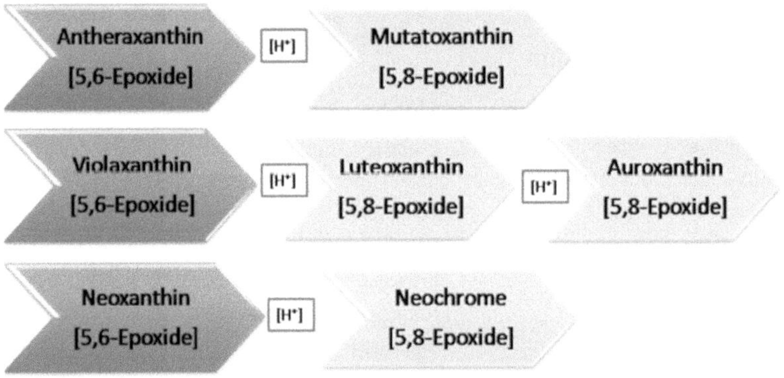

FIGURE 7.2 Conversion of 5,6-epoxides to 5,8-epoxides in photosynthetic organisms. The carotenoids vary in their structures when 5,6-epoxides are added by protons to form 5,8-epoxides, which are less stable.

TABLE 7.2
Epoxy carotenoid sources

S. no.	Source	Carotenoids	Reference
1.	Green leafy vegetables	Violaxanthin, neoxanthin, and neochrome	Meléndez-Martínez et al. (2007a)
2.	Processed food, juices	Luteoxanthin, auroxanthin, and mutatoxanthin	Meléndez-Martínez et al. (2007a)
3.	Brown algae, diatoms	Fucoxanthin and diadinoxanthin	Wang et al. (2014)
4.	Green algae, red algae	Antheraxanthin	Khalid et al. (2018)
5.	Lichens	Mutatochrome, aurochrome	Alós et al. (2019)

Common Sources of Epoxy Carotenoids

Epoxy carotenoids are mainly xanthophylls, which are abundant in flowers. Besides this, they are available in our everyday consumption in unknown quantities. Table 7.2 shows the availability of these valuable carotenoids.

Antheraxanthin

Antheraxanthin has been derived from a Greek name, wherein 'anthos' means flower and 'xantos' means yellow. It is a bright yellow coloured accessory pigment present in many photosynthetic organisms including green algae, red algae, and plants. Antheraxanthin, derived from zeaxanthin, is a light-yellow coloured pigment by the enzyme zeaxanthin epoxidase (Khalid et al., 2018). Antheraxanathin is also involved in synthesis of two other important carotenoids, namely capsanthin and mutatoxanthin. Capsanthin is mainly present in *Capsicum annum* varieties that are used as colouring compound in food industries. The conversion takes place through the enzyme capsanthin synthase (Ariizumi et al., 2014). Antheraxanthin plays an important role as an intermediate in plants' VAZ cycle as photoprotection mechanism (Table 7.3).

Violaxanthin

An orange coloured pigment and one of the predominant carotenoids present in the plant kingdom besides zeaxanthin and leutin. They are reported largely in pansy flowers as it was the main carotenoid observed (Rimkiene et al., 2003). It was noted that leaves turn yellow during the autumn season due to two main carotenoids: violaxanthin and lutein (Młodzińska, 2009). Violaxanthin

TABLE 7.3
Plants containing Antheraxanthin

S. no.	Botanical name	Common name	Family	Pigment presence	Reference
1.	*Triticum aestivum*	Wheat	Poaceae	Kernel	Paznocht et al. (2019)
2.	*Staphylea pinnata*	European bladdernut	Staphyleaceae	Kernel	Šircelj (2019)
3.	*Oryza sativa*	Rice	Poaceae	Kernel	Ha et al. (2019)
4.	*Hordeum vulgare*	Barley	Poaceae	Kernel	Janečková et al. (2019)
5.	*Lilium candidum*	Lilium	Liliaceae	Flowers	Wang et al. (2019)
6.	*Citrus sinensis*	Orange	Rutaceae	Fruit	Abdelaali et al. (2018)
7.	*Solanum betaceum*	Tamarillo	Solanaceae	Fruit	Giuffrida et al. (2018)
8.	*Pistacia vera*	Pistachio	Anacardiaceae	Oil	Rabadán et al. (2018)
9.	*Olea europaea*	Olive	Oleaceae	Fruits	Gandul et al. (2018)
10.	*Coriandrum sativum*	Coriander	Apiaceae	Leaves	Divya et al. (2018)
11.	*Antirrhinum majus*	Snapdragon	Plantaginaceae	Flowers	González et al. (2018)
12.	*Viola wittrockiana*	Pansy	Violaceae	Flowers	González et al. (2018)
13.	*Brassica oleracea*	Broccoli	Brassicaceae	Flowers	Zhang et al. (2018)
14.	*Capsicum annuum*	Yellow bell pepper	Solanaceae	Fruits	Zhang et al. (2018)
15.	*Ipomoea batatas*	Sweet potato	Convolvulaceae	Fruits	Zhang et al. (2018)

is synthesized from antheraxanthin through enzyme zeaxanthin epoxidase and ZEP gene. It is involved in synthesis of some important carotenoids, namely capsorubin, auroxanthin, and luteoxanthin. Capsorubin is present along with Capsanthin in *Capsicum annum* varieties and the enzyme responsible for the conversion is capsorubin synthase. Both capsathin and capsorubin are used in preparation of paprika oleoresin, which has wide application in food and feed sector (Pérez-Gálvez et al., 2003).

VAZ Cycle

In many photosynthetic organisms the VAZ (Violaxanthin Antheraxanthin Zeaxanthin) cycle occurs; also termed as xanthophyll cycle or violaxanthin cycle (Figure 7.3 and Table 7.4).

Luteoxanthin and Mutatoxanthin

Luteoxanthin is a light-yellow coloured pigment which is synthesized from violaxanthin (Table 7.5). It acts as an intermediate for the auroxanthin synthesis. The molecular formula and structural similarity have proved that it is a geometric isomer of violaxanthin. It was reported in *Solanum sessiliflorum* fruits (origin to amazon) as a main constituent with other two carotenoids leutin and β-carotene (Colodel and de Oliveira Petkowicz, 2019). For chemical synthesis, violaxanthin is treated with hydrochloric acid to get luteoxanthin and when luteoxanthin is treated with HCL auroxanthin is formed. In both the reactions the structure undergoes epoxy furanoid rearrangement (Figure 7.4).

Mutatoxanthin is also a yellow coloured pigment which is synthesized from antheraxanthin and antheraxanthin, mutatoxanthin have same molecular formulae (Table 7.6). Mostly, they are found in *Xanthora* sp. (lichens) and *Euglena* sp. along with auroxanthin (Czeczuga, 1983). There are also found in *Rhodophyta* sp. They have been reported to have an epoxy furanoid rearrangement from antheraxanthin.

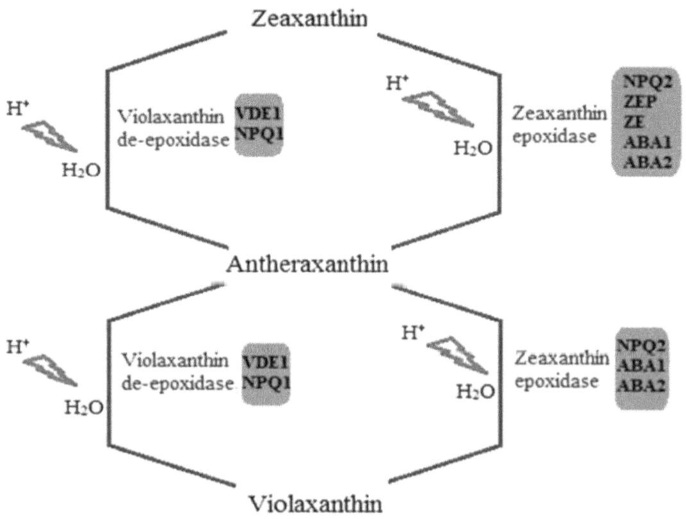

FIGURE 7.3 VAZ cycle. VAZ cycle, which is a common cycle, occurs in most of the photosynthetic organisms in their carotenoid biosynthesis pathway. In this cycle interconversion between the pigments takes place in order to protect the organisms by dissipating the excess heat. When the intensity of the light is higher, the violaxanthin in the organism is converted back to its precursor zeaxanthin through antheraxanthin by the enzyme violaxanthin de-epoxidase (VDE). When the condition turns back to normal, zeaxanthin again gets converted into violaxanthin by the enzyme zeaxanthin epoxidase (ZEP) (Badejo 2018). The key genes NPQ2, ZEP, ZE, ABA1 and ABA2 are involved in the conversion of zeaxanthin to antheraxanthin with respect to different organisms, and on the other hand, the genes responsible for the conversion from antheraxanthin to violaxanthin are as follows: NPQ2, ABA1 and ABA2. In contrast to that, the genes VDE1 and NPQ1 are involved in the conversion of violaxanthin to antheraxanthin and antheraxanthin to zeaxanthin. During these processes, the proton is converted into water as an energy conversion.

TABLE 7.4
Plants containing violaxanthin

S. no.	Botanical name	Common name	Family	Pigment presence	Reference
1.	*Acrocomia aculeate*	Palm	Arecaceae	Fruits	Schex et al. (2018)
2.	*Calendula officinalis*	Marigold	Asteraceae	Flowers	Rodrigues et al. (2018)
3.	*Arabidopsis thaliana*	Thale cress	Brassicaceae	Leaves	Morris et al. (2018)
4.	*Mangifera indica*	Mango	Anacardiaceae	Fruits	Rosalie et al. (2018)
5.	*Prunus armeniaca*	Armenian plum	Rosaceae	Fruits	Fratianni et al. (2017)
6.	*Pouteria sapota*	Mameysapote	Sapotaceae	Fruits	Chacón et al. (2017)
7.	*Capsicum annuum*	Pepper	Solanaceae	Fruits	Silveira et al. (2017)
8.	*Fragaria ananassa*	Strawberry	Rosaceae	Fruits	Delgado et al. (2016)
9.	*Cucumis sativus*	Cucumber	Cucurbitaceae	Leaves	Li et al. (2015)
10.	*Persea americana*	Avocado	Lauraceae	Leaves	Matsubara et al. (2011)
11.	*Citrus sinensis*	Orange	Rutaceae	Fruit	Meléndez et al. (2007b)
12.	*Hordeum vulgare*	Barley	Poaceae	Leaves	Phillip et al. (1999)
13.	*Spinacia oleracea*	Spinach	Amaranthaceae	Leaves	Havir et al. (1997)
14.	*Viola tricolor*	Johnny Jump up	Violaceae	Flowers	Hansmann et al. (1982)
15.	*Lactuca sativa*	Lettuce	Asteraceae	Leaves	Saini et al. (1975)

TABLE 7.5
Plants Containing Luteoxanthin

S. no.	Botanical name	Common name	Family	Pigment presence	Reference
1.	*Citrus sinensis*	Orange	Rutaceae	Fruit	Abdelaali et al. (2018)
2.	*Pistacia vera*	Pistachio	Anacardiaceae	Oil	Rabadán et al. (2018)
3.	*Malus pumila*	Apple	Rosaceae	Fruit juice	Galvis et al. (2018)
4.	*Prunus persica*	Peach	Rosaceae	Fruit juice	Galvis et al. (2018)
5.	*Pyrus amygdaliformis*	Pear	Rosaceae	Fruit juice	Galvis et al. (2018)
6.	*Olea europaea*	Olive	Oleaceae	Fruits	Gandul et al. (2018)
7.	*Passiflora caerulea*	Passion fruit	Passifloraceae	Fruits	Reis et al. (2018)
8.	*Eriobotrya japonica*	Loquat	Rosaceae	Fruits	Alos et al. (2018)
9.	*Lycium barbarum*	Goji berry	Solanaceae	Fruits	Bertoldi et al. (2018)
10.	*Moringa oleifera*	Moringa	Moringaceae	Fruits	Ma et al. (2018)
11.	*Solanum sessiliflorum*	Cocona	Solanaceae	Fruits	Colodel et al. (2018)
12.	*Mangifera indica*	Mango	Anacardiaceae	Fruits	Ntsoane et al. (2018)
13.	*Vitis vinifera*	Grape vine	Vitaceae	Fruits	Gutiérrez et al. (2018)
14.	*Malpighia emarginata*	Acerola	Malpighiaceae	Fruits	Belwal et al. (2018)
15.	*Pouteria sapota*	Mameysapote	Sapotaceae	Fruits	Chacón et al. (2017)

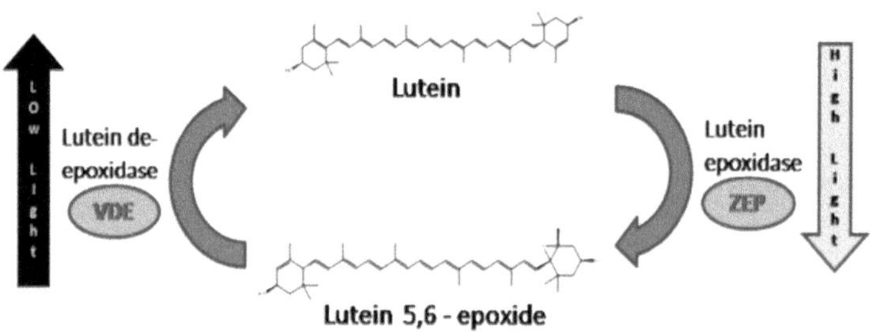

FIGURE 7.4 Lx cycle. Lx cycle, the photoprotection mechanisms in plants; many use VAZ cycle, but mostly woody plants use Lx cycle. Lx cycle is also a light driven epoxide rearrangement of pigment leutin. They particularly occur in leaves to protect the plants from photo damage. When plants face strong light, lutein in the leaves gets converted into lutein epoxidase through enzyme lutein epoxidase, and again in weak light, lutein epoxidase gets reverted to lutein through an enzyme lutein de-epoxidase (Esteban et al., 2009, Papageorgiou, 2014). The key genes involved in these epoxidation reactions were ZEP and VDE. However, the complete function and use remain unknown.

Neoxanthin and Neochrome

Neoxanthin is an important intermediate in the biosynthesis of ABA (abscisic acid) which is an important plant growth hormone (Table 7.7). ABA helps in plants senescence and ripening, which are necessary for plants' metabolisms. Neoxanthin is synthesized from violaxanthin through an enzyme neoxanthin synthase. They are present in green leafy vegetables; predominantly in spinach. In certain plants this neoxanthin, which is a 5,6 epoxide, gets converted into neochrome: a 5,8 epoxide (Figure 7.5). Neochrome can also be synthesized through HCL treatment with neoxanthin, and they have similar structures. Neochrome has one hydrogen molecule less when compared to neoxanthin, and it has been reported that both are water insoluble. They have been reported for their potential role against human cancer (Asai et al., 2004).

TABLE 7.6
Plants containing mutatoxanthin

S. no.	Botanical name	Common name	Family	Pigment presence	Reference
1.	*Pistacia vera*	Pistachio	Anacardiaceae	Oil	Rabadán et al. (2018)
2.	*Olea europaea*	Olive	Oleaceae	Fruits	Gandul et al. (2018)
3.	*Citrus sinensis*	Orange	Rutaceae	Fruits	Abdelaali et al. (2018)
4.	*Diospyros kaki*	Persimmon	Ebenaceae	Fruits	Bordiga et al. (2018)
5.	*Anacardium occidentale*	Caju	Anacardiaceae	Fruits	Brito et al. (2018)
6.	*Passiflora caerulea*	Passion fruit	Passifloraceae	Fruits	Reis et al. (2018)
7.	*Malus pumila*	Apple	Rosaceae	Fruit juice	Galvis et al. (2018)
8.	*Prunus persica*	Peach	Rosaceae	Fruit juice	Galvis et al. (2018)
9.	*Pyrus amygdaliformis*	Pear	Rosaceae	Fruit juice	Galvis et al. (2018)
10.	*Cocos nucifera*	Coconut	Arecaceae	Fruits	Silveira Agostini-Costa (2018)
11.	*Phoenix dactylifera*	Date palm	Arecaceae	Fruits	Silveira Agostini-Costa (2018)
12.	*Elaeis guineensis*	African oil palm	Arecaceae	Fruits	Silveira Agostini-Costa (2018)
13.	*Arbutus unedo*	Strawberry	Ericaceae	Fruits	Morgado et al. (2018)
14.	*Caryocar brasilense*	Pequi	Caryocaraceae	Fruits	Torres et al. (2018)
15.	*Solanum tuberosum*	Potato	Solanaceae	Tubers	Xie et al. (2018)

TABLE 7.7
Plants containing neoxanthin

S. no.	Botanical name	Common name	Family	Pigment presence	Reference
1.	*Eriobotrya japonica*	Loquat	Rosaceae	Fruits	Alos et al. (2018)
2.	*Lilium candidum*	Lilium	Liliaceae	Flowers	Wang et al. (2019)
3.	*Staphylea pinnata*	European bladdernut	Staphyleaceae	Kernel	Šircelj H (2019)
4.	*Triticum aestivum*	Wheat	Poaceae	Kernel	Paznocht et al. (2019)
5.	*Brassica rapa* ssp. *chinensis*	Pak choi	Brassicaceae	Leaves	Frede et al. (2019)
6.	*Medicago sativa*	Alfalfa	Fabaceae	Sprouts	Fiutak et al. (2019)
7.	*Ipomoea batatas*	Sweet potato	Convolvulaceae	Fruits	Ke et al. (2019)
8.	*Opuntia ficusindica*	Prickly pear	Cactaceae	Peel	Amaya et al. (2019)
9.	*Mangifera indica*	Mango	Anacardiaceae	Fruits	Ntsoane et al. (2019)
10.	*Cuscuta epithymum*	Dodder	Convolvulaceae	Leaves	Chabra et al. (2019)
11.	*Coriandrum sativum* L.	Coriander	Apiaceae	Leaves	Wei et al. (2019)
12.	*Vitis vinifera*	Grapevine	Vitaceae	Fruits	Sánchez et al. (2019)
13.	*Glycine max*	Soybean	Fabaceae	Sprouts	Jiao et al. (2019)
14.	*Solanum lycopersicum*	Tomatoe	Solanaceae	Fruits	Salehi et al. (2019)
15.	*Coffea Arabica*	Arabian coffee	Rubiaceae	Leaves	Marcheafave et al. (2019)

Mutatochrome and Aurochrome

Mutatochrome and aurochrome are synthesized from mutatoxanthin and auroxanthin, respectively. It is noted that the synthesized chrome compounds differ by absence of OH group in their furanoid ring. They were reported in many plant species along with their precursor molecules. Mutatochrome is also found in yellow flesh pear cultivars with luteoxanthin and mutatoxanthin (Alós et al., 2019).

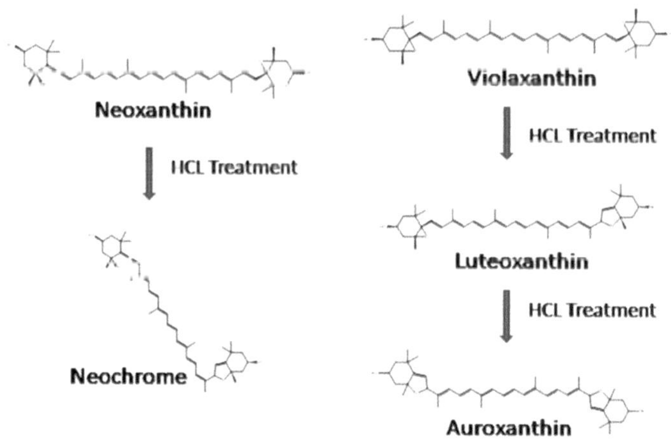

FIGURE 7.5 Conversion of 5,6 epoxides 5,8 epoxides. The figure depicts 5,6 epoxides (violaxanthin and neoxanthin) conversion to 5,8 epoxides (luteoxanthin, auroxanthin and neochrome) on HCL treatment. When HCL is added to 5,6 epoxides, the epoxy arrangement shifts and forms 5,8 epoxides.

Aurochrome, when treated with $CHCl_3$, gives rise to many meso forms and eight derivatives have been studied so far (Acemoglu and Eugster, 1984).

Diadinoxanthin

Diadinoxanthins are mostly present in marine algae sp. like heterokontophyta, haptophyta, dinophyta and euglenophyta (Table 7.8). They are synthesized through violaxanthin epoxide arrangement, which has two oxygen molecules and one hydrogen molecule lesser than violaxanthin.

TABLE 7.8
Some of the organisms containing diadinoxanthin

S. no.	Botanical name	Common name	Family	Pigment presence	Reference
1.	*Phaeodactylum tricornutum*	–	Phaeodactylaceae	Cells	Duarte et al. (2019)
2.	*Nannochloropsis oceanica*	Nannochloropsis	Eustigmataceae	Cells	Mitra et al. (2019)
3.	*Nanofrustulum shiloi*	–	Fragilariaceae	Cells	Sahin et al. (2019)
4.	*Noctiluca scintillans*	Sea sparkle	Noctilucaceae	Cells	Sarma et al. (2019)
5.	*Aureococcus anophagefferens*	–	Pelagomonadaceae	Cells	Sarkar SK (2019)
6.	*Phaeocystis antarctica*	–	Phaeocystaceae	Cells	Stamatakis et al. (2019)
7.	*Cyclotella meneghiniana*	–	Stephanodiscaceae	Cells	Elnour et al. (2018)
8.	*Chaetoceros calcitrans*	–	Chaetocerotaceae	Cells	Yamano et al. (2018)
9.	*Euglena gracilis*	–	Euglenaceae	Cells	Tamaki et al. (2018)
10.	*Tisochrysis lutea*	–	Isochrysidaceae	Cells	Balduyck et al. (2018)
11.	*Prorocentrum donghaiense*	–	Prorocentraceae	Cells	Shi et al. (2018)
12.	*Chaetoceros gracilis*	–	Chaetocerotaceae	Cells	Nagao et al. (2018)
13.	*Alexandrium minutum*	–	Goniodomaceae	Cells	Long et al. (2018)
14.	*Odontella aurita*	–	Eupodiscaceae	Cells	Xia et al. (2018)
15.	*Amphidinium carterae*	–	Gymnodiniaceae	Cells	Molina-Miras et al. (2018)

Table 7.9 Some of the organisms containing fucoxanthin

S. no.	Botanical name	Common name	Family	Pigment presence	Reference
1.	*Aureococcus anophagefferens*	–	Pelagomonadaceae	Cells	Sarkar SK (2019)
2.	*Phaeodactylum tricornutum*	–	Phaeodactylaceae	Cells	Duarte et al. (2019)
3.	*Trichodesmium erythraeum*	Trichodesmium	Microcoleaceae	Cells	Lima et al. (2019)
4.	*Nannochloropsis oceanica*	Nannochloropsis	Eustigmataceae	Cells	Mitra et al. (2019)
5.	*Laminaria ochroleuca*	Kombu	Laminariaceae	Cells	Olmo et al. (2019)
6.	*Thalassiosira pseudonana*	–	Thalassiosiraceae	Cells	Jäger et al. (2019)
7.	*Crassostrea hongkongensis*	Hong Kong oyster	Ostreidae	Shells	Cai et al. (2019)
8.	*Sargassum muticum*	Japanese wireweed	Sargassaceae	Cells	Flórez et al. (2019)
9.	*Macrocystis pyrifera*	Giant kelp	Laminariaceae	Cells	Camus et al. (2019)
10.	*Sargassum serratifolium*	–	Sargassaceae	Cells	Lim et al. (2019)
11.	*Noctiluca scintillans*	Sea sparkle	Noctilucaceae	Cells	Sarma et al. (2019)
12.	*Laminaria setchellii*	Kelp	Laminariaceae	Cells	Tedesco et al. (2019)
13.	*Nanofrustulum shiloi*	–	Fragilariaceae	Cells	Sahin et al. (2019)
14.	*Tetraselmis suecica*	–	Chlorodendraceae	Cells	Messina et al. (2019)
15.	*Nitzschia laevis*	–	Bacillariaceae	Cells	Sun et al. (2019)

It is responsible for synthesizing another two compounds, namely fucoxanthin and diatoxanthin, through diatoxanthin synthase (Wang et al., 2014). Diatoxanthin can be reverted back to diadinoxanthin through diadinoxanthin de-epoxidase, as reported in *Cyclotella meneghinaina* (Takaichi, 2011).

Fucoxanthin

These are classically found in marine algae, which are phototrophs. Fucoxanthin are predominantly found in brown algae, diatoms and dinoflagellates (Table 7.9). Fucoxanthin are a class-specific carotenoid that avail a unique structure (Peng et al., 2011). They are synthesized by two means: one is from neoxanthin and the other is from diadinoxanthin. They have two hydrogen molecules and two more oxygen molecules than neoxanthin. Its production was reported in a diatom *Phaeodactylum tricornutum* that was grown on palm oil mill (Nur et al., 2018).

Auroxanthin – An Unexplored Carotenoid

Auroxanthin ($C_{40}H_{56}O_4$) is a 5,8: 5′,8′-diepoxy-5,8,5′,8′-tetrahydro-beta,beta-carotene-3,3′-diol. It is a geometric isomer of violaxanthin (Meléndez-Martínez et al., 2008), which is often found in higher plants (Araki et al., 2016). The molecular weight of auroxanthin was 600.87, which is similar to violaxanthin, and has melting point of 203°C. Auroxanthin can be detected based on the presence of two furanoid rings. It has lambda max 40 nm lower than that of violaxanthin. Auroxanthin has three isomers, Auroxanthin A (8R,8′R), Auroxanthin B (8R,8′S), Auroxanthin C (8S,8′S) (Delpino-Rius et al., 2014). For HPLC determination of auroxanthin, we need t-BME, methanol and water as mobile phase in Silica C30 spheres (3 μm) column (Singh et al., 2015). In HPTLC, violaxanthin and auroxanthin have been reported to have same Rf values and appears yellow in colour (Gulyaeva, 2015). Auroxanthin is responsible for the last tinge of colour before the extract becomes colourless (Aparicio and Harwood, 2013).

As maximum carotenoids have their ester forms, auroxanthin likewise have ester forms. There are two major esters which have been reported, namely auroxanthin palmitate and auroxanthin stearate, which were found to be identical to auroxanthin and are constituted in marigold petals (Rodrigues et al., 2018).

Auroxanthin Sources

Auroxanthin content is reported to be more in pot marigold flowers when compared to their other sources, mainly in alice orange, yellow, double esterel orange and bonbon apricot varieties; the respective yields were 7.1%, 10%, 18.97%, and 15.43% of total carotenoids present (Kishimoto and Ohmiya, 2009; Khalid and da Silva, 2012) and their presence was observed in other varieties too (Sausserde and Kampuss, 2014; Singh and Sindhu, 2015). Violaxanthin, luteoxanthin and auroxanthin were found to be the main carotenoids in the mango pulp of about 100 mg/g fresh weight (Alós et al., 2019).

CHEMICAL SYNTHESIS

In many food processing industries like juice and oil processing, violaxanthin gets converted to auroxanthin during processing, and its content is found to be more after being processed (Polidori et al., 2018). It has been noted that during storage, fruits and vegetables exhibit an increase in auroxanthin levels. In a recent study, orange juice when stored at 42°C for 8 weeks showed 48% rise in the auroxanthin level (Wibowo et al., 2015). Therefore, storage plays a vital role in carotenoid levels and accumulation of auroxanthin (Table 7.10). Although violaxanthin was reported as a superior carotenoid in mango pulp, it was missing in processed juice. This was noted in citrus juices too, where complete desertion of violaxanthin and replacement with auroxanthin occurred when it was heated and processed (Figure 7.6). The auroxanthin occurrence proved such heating in acidified medium endorses furanoid formation over epoxy isomerization (Hadjal et al., 2013). In addition, Liu et al. found that there is an increase in the acidic environment when violaxanthin gets converted into auroxanthin (Liu et al., 2014). As per scientific literature, auroxanthin can be synthesized chemically by its precursor in acidic condition, but still cannot be used in the food industries due to chemical synthesis; therefore, we must look forward for their natural sources and their biosynthetic pathway.

Carotenoid Metabolism in Plants

Carotenoids are synthesized in the plastids, which are storage compartments in plant cells. There are many types of plastids, namely chloroplasts, amyloplasts, etioplasts, proplastids, and chromoplasts. Plastids play a major role in synthesizing and administering these carotenoids.

Gene Regulation in Epoxy Carotenoid Biosynthesis

In the chromoplasts there are several genes that regulate the formation of carotenoids. In the context of gene regulation, the genes that are involved are similar to the enzymes for carotenoid accumulation. Starting from phytoene to neoxanthin formation, the genes that are involved in the regulation will be discussed further. PSY1 and PSY2 are the key genes in the formation of phytoene. Many studies for the upregulation of PSY gene has been carried out to enhance provitamin A, which resulted in the production of golden cotton seeds (Yao et al., 2018). PDS gene helps in the regulation of ζ-carotene formation. Gene editing of this particular gene has helped in creating disease resistance in banana cultivars (Naim et al., 2018). LCYB1 and LCYB2 are the two genes responsible for the synthesis of lycopene. For the enhancement of carotenoid levels and to resist against abiotic stresses, LCYB gene was used in sweet potato transgenic study (Kang et al., 2018). CHXB gene was involved in the synthesis of zeaxanthin from β-carotene. ZEP gene regulates the pathway from zeaxanthin to violaxanthin and the role of the gene has been discussed in the VAZ cycle. ZEP gene was considered to be an important gene for the regulation of the carotenoid pathway (Ma et al., 2018), whereas NXS gene acts as key gene for neoxanthin formation (Figure 7.7).

TABLE 7.10
Some of the plant sources containing auroxanthin

S. no.	Scientific name	Common name	Family	Present part	Reference
1.	*Calendula officinalis*	Marigold	Asteraceae	Flowers	Bakó et al. (2002), Kishimoto Sanae et al. (2007), El-Gamal (2015)
2.	*Curcubita pepo*	Acorn Squash	Cucurbitaceae	Flowers	Duke (2004)
3.	*Delonix regia*	Flametree, Gulmohar	Fabaceae	Flowers, bark	Kamalambigeswari and Jeyanthi Rebecca (2016)
4.	*Lonicera japonica*	Japanese Honeysuckle	Caprifoliaceae	Fruits	Goodwin (1952)
5.	*Lycopersicon esculentum*	Tomato	Solanaceae	Fruits	Yahia et al. (2017)
6.	*Medicago sativa* subsp. *sativa*	Alfalfa blue variety	Fabaceae	Flowers, leaves	Duke (2004)
7.	*Viola tricolor*	Viola	Violoideae	Flowers	Duke (2004)
8.	*Capsicum annuum*	Bell peppers	Solanaceae	Fruits	Duke (2004)
9.	*Capsicum annuum*	Chilli pepper	Solanaceae	Fruits	Camara and Moneger (1978), Hornero-Méndez et al. (2000)
10.	*Citrus sinensis*	Sweet Orange	Rutaceae	Fruits	Duke (2004)
11.	*Vitis vinifera*	Grapes	Vitaceae	Fruits	Duke (2004)
12.	*Arum italicum*	Italian arum and Italian lords-and-ladies	Araceae	Fruits	Delgado-Vargas and Paredes-López (2002), Duke (2004)
13.	*Chrysanthemum carinathum*	Ismelia	Asteraceae	Flowers	Duke (2004)
14.	*Chrysanthemum coronarium*	Garland Chrysanthemum	Asteraceae	Flowers	Duke (2004)
15.	*Actinidia deliciosa*	Kiwi	Actinidiaceae	Fruits	Duke (2004)
16.	*Spinacia oleracea*	Spinach	Amaranthaceae	Leaves	Duke (2004)
17.	*Anacardium occidentale*	Cashew apple	Anacardiaceae	Processed juice	Duke (2004), de Brito et al. (2007), Talasila and Shaik (2015)
18.	*Ginkgo biloba*	Ginko, Maidenhair tree	Ginkgoaceae	Tissue culture, callus	Duke (2004)
19.	*Narcissus poeticus*	Narcissus	Amaryllidaceae	Flowers	Valadon and Mummery (1971)
20.	*Narcissus triandrus*	Daffodil	Amaryllidaceae	Flowers	Valadon and Mummery (1971)
21.	*Tulipa gesneriana*	Tulip	Liliaceae	Flowers	Valadon and Mummery (1971)
22.	*Gynostemma pentaphyllum*	Jiaogulan	Cucurbitaceae	Leaves	Liu et al. (2004)
23.	*Actinidia chinensis*	Kiwi	Actinidiaceae	Fruits	Besler (2000)
24.	*Magnifera indica*	Mango	Anacardiaceae	Fruit pulp	Mercadante (1998)
25.	*Brassica oleracea*	Brussels sprouts	Brassicaceae	Leaves	Giger (2002)
26.	*Pisum sativum*	Green peas	Fabaceae	Flowers	Giger (2002)

(*continued*)

TABLE 7.10 (CONTINUED)
Some of the plant sources containing auroxanthin

S. no.	Scientific name	Common name	Family	Present part	Reference
27.	*Rosa rugosa*	Rose hips	Rosaceae	Fruits	Giger (2002)
28.	*Cucumis sativus*	Cucumber	Cucurbitaceae	Petals	Neamtu et al. (1975), Zhu et al. (2010)
29.	*Juncus accutus*	Spiny rush	Juncaceae	Flowers	Anjum et al. (2016)
30.	*Halimione portulacoides*	Sea purslane	Amaranthaceae	Leaves	Duke (2004)
31.	*Zostera noltii*	Dwarf eelgrass	Zosteraceae	Leaves	Repolho et al. (2017)
32.	*Eschscholzia californica*	California poppy	Papaveraceae	Flowers	Zhou (2012)
33.	*Tagetes patula*	French Marigold	Asteraceae	Flowers	Duke (2004)
34.	*Solanum lycopersicum*	Tomato	Solanaceae	Flesh	Martin et al. (1974)
35.	*Citrus ichangensis Swingle*	Ichang papeda	Rutaceae	Fruit peel	Kudritskaya et al. (1983)
36.	*Arnica montana* L.	Mountain tobacco	Asteraceae	Flower	Azimova et al. (2011)
37.	*Bidens ferulifolia* DC.	Apache beggarticks	Asteraceae	Flower	Valadon and Mummery (1971)
38.	*Chrysanthemum carinatum* Schousb.	Tricolour chrysanthemum	Asteraceae	Flower	Valadon and Mummery (1971)
39.	*Chrysanthemum coronarium* L.	Crowndaisy chrysanthemum	Asteraceae	Flower	Valadon and Mummery (1971)
40.	*Coreopsis tinctoria* Nutt.	Plains coreopsis	Asteraceae	Flower	Valadon and Mummery (1971)
41.	*Gaillardia aristata* Pursh.	Common blanketflower	Asteraceae	Flower	Valadon and Mummery (1971)
42.	*Helianthus decapitatus* var. (Loddon Gold Hort.)	Thinleaf sunflower	Asteraceae	Flower	Valadon and Mummery (1971)
43.	*Layia elegans* Torr. et A. Gray	–	Asteraceae	Flower	Valadon and Mummery (1971)
44.	*Layia platyglossa* A. Gray	Coastal tidytips	Asteraceae	Flower	Valadon and Mummery (1971)
45.	*Rudbeckia speciosa* Hort. ex Link	Black-eyed-susans	Asteraceae	Flower	Valadon and Mummery (1971)
46.	*Santolina viridis* Hort.	Olive Herb	Asteraceae	Flower	Valadon and Mummery (1971)
47.	*Solidago canadensis* L.	Canadian goldenrod	Asteraceae	Flower	Valadon and Mummery (1971)
48.	*Tragopogon pratensis* L.	Meadow salsify	Asteraceae	Flower	Valadon and Mummery (1971)
49.	*Ursinia calenduliflora* Benth. et Hook. f.	Lace-leaf ursinia	Asteraceae	Flower	Valadon and Mummery (1971)
50.	*Theobroma grandiflorum*	Cupuassu	Malvaceae	Fruit	Cerón et al. (2015)
51.	*Spartina patens*	Saltmeadow cordgrass	Poaceae	Leaves	Duarte et al. (2016)
52.	*Olea europaea*	Olive	Oleaceae	Oil	Gandul-Rojas et al. (2013)

(continued)

TABLE 7.10 (CONTINUED)
Some of the plant sources containing auroxanthin

S. no.	Scientific name	Common name	Family	Present part	Reference
53.	*Dioscorea bulbifera* L.	Air potato	Dioscoreaceae	Bulbils	Martin et al. (1974), Guan et al. (2017)
54.	*Cucurbita pepo* subsp. *pepo*	Pumpkin	Cucurbitaceae	Seeds	Matus et al. (1993), Lelley et al. (2009)
55.	*Malpighia emarginata*	Barbados cherry	Malpighiaceae	Fruit	Mezadri et al. (2006)
56.	*Lycium barbarum*	Chinese wolfberry	Solanaceae	Seeds	Guoliang et al. (2011)
57.	*Citrus sinensis*	Orange	Rutaceae	Juice	Meléndez-Martínez et al. (2010)
58.	*Citrus reticulata*	Mandarin orange	Rutaceae	Fruits	Wagner et al. (2011)
59.	*Musa paradisiaca*	Banana	Musaceae	Fruits	Singh et al. (2016)
60.	*Arbutus unedo* L.	Strawberry tree	Ericaceae	Fruits	Delgado-Pelayo et al. (2016)
61.	*Aster tripolium*	Sea aster	Asteraceae	Flowers	Duarte et al. (2015a)
62.	*Calendula tripterocarpa* Rupr	Marigold	Asteraceae	Flowers	Al-Rifai (2018)
63.	*Cyperus longus* L.	Galingale, Sweet Cyperus	Cyperaceae	Flowers	Duarte et al. (2015c)
64.	*Spartina patens* Muhl	Saltmeadow cordgrass	Poaceae	Flowers	Duarte et al. (2015c)
65.	*Ipomoea aquatica* Forssk.	Water spinach	Convolvulaceae	Leaves	Kaur et al. (2016)
66.	*Scutellaria barbata* D. Don.	Barbed skullcap	Lamiaceae	Flowers	Liu et al. (2014)

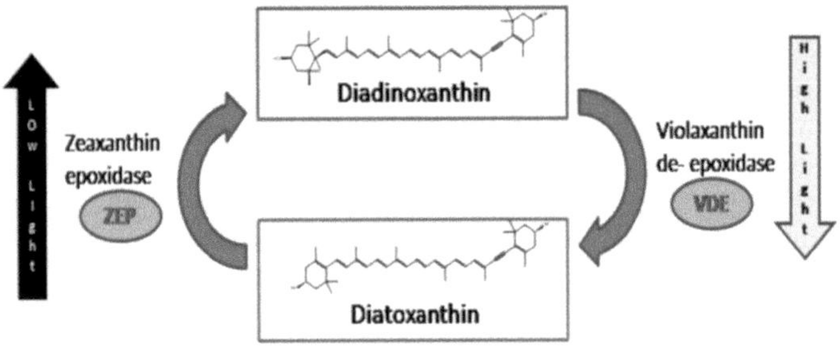

FIGURE 7.6 Dd-Dt cycle. Dd-Dt cycle, this cycle mainly occurs in diatoms to protect themselves from excess light. This cycle involves conversion of an epoxide diadinoxanthin to a non-epoxide diatoxanthin when the organism faces high light through the enzyme violaxanthin epoxidase regulated by the VDE gene. In low light, the reverse reaction occurs through enzyme zeaxanthin epoxidase regulated by the ZEP gene (Kuczynska et al., 2015).

Yellow-Flowered Cultivars

Auroxanthin is more often reported in the flowers of plants than other parts. In angiosperms, when carotenoids are present, the colours in the flower petals are in accordance with which carotenoid is present. As the colour of the carotenoids vary with their structure, their presence in the plants draws out the colour. Many flowered cultivars are available in nature but orange and yellow flowers specifically represent the carotenoids. The yellow-flowered cultivars differ from the orange ones when the carotenoids like lutein, lutein-5, 6-epoxide, zeaxanthin, violaxanthin, auroxanthin, luteoxanthin occur (Figure 7.8).

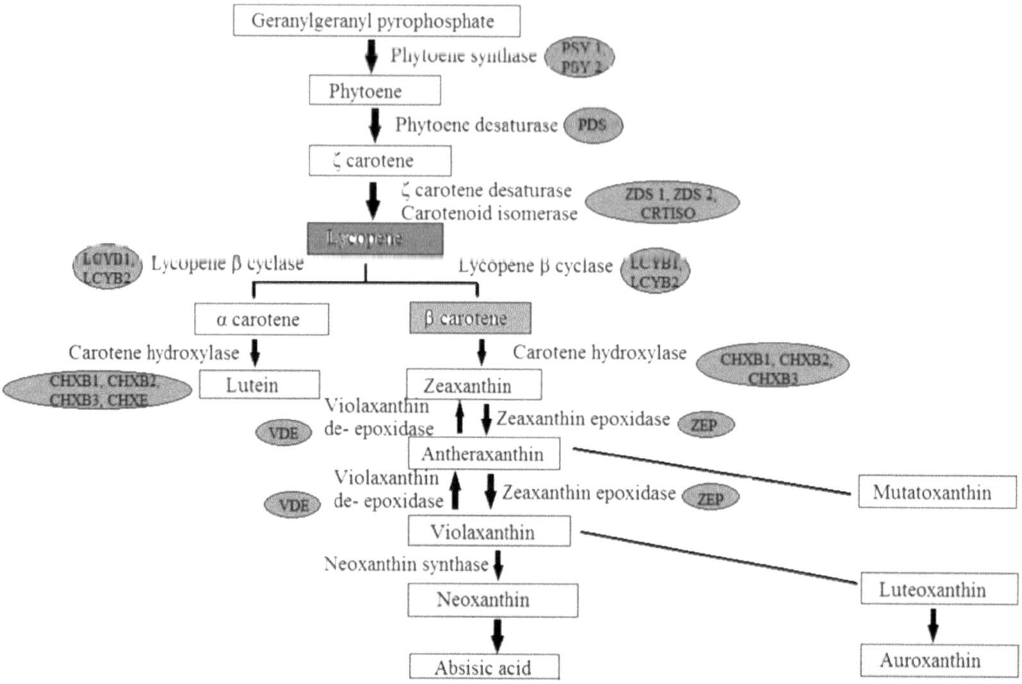

FIGURE 7.7 Biosynthesis of carotenoids. Carotenoid biosynthesis starts from isopentyl disphosphate and DMAPP catalysed by isopentyl disphosphate isomerases through the MEP pathway. Further Geranylgeranyl disphosphate (GGPP) is formed by interaction with Geranylgeranyl disphosphate synthase (GGPS). GGPP acts as a centre for chlorophyll and carotenoid synthesis. Phytoene synthase interacts with GGPP to form Phytoene, a colourless carotenoid. Phytoene interacts with phytoene desaturase to form ζ-carotene. Now ζ-carotene combines with the enzyme to form lycopene. Lycopene cyclization through the enzyme lycopene β-carotene cyclase results in the formation of two carotenoids: α-carotene and β-carotene (Majer et al., 2017). Through α-carotene, lutein is formed which interacts with carotene hydroxylase and via β-carotene; zeaxanthin, a light-yellow coloured xanthophyll, is formed via enzyme carotene hydroxylase (Sun et al., 2018). Through zeaxanthin, many xanthopyll compounds are formed. With zeaxanthin epoxidase interaction antheraxanthin has been reported to be synthesized from zeaxanthin. This antheraxanthin on interaction with zeaxanthin epoxidase again leads to the formation of violaxanthin taking place. From violaxanthin, a few unexplored carotenoids, like leuteoxanthin and auroxanthin, are formed, along with neoxanthin. Neoxathin further leads to the formation of Abscisic acid, which is an important plant hormone.

Auroxanthin in Storage

In canned mango products, the violaxanthin showed significantly decreased levels, and the amount of auroxanthin constantly increased. During a period of time, the mango puree and sliced mango pieces, when kept for storage, showed decreased violaxanthin and luteoxanthin levels due to biochemical reactions involved when stored; however, the level of auroxanthin remained constant (Solovchenko et al., 2019).

Auroxanthin Bioactivity – Role in Photoprotection

Non-photochemical quenching was observed when the light harvesting complexes were harvested in spinach leaves by auroxanthin. The quenching ability of violaxanthin and leutin was less, while that in zeaxanthin and auroxanthin was more. In the CP26 and LHC2 light harvesting complexes, auroxanthin showed the maximum quenching of 1.673 and 0.570. It was found to be a stronger

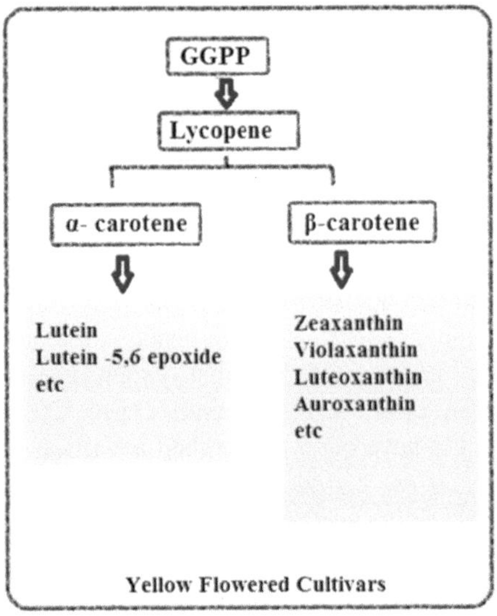

FIGURE 7.8 Carotenoid biosynthesis in yellow flower cultivars. The carotenoid biosynthesis in yellow flower cultivars is from Geranylgeranyl pyrophosphate to lycopene. They vary when lycopene gets converted to α-carotene and β-carotene by lycopene cyclase. From α-carotene, two carotenoids: lutein and lutein-5,6-epoxide; and from β-carotene; four carotenoids: zeaxanthin, violaxanthin, auroxanthin, and luteoxanthin, give the yellow colour (Kishimoto, 2006).

quencher than zeaxanthin, as it has C5,8 epoxide in its structure and its S1 energy is also higher than violaxanthin in green peas (Wentworth et al., 2000), which further enhances aggregation associated quenching (Ruban et al., 1998). Even though, when energy dissipation occurs in plants, violaxanthin is readily converted to zeaxanthin; on the other side there is an increase in the auroxanthin level (Anjum et al., 2016). The effectiveness of auroxanthin came to be known from the fact that the S1 energy level was higher than that of violaxanthin in halophytes. This effect is explained by its structure where the end groups of auroxanthin lie in the plane of the conjugated carbon double bond chain, as in zeaxanthin, whereas they are expected to be out-of-plane in violaxanthin. The coupled activation of the xanthophyll cycle along with a higher auroxanthin synthesis suggests that the heat-treated individuals had higher needs to dissipate excessive energy than the cells exposed to cold treatment. Thus, auroxanthin is a strong quencher and its content gets increased when plants are exposed to temperature stress (Duarte et al., 2015b, 2015c). Similarly, in *Zostera noltii* there was increase in the auroxanthin and antheraxanthin levels under warming conditions. The coupled activation of the xanthophyll cycle along with a higher auroxanthin synthesis suggests that warming-exposed individuals had higher needs to dissipate excessive energy than seagrass under control temperature treatments. It is worth noting that the increase in antioxidant pigments (e.g. β-carotene) in seagrass exposed to warming and acidification conditions, reinforces the need to counteract the effects of excessive redox potential accumulated inside its chloroplasts and subsequent generations of reactive oxygen species (ROS) (Repolho et al., 2017).

Role of Epoxy Carotenoids in Nutrition and Health

Many carotenoids play different roles in human health and nutrition, which vary with their structures and their metabolism. As β-carotene, an important provitamin A precursor, lutein and zeaxanthin play an important role in human eye macular degeneration (Eggersdorfer and Wyss, 2018), other carotenoids have potential applications. An intake of carotenoids boosts immunity, helps lower diabetes, fights against cancer and heart diseases, and plays a vital role in the diet as an antioxidant.

A study showed that violaxanthin from *Nannochloropsis oceanica* had a protective role on human dermal fibroplasts that were exposed to UV-B light. They also reported that they are not cytotoxic to the fibroplast cells (Kim et al., 2018b). Luteoxanthin from red and green lettuce cultivars showed a potential scavenging role on ABTS and DPPH radicals. These cultivars had many compounds with proven health benefits including luein and violaxanthin (Kim et al., 2018a). Mutatoxanthin, neochrome and auroxanthin which were found in fermented orange beverage when consumed reduced lipid peroxidation and inflammation in humans (Escudero-López et al., 2018). In yellow potatoes, the presence of neoxanthin and lutein has been determined. Neoxanthin has shown activity against obesity, which would stop the cascade of diseases from it (Chaparro et al., 2018). Fucoxanthin, which is mainly present in algae, has various health benefits, like anti-ageing, anti-angiogenic, and anti-metastatic activities. They are also used against human diseases as a bioactive molecule for prevention and treatment (Irvani et al., 2018; Zarekarizi et al., 2018). Diadinoxanthin, along with fucoxanthin, has been reported to have other antioxidant activities.

CONCLUSION

We infer that carotenoids are not synthesized in the human body, but they are supplemented by the food we eat. Although some 40 carotenoids have been produced industrially, many other carotenoids are not given importance as their metabolism, function and importance have not been well studied. In the category of epoxy carotenoids, there are many still to be explored which play an important role in plant metabolism, function and protection, like the known carotenoids. While some of the roles of epoxy carotenoids in the gene regulation have been reported, many of their important functions are yet to be explored. So, these unexplored carotenoids should be given importance by finding research gaps and untying the research knots. Still many of the carotenoid sources (and their bioavailability) remain unknown. Studies on lutein and zeaxanthin in human retina showed their importance in macular degeneration. One of the main objectives was their non-photochemical quenching; similarly, auroxanthin and luteoxanthin can be tested in human eye protection as they also had non-photochemical quenching ability. Many carotenoids, as mentioned, are food colourants and additives, which suggests promising roles of epoxy carotenoids in the industry as well. As there is a great need for natural colourants and necessity for production of carotenoids, we need to find the natural sources where these unexplored carotenoids are present on one hand, and their production can be done on the other. There is a huge demand for carotenoids in the market, and because of that, they are costlier too; so cost-effective production of these compounds will help in meeting the demand through tissue culture and bioreactors, which can be targeted with mutatoxanthin, luteoxanthin, auroxanthin, mutatochrome, aurochrome, and neochrome when they are explored. Therefore, in a nutshell, a lot more has to be discovered in the family of carotenoids, which could prove to be beneficial for the future of mankind.

REFERENCES

Abdelaali SB, Rodrigo MJ, Saddoud O, Zacarías L, Hajlaoui MR, Mars M. 2018. Carotenoids and colour diversity of traditional and emerging Tunisian orange cultivars (*Citrus sinensis* (L.) Osbeck). *Scientiahorticulturae* 227: 296–304.

Acemoglu M, Eugster CH. 1984. Die diasteromeren Aurochrome: Synthese, Analytik und Chiroptische Eigenschaften. *Helvetica Chimica Acta* 67: 471–487.

Agócs A, Murillo E, Turcsi E, Béni S, Darcsi A, Szappanos Á, Kurtán T, Deli J. 2018. Isolation of allene carotenoids from mamey. *Journal of Food Composition and Analysis* 65: 1–5.

Alós E, Rodrigo MJ, Zacarias L. 2019 Jan 1. Ripening and senescence. In: *Postharvest Physiology and Biochemistry of Fruits and Vegetables* (pp. 131–155). Woodhead Publishing.

Alos E, Martinez-Fuentes A, Reig C, Mesejo C, Zacarías L, Agustí M, Rodrigo MJ. 2019 Mar 1. Involvement of ethylene in color changes and carotenoid biosynthesis in loquat fruit (Eriobotrya japonica Lindl. cv. Algerie). *Postharvest Biology and Technology* 149: 129–138.

Al-Rifai A. 2018. Identification and evaluation of in-vitro antioxidant phenolic compounds from the *Calendula tripterocarpa* Rupr. *South African Journal of Botany* 116: 238–244.
Amaya-Cruz DM, Pérez-Ramírez IF, Delgado-García J, Mondragón-Jacobo C, Dector-Espinoza A, Reynoso-Camacho R. 2019 Apr 25. An integral profile of bioactive compounds and functional properties of prickly pear (Opuntiaficusindica L.) peel with different tonalities. *Food Chemistry* 278: 568–578.
Anantharaman A, Subramanian B, Chandrasekaran R, Seenivasan R, Siva R. 2014. Colorants and cancer: A review. *Industrial Crops and Products* 53: 167–186.
Anantharaman A, Priya RR, Hemachandran H, Sivaramakrishna A, Babu S, Siva R. 2015. Studies on interaction of norbixin with DNA: Multispectroscopic and in silico analysis. *Spectrochimica Acta Part A: Molecular and Biomolecular Spectroscopy* 144: 163–169.
Anantharaman A, Hemachandran H, Mohan S, Ayyathan DM, Siva R. 2016a. Induction of apoptosis by apocarotenoids in B16 melanoma cells through ROS-mediated mitochondrial-dependent pathway. *Journal of Functional Foods* 20: 346–357.
Anantharaman A, Hemachandran H, Priya RR, Sankari M, Gopalakrishnan M, Palanisami N, Siva R. 2016b. Inhibitory effect of apocarotenoids on the activity of tyrosinase: Multi-spectroscopic and docking studies. *Journal of Bioscience and Bioengineering* 121: 13–20.
Anjum NA, Duarte B, Caçador I, Sleimi N, Duarte AC, Pereira E. 2016. Biophysical and biochemical markers of metal/metalloid-impacts in salt marsh halophytes and their implications. *Frontiers in Environmental Science* 4: 24.
Aparicio R, Harwood J. 2013. *Handbook of Olive Oil. Analysis and Properties*, 2nd edn. Springer, New York.
Araki M, Kaku N, Harada M, Ando Y, Yamaguchi R, Shindo K. 2016. Production of auroxanthins from violaxanthin and 9-*cis*-violaxanthin by acidic treatment and the antioxidant activities of violaxanthin, 9-*cis*-violaxanthin, and auroxanthins. *Journal of Agricultural and Food Chemistry* 64: 9352–9355.
Ariizumi T, Kishimoto S, Kakami R, Maoka T, Hirakawa H, Suzuki Y, Ozeki Y, Shirasawa K, Bernillon S, Okabe Y. 2014. Identification of the carotenoid modifying gene PALE YELLOW PETAL 1 as an essential factor in xanthophyll esterification and yellow flower pigmentation in tomato (*S. olanum lycopersicum*). *The Plant Journal* 79: 453–465.
Asai A, Terasaki M, Nagao A. 2004. An epoxide–furanoid rearrangement of spinach neoxanthin occurs in the gastrointestinal tract of mice and *in vitro*: Formation and cytostatic activity of neochrome stereoisomers. *The Journal of Nutrition* 134: 2237–2243.
Azimova SS, Glushenkova AI, Vinogradova VI. 2011. *Lipids, Lipophilic Components and Essential Oils from Plant Sources*. Springer Science & Business Media, New York.
Badejo AA. 2018. Elevated carotenoids in staple crops: The biosynthesis, challenges and measures for target delivery. *Journal of Genetic Engineering and Biotechnology*
Bakó E, Deli J, Tóth G. 2002. HPLC study on the carotenoid composition of Calendula products. *Journal of Biochemical and Biophysical Methods* 53: 241–250.
Balduyck L, Dejonghe C, Goos P, Jooken E, Muylaert K, Foubert I. 2019 Mar 1. Inhibition of lipolytic reactions during wet storage of T-Isochrysislutea biomass by heat treatment. *Algal Research* 38: 101388.
Belwal T, Devkota HP, Hassan HA, Ahluwalia S, Ramadan MF, Mocan A, Atanasov AG. 2018 Apr 1. Phytopharmacology of Acerola (Malpighia spp.) and its potential as functional food. *Trends in Food Science & Technology* 74: 99–106.
Bertoldi D, Cossignani L, Blasi F, Perini M, Barbero A, Pianezze S, Montesano D. 2019 Mar 1. Characterisation and geographical traceability of Italian goji berries. *Food Chemistry* 275: 585–593.
Besler, M, Fahlbusch B, Santiago Q. 2000. Kiwi Fruit (Actinidia chinensis). http://www.food-allergens.de
Bordiga M, Travaglia F, Giuffrida D, Mangraviti D, Rigano F, Mondello L, Arlorio M, Coïsson JD. 2018 Nov 18. Characterization of peel and pulp proanthocyanidins and carotenoids during ripening in persimmon "Kaki Tipo" cv, cultivated in Italy. *Food Research International.*
de Brito ES, de Araújo MCP, Lin L-Z, Harnly J. 2007. Determination of the flavonoid components of cashew apple (*Anacardium occidentale*) by LC-DAD-ESI/MS. *Food Chemistry* 105: 1112–1118.
Brito ES, de Oliveira Silva E, Rodrigues S. 2018 Jan 1. Caju—Anacardiumoccidentale. *InExotic Fruits* 85–89.
Cai B, Wan P, Chen H, Chen D, Chen X, Sun H, Pan J. 2019 Mar 1. Composition characterization of oyster polysaccharides from Crassostreahongkongensis and their protective effect against H_2O_2-induced oxidative damage in IEC-6 cells. *International Journal of Biological Macromolecules* 124: 246–254.
Camara B, Moneger R. 1978. Free and esterified carotenoids in green and red fruits of *Capsicum annuum. Phytochemistry* 17: 91–93.
Camus C, Infante J, Buschmann AH. 2019 Mar 15. Revisiting the economic profitability of giant kelp Macrocystispyrifera (Ochrophyta) cultivation in Chile. *Aquaculture* 502: 80–86.

Cerón I, Higuita J, Cardona C. 2015. Analysis of a biorefinery based on *Theobroma grandiflorum* (copoazu) fruit. *Biomass Conversion and Biorefinery* 5: 183–194.

Chabra A, Monadi T, Azadbakht M, Haerizadeh SI. 2018 Dec 3. Ethnopharmacology of Cuscuta epithymum: A comprehensive review on ethnobotany, phytochemistry, pharmacology and toxicity. *Journal of Ethnopharmacology*.

Chacón-Ordóñez T, Schweiggert RM, Bosy-Westphal A, Jiménez VM, Carle R, Esquivel P. 2017. Carotenoids and carotenoid esters of orange-and yellow-fleshed mamey sapote (*Pouteria sapota* (Jacq.) HE Moore & Stearn) fruit and their post-prandial absorption in humans. *Food Chemistry* 221: 673–682.

Chaparro JM, Holm DG, Broeckling CD, Prenni JE, Heuberger AL. 2018. Metabolomics and ionomics of potato tuber reveals an influence of cultivar and market class on human nutrients and bioactive compounds. *Frontiers in Nutrition* 5: 36.

Colodel C, de Oliveira Petkowicz CL. 2019. Acid extraction and physicochemical characterization of pectin from cubiu (*Solanum sessiliflorum* D.) fruit peel. *Food Hydrocolloids* 86: 193–200.

Colodel C. and de Oliveira Petkowicz, CL 2019. Acid extraction and physicochemical characterization of pectin from cubiu (Solanum sessiliflorum D.) fruit peel. *Food Hydrocolloids* 86: 193–200.

Czeczuga B. 1983. Mutatoxanthin, the dominant carotenoid in lichens of the *Xanthoria* genus. *Biochemical Systematics and Ecology* 11: 329–331.

Delgado-Pelayo R, Gallardo-Guerrero L, Hornero-Méndez D. 2016. Carotenoid composition of strawberry tree (*Arbutus unedo* L.) fruits. *Food Chemistry* 199: 165–175.

Delgado-Vargas F, Paredes-López O. 2002. *Natural Colorants for Food and Nutraceutical Uses*. CRC Press, Boca Raton, FL.

Deli J, Osz E. 2004. Carotenoid 5, 6-, 5, 8-and 3, 6-epoxides. *Arkivoc* 2004: 150–168.

Delpino-Rius A, Eras J, Marsol-Vall A, Vilaró F, Balcells M, Canela-Garayoa R. 2014. Ultra performance liquid chromatography analysis to study the changes in the carotenoid profile of commercial monovarietal fruit juices. *Journal of Chromatography A* 1331: 90–99.

Divya P, Puthusseri B, Savanur MA, Lokesh V, Neelwarne B. 2018 Sep 1. Effects of methyl jasmonate and carotenogenic inhibitors on gene expression and carotenoid accumulation in coriander (Coriandrumsativum L.) foliage. *Food Research International* 111: 11–19.

Duare B, et al. 2015. Ecophysiological constraints of Aster tripolium under extreme thermal events impacts: Merging biophysical, biochemical and genetic insights. *Plant Physiology and Biochemistry* 97: 217–228.

Duarte B, Goessling JW, Marques J, Caçador I. 2015a. Ecophysiological constraints of *Aster tripolium* under extreme thermal events impacts: Merging biophysical, biochemical and genetic insights. *Plant Physiology and Biochemistry* 97: 217–228.

Duarte B, Santos D, Marques J, Caçador I. 2015b. Impact of heat and cold events on the energetic metabolism of the C3 halophyte *Halimione portulacoides. Estuarine, Coastal and Shelf Science* 167: 166–177.

Duarte B, Marques J, Caçador I. 2016. Ecophysiological response of native and invasive *Spartina* species to extreme temperature events in Mediterranean marshes. *Biological Invasions* 18: 2189–2205.

Duke J. 2004. Dr. Duke's Phytochemical and Ethnobotanical Databases. United States Department of Agriculture, Agricultural Research Service, accessed April 27. https://phytochem.nal.usda.gov/phytochem/search/list

Eggersdorfer M, Wyss A. 2018. Carotenoids in human nutrition and health. *Archives of Biochemistry and Biophysics* 652: 18–26.

El-Gamal SM. 2015. Influence of pinching and potassium sources on growth and flowers yield of *Calendula officinalis. Egyptian Journal of Horticulture* 42(1): 639–654.

Elnour HM, Dietzel L, Ramanan C, Büchel C, van Grondelle R, Krüger TP. 2018 Oct 1. Energy dissipation mechanisms in the FCPb light-harvesting complex of the diatom Cyclotella meneghiniana. *BiochimicaetBiophysicaActa (BBA)-Bioenergetics* 1859(10):1151–1160.

Escudero-López B, Ortega Á, Cerrillo I, Rodríguez-Griñolo MR, Muñoz-Hernández R, Macher HC, Martín F, Hornero-Méndez D, Mena P, Del Rio D. 2018. Consumption of orange fermented beverage improves antioxidant status and reduces peroxidation lipid and inflammatory markers in healthy humans. *Journal of the Science of Food and Agriculture* 98: 2777–2786.

Esteban R, Becerril JM, García-Plazaola JI. 2009. Lutein epoxide cycle, more than just a forest tale. *Plant Signaling & Behavior* 4: 342–344.

Fernandez-Orozco R, Gallardo-Guerrero L, Hornero-Méndez D. 2013. Carotenoid profiling in tubers of different potato (*Solanum* sp) cultivars: Accumulation of carotenoids mediated by xanthophyll esterification. *Food Chemistry* 141: 2864–2872.

Fiutak G, Michalczyk M, Filipczak-Fiutak M, Fiedor L, Surówka K.2019 Jan 19. The impact of LED lighting on the yield, morphological structure and some bioactive components in alfalfa (Medicagosativa L.) sprouts. *Food Chemistry*.

Flórez-Fernández N, Domínguez H, Torres MD. 2019 Mar 1. A green approach for alginate extraction from Sargassummuticum brown seaweed using ultrasound-assisted technique. *International Journal of Biological Macromolecules* 124 :451–459.
Fratianni A, Niro S, Messia MC, Cinquanta L, Panfili G, Albanese D, Di Matteo M. 2017 Sep 1. Kinetics of carotenoids degradation and furosine formation in dried apricots (Prunus armeniaca L.). *Food Research International* 99: 862–867.
Frede K, Schreiner M, Baldermann S. 2019 Feb 13. Light quality-induced changes of carotenoid composition in pakchoi Brassica rapa ssp. chinensis. *Journal of Photochemistry and Photobiology B: Biology.*
Galvis-Sánchez AC, Vinholes J. 2018 Jan 1. Fruit Juices (Apple, Peach, and Pear) and Changes in the Carotenoid Profile. *InFruit Juices* 59–74.
Gandul-Rojas B, Gallardo-Guerrero L. 2018 Jun 1. Pigment changes during preservation of green table olive specialities treated with alkali and without fermentation: Effect of thermal treatments and storage conditions. *Food Research International* 108: 57–67.
Gandul-Rojas B., Gallardo-Guerrero L., Roca M., Aparicio-Ruiz R. 2013 Chromatographic methodologies: Compounds for olive oil color issues. In: Aparicio R., Harwood J. (eds) *Handbook of Olive Oil.* Springer, Boston, MA. https://doi.org/10.1007/978-1-4614-7777-8_7.
Giger A. 2002. Chemical synthesis project. A new yellow carotenoid. *Pure and Applied Chemistry* 74: 1383–1390.
Giuffrida D, Zoccali M, Arigò A, Cacciola F, Roa CO, Dugo P, Mondello L. 2018 May 15. Comparison of different analytical techniques for the analysis of carotenoids in tamarillo (Solanumbetaceum Cav.). *Archives of Biochemistry and Biophysics* 646: 161–167.
González-Barrio R, Periago MJ, Luna-Recio C, Garcia-Alonso FJ, Navarro-González I. 2018 Jun 30. Chemical composition of the edible flowers, pansy (Viola wittrockiana) and snapdragon (Antirrhinum majus) as new sources of bioactive compounds. *Food Chemistry* 252: 373–380.
Goodwin T. 1952. The carotenoids of the berries of *Lonicera japonica. Biochemical Journal* 51: 458.
Guan X-R, Zhu L, Xiao Z-G, Zhang Y-L, Chen H-B, Yi T. 2017. Bioactivity, toxicity and detoxification assessment of *Dioscorea bulbifera* L.: A comprehensive review. *Phytochemistry Reviews* 16: 573–601.
Gulyaeva K. 2015. Development of protocols for the extraction, separation and characterization of lipids-determination of photosynthetic lipid composition under different light environments in the microalga *Nannochloropsis oceanica.* Masters thesis, NTNU.
Guoliang L, Junyou S, Yourui S, Zhiwei S, Lian X, Jie Z, Jinmao Y, Yongjun L. 2011. Supercritical CO_2 cell breaking extraction of *Lycium barbarum* seed oil and determination of its chemical composition by HPLC/APCI/MS and antioxidant activity. *LWT-Food Science and Technology* 44: 1172–1178.
Gutiérrez-Gamboa G, Marín-San Román S, Jofré V, Rubio-Bretón P, Pérez-Álvarez EP, Garde-Cerdán T. 2018. Effects on chlorophyll and carotenoid contents in different grape varieties (Vitisvinifera L.) after nitrogen and elicitor foliar applications to the vineyard. *Food Chemistry* 269: 380–386.
Ha SH, Kim JK, Jeong YS, You MK, Lim SH, Kim JK. 2019 Mar 1. Stepwise pathway engineering to the biosynthesis of zeaxanthin, astaxanthin and capsanthin in rice endosperm. *Metabolic Engineering* 52: 178–189.
Hadjal T, Dhuique-Mayer C, Madani K, Dornier M, Achir N. 2013. Thermal degradation kinetics of xanthophylls from blood orange in model and real food systems. *Food Chemistry* 138: 2442–2450.
Hadjipieri M, Georgiadou EC, Marin A, Diaz-Mula HM, Goulas V, Fotopoulos V, Tomás-Barberán FA, Manganaris GA. 2017. Metabolic and transcriptional elucidation of the carotenoid biosynthesis pathway in peel and flesh tissue of loquat fruit during on-tree development. *BMC Plant Biology* 17: 102.
Hansmann P, Kleinig H. 1982 Jan 1. Violaxanthin esters from Viola tricolor flowers. *Phytochemistry* 21(1):238–239.
Havir EA, Tausta SL, Peterson RB. 1997 Mar 28. Purification and properties of violaxanthin de-epoxidase from spinach. *Plant Science* 123(1–2):57–66.
Hornero-Méndez D, Gómez-Ladrón de Guevara R, Minguez-Mosquera MI. 2000. Carotenoid biosynthesis changes in five red pepper (*Capsicum annuum* L.) cultivars during ripening. Cultivar selection for breeding. *Journal of Agricultural and Food Chemistry* 48: 3857–3864.
Irvani N, Hajiaghaee R, Zarekarizi A. 2018. A review on biosynthesis, health benefits and extraction methods of fucoxanthin, particular marine carotenoids in algae. *Journal of Medicinal Plants* 3: 6–30.
Jäger S, Büchel C. 2019 Jan 1. Cation-dependent changes in the thylakoid membrane appression of the diatom Thalassiosirapseudonana. *Biochimica et Biophysica Acta (BBA)-Bioenergetics* 1860(1):41–51.
Janečková H, Husičková A, Lazár D, Ferretti U, Pospíšil P, Špundová M. 2019 Mar 1. Exogenous application of cytokinin during dark senescence eliminates the acceleration of photosystem II impairment caused by chlorophyll b deficiency in barley. *Plant Physiology and Biochemistry* 136: 43–51.

Jeknić Z, Morré JT, Jeknić S, Jevremović S, Subotić A, Chen TH. 2012. Cloning and functional characterization of a gene for capsanthin-capsorubin synthase from tiger lily (*Lilium lancifolium* Thunb.'Splendens'). *Plant and Cell Physiology* 53: 1899–1912.

Jiao C, Gu Z. 2019 Mar 1. iTRAQ-based proteomic analysis reveals changes in response to UV-B treatment in soybean sprouts. *Food Chemistry* 275: 467–473.

Kamalambigeswari R, Jeyanthi Rebecca L. 2016. Extraction of antioxidant lutein from various flowers. *International Journal of Pharmaceutical Sciences Review and Research* 39(1): 122–124

Kang C, Zhai H, Xue L, Zhao N, He S, Liu Q. 2018. A lycopene β-cyclase gene, IbLCYB2, enhances carotenoid contents and abiotic stress tolerance in transgenic sweetpotato. *Plant Science* 272: 243–254.

Kaur J, Rawat A, Kumar S, Narain S. 2016. Taxonomy, phytochemistry, traditional uses and cultivation of Ipomoea Aquatica Forsk. *Imperial Journal of Interdisciplinary Research* 2: 408–412.

Ke Q, Kang L, Kim HS, Xie T, Liu C, Ji CY, Kim SH, Park WS, Ahn MJ, Wang S, Li H. 2019 Apr 1. Downregulation of lycopene ε-cyclase expression in transgenic sweetpotato plants increases the carotenoid content and tolerance to abiotic stress. *Plant Science* 281: 52–60.

Khalid KA, da Silva JT. 2012. Biology of *Calendula officinalis* Linn.: focus on pharmacology, biological activities and agronomic practices. *Medicinal and Aromatic Plant Science and Biotechnology* 6: 12–27.

Khalid M, Bilal M, Iqbal HM, Huang D. 2018. Biosynthesis and biomedical perspectives of carotenoids with special reference to human health-related applications. *Biocatalysis and Agricultural Biotechnology* 17: 399–407.

Kim D-E, Shang X, Assefa AD, Keum Y-S, Saini RK. 2018a. Metabolite profiling of green, green/red, and red lettuce cultivars: Variation in health beneficial compounds and antioxidant potential. *Food Research International* 105: 361–370.

Kim HM, Jung JH, Kim JY, Heo J, Cho DH, Kim HS, An S, An IS, Bae S. 2018b. The protective effect of violaxanthin from *Nannochloropsis oceanica* against ultraviolet B-induced damage in normal human dermal fibroblasts. *Photochemistry and Photobiology* 95: 595–604.

Kiokias S, Proestos C, Varzakas T. 2016. A review of the structure, biosynthesis, absorption of carotenoids-analysis and properties of their common natural extracts. *Current Research in Nutrition and Food Science Journal* 4: 25–37.

Kishimoto S. 2006. Studies on carotenoids in petals of compositae plants. *Bulletin of the National Institute of Floricultural Science (Japan)* 78: 263–272.

Kishimoto S, Ohmiya A. 2009. Studies on carotenoids in the petals of Compositae plants. *Journal of the Japanese Society for Horticultural Science* 78: 263–272.

Kuczynska P, Jemiola-Rzeminska M, Strzalka K. 2015. Photosynthetic pigments in diatoms. *Marine Drugs* 13: 5847–5881.

Kudritskaya S, Fishman G, Zagorodskaya L, Chikovani D. 1983. Carotenoids of promising citrus hybrids. *Chemistry of Natural Compounds* 19: 112–112.

Lelley T., Loy B., Murkovic M. 2009. Hull-less oil seed pumpkin. In: Vollmann J, Rajcan I (eds) *Oil Crops. Handbook of Plant Breeding*, Vol 4. Springer, New York, NY. https://doi.org/10.1007/978-0-387-77594-4_16.

Li X, Sui X, Zhao W, Huang H, Chen Y, Zhang Z. 2015 Apr 1. Characterization of cucumber violaxanthin de-epoxidase gene promoter in Arabidopsis. *Journal of Bioscience and Bioengineering* 119(4):470–477.

Lim S, Choi AH, Kwon M, Joung EJ, Shin T, Lee SG, Kim NG, Kim HR. 2019 Apr 25. Evaluation of antioxidant activities of various solvent extract from Sargassumserratifolium and its major antioxidant components. *Food Chemistry* 278: 178–184.

Lima CR, Mendes CR, Tavano VM, Detoni AM, Secchi ER. 2019 Jan 21. Chemotaxonomy-based mapping of phytoplankton communities in the subtropical Southwestern Atlantic Ocean, with emphasis on the marine cyanobacterium Trichodesmium. *Progress in Oceanography*.

Liu H, Kao T, Chen B. 2004. Determination of carotenoids in the chinese medical herb Jiao-Gu-Lan (Gynostemma Pentaphyllum MAKINO) by liquid chromatography. *Chromatographia* 60: 411–417.

Liu H-L, Chen B-H, Kao T-H, Shiau C-Y. 2014. Carotenoids composition in *Scutellaria barbata* D. Don as detected by high performance liquid chromatography-diode array detection-mass spectrometry-atmospheric pressure chemical ionization. *Journal of Functional Foods* 8: 100–110.

Long M, Tallec K, Soudant P, Le Grand F, Donval A, Lambert C, Sarthou G, Jolley DF, Hégaret H. 2018 Nov 1. Allelochemicals from Alexandriumminutum induce rapid inhibition of metabolism and modify the membranes from Chaetocerosmuelleri. *Algal Research* 35: 508–518.

Ma X, Zheng B, Ma Y, Xu W, Wu H, Wang S. 2018. Carotenoid accumulation and expression of carotenoid biosynthesis genes in mango flesh during fruit development and ripening. *Scientia Horticulturae* 237: 201–206.

Majer E, Llorente B, Rodríguez-Concepción M, Daròs J-A. 2017. Rewiring carotenoid biosynthesis in plants using a viral vector. *Scientific Reports* 7: 41645.

Marcheafave GG, Tormena CD, Pauli ED, Rakocevic M, Bruns RE, Scarminio IS. 2019 Jan 29. Experimental mixture design solvent effects on pigment extraction and antioxidant activity from Coffea arabica L. leaves. *Microchemical Journal.*

Martin FW, Telek L, Ruberte RM. 1974. Yellow pigments of *Dioscorea bulbifera. Journal of Agricultural and Food Chemistry* 22: 335–337.

Matsubara S, Chen YC, Caliandro R, Clegg RM. 2011 Jul 1. Photosystem II fluorescence lifetime imaging in avocado leaves: contributions of the lutein-epoxide and violaxanthin cycles to fluorescence quenching. *Journal of Photochemistry and Photobiology B: Biology* 104(1–2):271–284.

Matus Z, Molnár P, Szabó LG. 1993. Main carotenoids in pressed seeds (*Cucurbitae semen*) of oil pumpkin (*Cucurbita pepo* convar. *pepo* var. *styriaca*). *Acta Pharmaceutica Hungarica* 63: 247–256.

Meléndez-Martínez AJ, Britton G, Vicario IM, Heredia FJ. 2007a. Relationship between the colour and the chemical structure of carotenoid pigments. *Food Chemistry* 101: 1145–1150.

Meléndez-Martínez AJ, Vicario IM, Heredia FJ. 2007b Jan 1. Geometrical isomers of violaxanthin in orange juice. *Food Chemistry* 104(1):169–175.

Meléndez-Martínez AJ, et al. 2008. The complex carotenoid pattern of orange juices from concentrate. *Food Chemistry* 109(3): 546–553.

Meléndez-Martínez AJ, Escudero-Gilete ML, Vicario IM, Heredia FJ. 2010. Effect of increased acidity on the carotenoid pattern and colour of orange juice. *European Food Research and Technology* 230: 527.

Mercadante AZ, Rodriguez-Amaya DB. 1998 Jan 19. Effects of ripening, cultivar differences, and processing on the carotenoid composition of mango. *Journal of Agricultural and Food Chemistry* 46(1):128–130.

Messina M, Bulfon C, Beraldo P, Tibaldi E, Cardinaletti G. 2019 Feb 1. Intestinal morpho-physiology and innate immune status of European sea bass (Dicentrarchuslabrax) in response to diets including a blend of two marine microalgae, Tisochrysislutea and Tetraselmissuecica. *Aquaculture* 500: 660–669.

Mezadri T, Fernández-Pachón MS, Villano D, García-Parrilla MC, Troncoso AM. 2006. The acerola fruit: Composition, productive characteristics and economic importance. *Archivos latinoamericanos de nutricion* 56: 101–109.

Mitra M, Mishra S. 2019 Mar 1. A comparative analysis of different extraction solvent systems on the extractability of eicosapentaenoic acid from the marine eustigmatophyte Nannochloropsis oceanica. *Algal Research* 38: 101387.

Miyashita K, Nishikawa S, Beppu F, Tsukui T, Abe M, Hosokawa M. 2011. The allenic carotenoid fucoxanthin, a novel marine nutraceutical from brown seaweeds. *Journal of the Science of Food and Agriculture* 91: 1166–1174.

Młodzińska E. 2009. Survey of plant pigments: Molecular and environmental determinants of plant colors. *Acta Biologica Cracovienca Series Botanica* 51: 7–16.

Mohan S, Hemachandran H, Sneha P, George Priya Doss C, Godwin Christopher J, Jayaraman G, Ramamoorthy S. 2018. Structural insights into the binding mode and conformational changes of BSA induced by bixin and crocin. *Journal of Biomolecular Structure and Dynamics* 36: 2085–2098.

Molina-Miras A, López-Rosales L, Sánchez-Mirón A, Cerón-García MC, Seoane-Parra S, García-Camacho F, Molina-Grima E. 2018 Oct 1. Long-term culture of the marine dinoflagellate microalga Amphidiniumcarterae in an indoor LED-lighted raceway photobioreactor: production of carotenoids and fatty acids. *Bioresource Technology* 265: 257–267.

Morris JM, Fleming GR. 2018 Apr 4. Quantitative modeling of energy dissipation in Arabidopsis thaliana. *Environmental and Experimental Botany.*

Nagao R, Ueno Y, Yokono M, Shen JR, Akimoto S. 2018 Jul 1. Alterations of pigment composition and their interactions in response to different light conditions in the diatom Chaetocerosgracilis probed by time-resolved fluorescence spectroscopy. *Biochimica et Biophysica Acta (BBA)-Bioenergetics* 1859(7):524–530.

Naim F, Dugdale B, Kleidon J, Brinin A, Shand K, Waterhouse P, Dale J. 2018. Gene editing the phytoene desaturase alleles of Cavendish banana using CRISPR/Cas9. *Transgenic Research* 27: 451–460.

Neamtu G, Zoltan N, Bilasus C, Laszlo T. 1975. Chemotaxonomical researches in higher plants. IX. Carotenoid pigments in some species of the genus *Cucumis. Notulae Botanicae Horti Agrobotanici Cluj-Napoca* 8: 53–58.

Ntsoane ML, Jalali A, Römer J, Duewel K, Göller C, Kühn R, Mähne K, Geyer M, Sivakumar D, Mahajan PV. 2019a Mar 1. Performance evaluation of silica gel-water adsorption based cooling system for mango fruit storage in Sub-Saharan Africa. *Postharvest Biology and Technology* 149: 195–199.

Ntsoane ML, Zude-Sasse M, Mahajan P, Sivakumar D. 2019b Apr 30. Quality assessment and postharvest technology of mango: A review of its current status and future perspectives. *ScientiaHorticulturae* 249: 77–85.

Nur MMA, Muizelaar W, Boelen P, Buma A. 2018. Environmental and nutrient conditions influence fucoxanthin productivity of the marine diatom *Phaeodactylum tricornutum* grown on palm oil mill effluent. *Journal of Applied Phycology* 31: 111–122.

Olmo A, Picon A, Nunez M. 2019 Feb 13. High pressure processing for the extension of Laminariaochroleuca (kombu) shelf-life: A comparative study with seaweed salting and freezing. *Innovative Food Science & Emerging Technologies.*

Papageorgiou G.C. 2014. The non-photochemical quenching of the electronically excited state of chlorophyll a in plants: Definitions, timelines, viewpoints, open questions. In: Demmig-Adams B, Garab G, Adams III W, Govindjee (eds) *Non-Photochemical Quenching and Energy Dissipation in Plants, Algae and Cyanobacteria. Advances in Photosynthesis and Respiration (Including Bioenergy and Related Processes)*, Vol 40. Springer, Dordrecht. https://doi.org/10.1007/978-94-017-9032-1_1.

Paznocht L, Kotíková Z, Orsák M, Lachman J, Martinek P. 2019 Mar 30. Carotenoid changes of colored-grain wheat flours during bun-making. *Food Chemistry* 277: 725–734.

Peng J, Yuan J-P, Wu C-F, Wang J-H. 2011. Fucoxanthin, a marine carotenoid present in brown seaweeds and diatoms: Metabolism and bioactivities relevant to human health. *Marine Drugs* 9: 1806–1828.

Pérez-Gálvez A, Martin HD, Sies H, Stahl W. 2003. Incorporation of carotenoids from paprika oleoresin into human chylomicrons. *British Journal of Nutrition* 89: 787–793.

Phillip D, Molnar P, Toth G, Young AJ. 1999 Apr 1. Light-induced formation of 13-cis violaxanthin in leaves of Hordeumvulgare. *Journal of Photochemistry and Photobiology B: Biology* 49(2–3):89–95.

Polidori J, Dhuique-Mayer C, Dornier M. 2018. Crossflow microfiltration coupled with diafiltration to concentrate and purify carotenoids and flavonoids from citrus juices. *Innovative Food Science & Emerging Technologies* 45: 320–329.

Price EJ, Bhattacharjee R, Lopez-Montes A, Fraser PD. 2018. Carotenoid profiling of yams: Clarity, comparisons and diversity. *Food Chemistry* 259: 130–138.

Priya R, Sneha P, Rivera Madrid R, Doss CGP, Singh P, Siva R. 2017. Molecular modeling and dynamic simulation of *Arabidopsis thaliana* carotenoid cleavage dioxygenase gene: A comparison with *Bixa orellana* and *Crocus sativus*. *Journal of Cellular Biochemistry* 118: 2712–2721.

Rabadán A, Gallardo-Guerrero L, Gandul-Rojas B, Álvarez-Ortí M, Pardo JE. 2018 Oct 30. Effect of roasting conditions on pigment composition and some quality parameters of pistachio oil. *Food Chemistry* 264: 49–57.

Rao AV, Rao AV. 2018. Chapter 6: Lycopene, tomatoes, and bone health, p. 129. In *Lycopene and Tomatoes in Human Nutrition and Health*, Taylor & Francis Group, Boca Raton, FL.

Rao AV, Young GL, Rao LG, editors. ; 2018 Mar 22. *Lycopene and tomatoes in human nutrition and health.* CRC Press (Vol 6 129).

Reis LC, Facco EM, Flôres SH, de Oliveira Rios A. 2018 Apr 1. Stability of functional compounds and antioxidant activity of fresh and pasteurized orange passion fruit (Passifloracaerulea) during cold storage. *Food Research International* 106: 481–486.

Repolho T, Duarte B, Dionísio G, Paula JR, Lopes AR, Rosa IC, Grilo TF, Caçador I, Calado R, Rosa R. 2017. Seagrass ecophysiological performance under ocean warming and acidification. *Scientific Reports* 7: 41443.

Rimkiene S, Ragazinskiene O, Savickiene N. 2003. The cumulation of Wild pansy (*Viola tricolor* L.) accessions: The possibility of species preservation and usage in medicine. *Medicina (Kaunas)* 39: 411–416.

Rodrigues DB, Mercadante AZ, Mariutti LRB. 2018. Marigold carotenoids: Much more than lutein esters. *Food Research International* 119: 653–664.

Rodriguez-Amaya DB. 2018. Update on natural food pigments – A mini-review on carotenoids, anthocyanins, and betalains. *Food Research International* 124: 200–205.

Rosalie R, Léchaudel M, Dhuique-Mayer C, Dufossé L, Joas J. 2018 May 1. Antioxidant and enzymatic responses to oxidative stress induced by cold temperature storage and ripening in mango (Mangiferaindica L. cv.'Cogshall') in relation to carotenoid content. *Journal of Plant Physiology* 224: 75–85.

Ruban AV, Phillip D, Young AJ, Horton P. 1998. Excited-state energy level does not determine the differential effect of violaxanthin and zeaxanthin on chlorophyll fluorescence quenching in the isolated light-harvesting complex of photosystem II. *Photochemistry and Photobiology* 68: 829–834.

Sahin MS, Khazi MI, Demirel Z, Dalay MC. 2019 Jan 1. Variation in growth, fucoxanthin, fatty acids profile and lipid content of marine diatoms Nitzschia sp. and Nanofrustulumshiloi in response to nitrogen and iron. *Biocatalysis and Agricultural Biotechnology* 17: 390–398.

Saini RK, Moon SH, Gansukh E, Keum YS. 2018 May 24. An efficient one-step scheme for the purification of major xanthophyll carotenoids from lettuce, and assessment of their comparative anticancer potential. *Food Chemistry.*

Salehi B, Sharifi-Rad R, Sharopov F, Namiesnik J, Roointan A, Kamle M, Kumar P, Martins N, Sharifi-Rad J. 2019 Jan 25. Beneficial Effects and Potential Risks of Tomatoes Consumption for human health: An overview. *Nutrition.*

Sánchez-Gómez R, Torregrosa L, Zalacain A, Ojeda H, Bouckenooghe V, Schneider R, Alonso GL, Salinas MR. 2019 Feb 27. Behavior of glycosylated aroma precursors in Microvine fruits after guaiacol foliar application. *ScientiaHorticulturae* 246:e1–e8.

Sarialtin SY, Coban T. 2018. An overview on the role of macular xanthophylls in ocular diseases. *Records of Natural Products* 12: 107–120.

Sarkar SK. 2018. Harmful algal blooms (HABs). In *Marine Algal Bloom: Characteristics, Causes and Climate Change Impacts*, pp. 111–170. Springer, Singapore.

Sarma VV, Patil JS, Shankar D, Anil AC. 2019 Jan 1. Shallow convective mixing promotes massive Noctilucascintillans bloom in the northeastern Arabian Sea. *Marine Pollution Bulletin* 138: 428–436.

Sausscrdc R, Kampuss K. 2014. Composition of carotenoids in calendula (*Calendula officinalis* L.) flowers, p. 13. In *9th Baltic Conference on Food Science and Technology 'Food for Consumer Well-Being'*, Jelgava, Lativia.

Schex R, Lieb VM, Jiménez VM, Esquivel P, Schweiggert RM, Carle R, Steingass CB. 2018 Mar 1. HPLC-DAD-APCI/ESI-MSn analysis of carotenoids and α-tocopherol in Costa Rican Acrocomiaaculeata fruits of varying maturity stages. *Food Research International* 105: 645–653.

Shi X, Li L, Lin S. 2018 May 31. Circadian and irradiance effects on expression of antenna protein genes and pigment contents in dinoflagellateProrocentrumdonghaiense (Dinophycae). *Harmful Algae* 75: 27–34.

Shimode S, Miyata K, Araki M, Shindo K. 2018. Antioxidant activities of the antheraxanthin-related carotenoids, antheraxanthin, 9-*cis*-antheraxanthin, and mutatoxanthins. *Journal of Oleo Science* 67: 977–981.

Silveira Agostini-Costa T. 2018 May 26. Bioactive compounds and health benefits of some palm species traditionally used in Africa and the Americas–a review. *Journal of Ethnopharmacology.*

Silveira Agostini-Costa T, da Silva Gomes I, de Melo LA, Reifschneider FJ, da Costa Ribeiro CS. 2017 Apr 1. Carotenoid and total vitamin C content of peppers from selected Brazilian cultivars. *Journal of Food Composition and Analysis* 57: 73–79.

Singh KP, Sindhu S. 2015. Nutraceutical pigments sourced from flower crops – An appraisal. *Journal of Ornamental Horticulture* 18: 1–16.

Singh A, Ahmad S, Ahmad A. 2015. Green extraction methods and environmental applications of carotenoids-a review. *RSC Advances* 5: 62358–62393.

Singh B, Singh JP, Kaur A, Singh N. 2016. Bioactive compounds in banana and their associated health benefits – A review. *Food Chemistry* 206: 1–11.

Šircelj H. 2019 Jan 28. Evaluation of bioactive constituents in European bladdernut (Staphyleapinnata L.) seed kernels. *Journal of Food Composition and Analysis.*

Solovchenko A, Yahia EM, Chen C. 2019 Jan 1. Pigments. In: *Postharvest Physiology and Biochemistry of Fruits and Vegetables* (pp. 225–252). Woodhead Publishing.

Stamatakis K, Broussos PI, Panagiotopoulou A, Gast RJ, Pelecanou M, Papageorgiou GC. 2019 Jan 1. Light-adaptive state transitions in the Ross Sea haptophyte Phaeocystisantarctica and in dinoflagellate cells hosting kleptoplasts derived from it. *BiochimicaetBiophysicaActa (BBA)-Bioenergetics* 1860(1):102–110.

Sun T, Yuan H, Cao H, Yazdani M, Tadmor Y, Li L. 2018. Carotenoid metabolism in plants: The role of plastids. *Molecular Plant* 11: 58–74.

Sun P, Wong CC, Li Y, He Y, Mao X, Wu T, Ren Y, Chen F. 2019 Mar 30. A novel strategy for isolation and purification of fucoxanthinol and fucoxanthin from the diatom Nitzschialaevis. *Food Chemistry* 277: 566–572.

Takaichi S. 2011. Carotenoids in algae: Distributions, biosyntheses and functions. *Marine Drugs* 9: 1101–1118.

Talasila U, Shaik KB. 2015. Quality, spoilage and preservation of cashew apple juice: A review. *Journal of Food Science and Technology* 52: 54–62.

Tamaki S, Kato S, Shinomura T, Ishikawa T, Imaishi H. 2019 Jan 1. Physiological role of β-carotene monohydroxylase (CYP97H1) in carotenoid biosynthesis in Euglena gracilis. *Plant Science* 278: 80–87.

Tedesco S, Daniels S. 2019 Feb 14. Evaluation of inoculum acclimatation and biochemical seasonal variation for the production of renewable gaseous fuel from biorefinedLaminaria sp. waste streams. *Renewable Energy.*

Torregrosa-Crespo J, Montero Z, Fuentes J, Reig García-Galbis M, Garbayo I, Vílchez C, Martínez-Espinosa R. 2018. Exploring the valuable carotenoids for the large-scale production by marine microorganisms. *Marine Drugs* 16: 203.

Torres LR, Santana FC, Shinagawa FB, Mancini-Filho J. 2018 Jun 4. Bioactive compounds and functional potential of pequi (Caryocar spp.), a native Brazilian fruit: a review. *Grasas y Aceites* 69(2):257.

Turcsi E, Nagy V, Deli J. 2016. Study on the elution order of carotenoids on endcapped C18 and C30 reverse silica stationary phases. A review of the database. *Journal of Food Composition and Analysis* 47: 101–112.

Valadon L, Mummery RS. 1971. Carotenoids of compositae flowers. *Phytochemistry* 10: 2349–2353.

Wagner H, Bauer R, Melchart D, Xiao PG, Staudinger A. 2011. Pericarpium Citri Reticulatae — *Chenpi* Pericarpium Citri Reticulatae Viride — *Qingpi*. In: Wagner H., Bauer R, Melchart D, Xiao PG, Staudinger A (eds) *Chromatographic Fingerprint Analysis of Herbal Medicines*. Springer, Vienna. https://doi.org/10.1007/978-3-7091-0763-8_55

Wang X, Yamagishi M. 2019 Jan 3. Mechanisms suppressing carotenoid accumulation in flowers differ depending on the hybrid groups of lilies (Lilium spp.). *ScientiaHorticulturae* 243: 159–168.

Wang C, Kim J-H, Kim S-W. 2014. Synthetic biology and metabolic engineering for marine carotenoids: New opportunities and future prospects. *Marine Drugs* 12: 4810–4832.

Wang K, Tu W, Liu C, Rao Y, Gao Z, Yang C. 2017. 9-*Cis*-neoxanthin in LHCII regulates the binding of violaxanthin and xanthophyll cycle. *Plant Physiology* 174: 86–96.

Wei JN, Liu ZH, Zhao YP, Zhao LL, Xue TK, Lan QK. 2019 Feb 1. Phytochemical and Bioactive Profile of Coriandrumsativum L. *Food Chemistry*.

Wentworth M, Ruban AV, Horton P. 2000. Chlorophyll fluorescence quenching in isolated light harvesting complexes induced by zeaxanthin. *FEBS Letters* 471: 71–74.

Wibowo S, Grauwet T, Santiago JS, Tomic J, Vervoort L, Hendrickx M, Van Loey A. 2015. Quality changes of pasteurised orange juice during storage: A kinetic study of specific parameters and their relation to colour instability. *Food Chemistry* 187: 140–151.

Xia S, Gao B, Fu J, Xiong J, Zhang C. 2018 Dec 1. Production of fucoxanthin, chrysolaminarin, and eicosapentaenoic acid by Odontellaaurita under different nitrogen supply regimes. *Journal of Bioscience and Bioengineering* 126(6):723–729.

Yahia N, Marniche F, Bellache M, Benslimane W, Mouheb S. 2017. Faunal biodiversity associated to the tomato culture (*Lycopersicon esculentum* Mill., Tavera variety) in Metijda, pp. 1108–1113. In *VIII International Scientific Agriculture Symposium, 'Agrosym 2017'*, Jahorina, Bosnia and Herzegovina, October 2017. Book of Proceedings: Faculty of Agriculture, University of East Sarajevo.

Yamano N, Mizoguchi T, Fujii R. 2018 May 1. The pH-dependent photophysical properties of chlorophyll-c bound to the light-harvesting complex from a diatom, Chaetoceroscalcitrans. *Journal of Photochemistry and Photobiology A: Chemistry* 358: 379–385.

Yao D, Wang Y, Li Q, Ouyang X, Li Y, Wang C, Ding L, Hou L, Luo M, Xiao Y. 2018. Specific upregulation of a cotton phytoene synthase gene produces golden cottonseeds with enhanced provitamin A. *Scientific Reports* 8: 1348.

Zarekarizi A, Hoffmann L, Burritt D. 2018. Approaches for the sustainable production of fucoxanthin, a xanthophyll with potential health benefits. *Journal of Applied Phycology* 31: 281–299.

Zhang Z, Wei Q, Nie M, Jiang N, Liu C, Liu C, Li D, Xu L. 2018 Oct 1. Microstructure and bioaccessibility of different carotenoid species as affected by hot air drying: Study on carrot, sweet potato, yellow bell pepper and broccoli. *LWT* 96: 357–363.

Zhou J. 2012. Regulation of the carotenoid biosynthetic pathway in petals of California poppy (*Eschscholzia californica*). A thesis presented in partial fulfilment of the requirements for the degree of Master of Science in Plant Biology at Massey University, Palmerston North, New ZealandMassey University.

Zhu C, Bai C, Sanahuja G, Yuan D, Farré G, Naqvi S, Shi L, Capell T, Christou P. 2010. The regulation of carotenoid pigmentation in flowers. *Archives of Biochemistry and Biophysics* 504: 132–141.

8 Anticarcinogenic Effect of Apocarotenoids

Yair Cárdenas-Conejo and Edith E. Uresti-Rivera

CONTENTS

INTRODUCTION

Apocarotenoids are cleavage products derived from carotenoids. These compounds have essential functions in plants and mammals. The breakdown products of carotenoids are produced by enzymatic and non-enzymatic reactions (Hou et al., 2016). The enzymes that catalyze the oxidative cleavage in carotenoids at specific double bonds are named carotenoid cleavage dioxygenases (CCDs) (Harrison and Bugg, 2014). On the other hand, apocarotenoids generated by non-enzymatic reactions come from the oxidation of carotenoids by reactive oxygen species (ROS) (Havaux, 2014).

The most well-known apocarotenoids are those belonging to vitamin A (retinol, retinal, retinoic acid (RA)). Vitamin A is a cleavage product of β-carotene at the 15,15′-double bond by β-carotene oxygenase 1 (BCO1) enzyme (Wyss et al., 2000). Vitamin A is essential in several physiological processes in humans, and it regulates the expression of several genes through binding to transcription factors (Blomhoff and Blomhoff, 2006; Harrison and Quadro, 2018). In plants, apocarotenoids are responsible for the pleasant aromas and flavours of many flowers and fruits; they have the function to attract pollinators (McQuate and Peck, 2001). Some apocarotenoids from plants are industrially important, for instance, saffron from *Crocus sativus* and annatto from *Bixa orellana* are the most valuable (Walter and Strack, 2011). Physiologically, apocarotenoids worked as phytohormones (abscisic acid and strigolactone) and recently it has been suggested that they can play signalling roles (Hou et al., 2016).

A large number of apocarotenoids are reactive electrophile species (RES), such as β-cyclocitral and β-ionone. RES are electrophilic compounds containing α- and β-unsaturated carbonyl groups, able to react with nucleophilic compounds (Farmer and Davoine, 2007). RES reacts with nucleophilic thiols, so they can modify transcription factors and induce gene expression (Levonen et al., 2004). Some apocarotenoids such as crocin (apocarotenoid from saffron) and bixin (apocarotenoid found in annatto) can regulate gene expression; for instance, crocin, reduces the mRNA of lipogenesis-related regulators (Luoet al., 2019), and bixin activates as the nuclear factor erythroid 2-related factor 2 (Nrf2) signals and regulates the expression of the transforming growth factor-beta, matrix metallopeptidase 9 and fibronectin (Zhang et al., 2018a).

Several evidence lines have been proved that apocarotenoids have anticancer activity; for example, the apocarotenoid all-trans-retinoic acid (ATRA), an active metabolite of vitamin A, has anticancer effects, including apoptosis and differentiation of cancer cells (Siddikuzzaman and Berlin Grace, 2011). This apocarotenoid, under the trade name of Vesanoid, is used as a drug in the treatment of acute promyelocytic leukaemia (Degos and Wang, 2001). Most of the anticarcinogenic properties of apocarotenoids are due to the fact that they can modulate the activity of transcription factors involved in cancer development and cancer progression, such as Nrf2, the nuclear factor κB (NF-κB), the retinoic acid receptor/retinoid X receptor (RAR/RXR), hypoxia-inducible factor-1 (HIF-1), signal transducers and activators of transcription (STATs), and oestrogen receptors (OR) (Table 8.1). Additionally, apocarotenoids enhance the gap junction intercellular communication (GJC) and modulate apoptosis (Table 8.1).

MODULATION OF THE NRF2 PATHWAY BY APOCAROTENOIDS

Apocarotenoids can interact with the Nrf2, which is involved in the redox homeostasis, these interactions allow apocarotenoids to perform prophylactic anticancer activities (Sharoni et al., 2012). The cell transcription factor Nrf2 controls over 200 genes, that protect cells from ROS and other electrophilic molecules, such as GSTA2 (glutathione S-transferase A2), NQO1 (NADPH: quinone oxidoreductase 1) and superoxide dismutases (SODs) (Nguyenet al., 2009; Jaramillo and Zhang, 2013). Compounds that modulate this system can help to prevent cancer (Lian and Wang, 2008) or may allow cancer progression and resistance to therapy (Wang et al., 2008) by inhibiting the endogenous generations of ROS.

Normally, in unstressed conditions, Nrf2 is located in the cytoplasm associated with Kelch-like ECH-associated protein 1 (Keap 1). The association Nrf2-Keap 1 maintains low protein levels in the cell since Keap 1 induces the Nrf2 ubiquitination and subsequent degradation. In oxidative stress or electrophilic stress conditions, RES or ROS react with cysteines 151, causing a conformational change in Keap 1, and the Nrf2 ubiquitination is inhibited. Therefore, Nrf2 accumulates in the cytosol and then translocates into the cell nucleus. In the nucleus, Nrf2 forms a protein complex with members of the masculoaponeurotic fibrosarcoma (Maf) protein family. The complex Maf:Nrf2 activated cell defence genes by the Nrf2 binding to the antioxidant response elements (ARE; 5′-RTGABNNNGCR-3′) (Figure 8.1). The signal is turned off when Keap 1 brings Nrf2 back into the cytosol (reviewed from (Jaramillo and Zhang, 2013; Rojo de la Vega et al., 2018a)).

Some apocarotenoids have been associated with the prevention of cancer by inhibiting ROS generation via Nrf2 activation in normal cells (Figure 8.1 and Table 8.1). For example, apocarotenoid apo-10′-lycopenoic acid, derived from lycopene, suppresses lung tumorigenesis in the A/J mouse model (Lian et al., 2007). The suppressing of tumorigenesis was associated with the activation of Nrf2 through interrupting Keap 1-mediated Nrf2 ubiquitination by aldehyde groups of apo-10′-lycopenoids (Lian and Wang, 2008; Linnewiel et al., 2009a). Apo-10′-lycopenoic acid increase by 50% Nrf2 in normal lung cells (BEAS-2B) and the expression of genes that codify detoxifying/antioxidant enzymes were upregulated, therefore ROS level was decreased (Lian and Wang 2008). Apparently, aldehyde groups of apo-10′-lycopenoids are involved in the interruption of Keap 1–Nrf2 association (Linnewiel et al., 2009b). Another example of cancer prevention by Nrf2 activation was described for bixin, orange pigment derived from lycopene. Activation of Nrf2 protects hairless mice (SKH-1) against UV radiation, which causes photodamage and carcinogenesis (Knatko et al., 2015). Bixin is able to increase levels of Nrf2 and Nrf2 target genes in primary human epidermal keratinocytes (HEKs) through interaction with Cys-151 residue in Keap 1 (Tao et al., 2015). Moreover, bixin suppresses skin photodamage induced by UV to mice treated with bixin through topical formulation (Rojo de la Vega et al., 2018b) or administrated via intraperitoneal (Tao et al., 2015). In both cases, the protection was related to the activation of cutaneous Nrf2 (Tao et al., 2015; Rojo de la Vega et al., 2018b).

TABLE 8.1
Anticarcinogenic effects of apocarotenoids on different cancers

Apocarotenoid	Mechanism	Cancer/model	Findings	References
apo-10′-lycopenoic acid	Activation of the Nrf2 system	Human Bronchial Epithelial Cell Line (BEAS-2B)	Nuclear Nrf2 protein was increased by 50% in BEAS-2B cells treated with 10 μM of apo-10′-lycopenoic acid after 3 h. Apo-10′-lycopenoic acid also increased the expression of ARE genes and inhibited the endogenous generations of ROS.	Lian and Wang (2008)
Bixin	Activation of the Nrf2 system	Hairless mice (SKH-1) and human cell HEKs	Bixin at 20 μM for 24 h increase levels of Nrf2 and Nrf2 target genes in HEKs through interaction with Cys-151 residue of Keap 1. Bixin protects skin against photodamage induced by UV to SKH-1 mice treated with topical formulation at 1% of bixin or administrated via intraperitoneal at 200 mg/kg.	Tao et al. (2015); Rojo de la Vega, Zhang et al. (2018b)
Lycopene oxidized derivatives by ethanolic extraction	Activation of the Nrf2 system	Liver cancer/ HepG2 cells	Lycopene oxidized derivatives at 5–25 μM induce the expression of ARE genes, such as NQO1, in a dose-dependent manner.	Ben-Dor et al. (2005)
Lycopene derivate, 10,10′-diapocarotene-10,10′-dial (10,10′)	Activation of Nrf2 system	Prostate cancer/ LNCaP cells. Breast cancer/ MCF-7 and T47D cells	10,10′ is a potent inducer of electrophile/antioxidant response elements (EpRE/ARE) activity at 10 μM. Activation was threefold higher in LNCaP and sevenfold higher in MCF-7. Apocarotenoids with aldehyde groups show better activation. Apocarotenoids at 10 μM attenuate breast and prostate cancer cell growth.	Linnewiel et al. (2009b)
Lycopene derivatives, 10,10'-diapocarotene-10,10′-dial (10,10′) 6,14′ diapocarotene-6,14′-dial (6,14′)	Inhibition of NF-κB	Breast cancer/ T47D cells	Lycopene derivates inhibited the induction of NF-κB via. 6.7 μM of 10,10′ reduce 52% ± 0.03 the NF-κB activation. 15 μM of 6,14′ reduce 90% the NF-κB activation. Apocarotenoids probably interfere with thiol groups of both IKK and p65.	Linnewiel-Hermoni et al. (2014)
apo-10-lycopenal	Inhibition of NF-κB	Prostate cancer/ LNCaP cells	apo-10-lycopenal at 1 μM significantly ($P < 0.05$, *t*-tests) reduced cell growth in LNCaP cells, but neither affected cell growth in DU145, C4-2 or PC3 cells. Relative quantities of NF-κB also were reduced in the nucleus ($P < 0.05$). apo-10-lycopenal increased expression of the BCO2 gene, which reduced NF-κB activity and decreased NF-κB nuclear translocation and DNA binding. The NF-κB blocking of BCO2 was independent of its enzymatic role.	Gong et al. (2016)

(*continued*)

TABLE 8.1 (CONTINUED)
Anticarcinogenic effects of apocarotenoids on different cancers

Apocarotenoid	Mechanism	Cancer/model	Findings	References
Crocin	Inhibition of NF-κB / Activation of Nrf2	Colon cancer/ICR mice	Crocin at 50 ppm, 100 ppm, and 200 ppm reduce the development of colonic adenocarcinomas ($P < 0.01$ at 200 ppm, *Bonferroni test*). Crocin at 200 ppm, reduce the NF-κB protein and mRNA. Crocin increases the mRNA of the Nrf2 expression and decreases the mRNA of inducible inflammatory enzymes (TNF-α IL-1β, IL-6, INF-γ, NF-κB, COX-2, and iNOS) in colorectal mucose ($P < 0.05$, *Kruskal-Wallis*).	Kawabata et al. (2012)
Apocarotenoids from saffron	Inhibition of NF-κB/ Proapoptotic	Liver cancer/Rats and HepG2 cells	In rats, saffron at doses of 300 mg/kg/day reduced the incidence of hepatic dyschromatic nodules induced by the inductor of liver cancer diethylnitrosamine (DEN). Saffron significantly ($P < 0.001$, ANOVA) inhibited the DEN-mediated elevations in numbers of cells positive for Ki-67, COX-2, iNOS, NF-κB p-65, and phosphorylated TNF receptor (TNFR). In HepG2 cells, saffron inhibited NF-κB p-65 and increased cleavage of caspase 3. Saffron increased the number of cells positive for TUNEL and M30 CytoDeath.	Amin et al. (2011)
Crocetin	Inhibition of NF-κB	Cervical cancer/ Kunming strain mice and HeLa cells	In Kunming strain mice treated with Methylcholanthrene, uterine cervical cancer inducer, crocetin significantly ($P < 0.05$, ANOVA) reduced polymorphonuclear cells activity, levels of cytokines (IL-1β and TNF-α) and levels of pro-inflammatory protein iNOS at 40 mg/kg. In HeLa cells, crocetin at 50 and 100 µmol/L significantly ($P < 0.05$, ANOVA) inhibited COX-2 production and expression.	Chen, Hou et al. (2015a)
Apo-10′-lycopenoic acid	Activation of RAR	NHBE, BEAS-2B and A549 cells	Apo-10'-lycopenoic acid to transactivate the promoter of the RARβ gene and to induce the expression of this receptor. The treatment with apo-10′-lycopenoic acid increased the mRNA levels of RARβ in NHBE, BEAS-2B and A549 cells.	Lian et al. (2007)

(continued)

TABLE 8.1 (CONTINUED)
Anticarcinogenic effects of apocarotenoids on different cancers

Apocarotenoid	Mechanism	Cancer/model	Findings	References
beta-apo-14′-carotenoic acid (14′CA)	Transactivation of RAR/ inhibition of cell growth	NHBE cells	14′CA inhibited the growth of NHBE cells (P < 0.05, ANOVA). The level of RARβ, a tumour suppressor, was reduced by 50% in the NHBE cells treated with BP (a primary lung carcinogen from cigarette smoke). However, treatment with 14′CA prevented the reduction of RARβ by BP. Furthermore, 14′CA transactivated the RARβ promoter primarily via its conversion to RA.	Prakash et al. (2004)
β-apo-13-carotenone	Antagonism of RXRα	Cos-1 cells	β-apo-13-carotenone was found to antagonize the activation of RXRα by 9-cis-RA and was effective at concentrations as low as 1nM. Molecular modelling studies revealed that β-apo-13-carotenone makes molecular interactions like an antagonist of RXRα.	Eroglu et al. (2010)
β-apo-13-carotenone	Antagonism of RXRα	Cos-7 cells Breast cancer/ MCF-7 cells	β-apo-13-carotenone inhibited 9cRA-induced transactivation of full-length RXRα. β-Apo-13-carotenone antagonizes RXRα through the induction of 'transcriptionally silent' RXRα tetramer.	Sun et al. (2014)
β-ionone	Inhibition of cell growth/Induction of apoptosis	Human gastric adenocarcinoma/ SGC-7901 cells	The growth of SGC-7901 cells was inhibited by β-ionone. Seven days after treatment with β-ionone at four concentrations, the inhibition rates were 12.04%, 30.59%, 78.25% and 94.15%, respectively. β-ionone also induced apoptosis of SGC-7901 cells.	Liu, Chen, et al. (2004a)
β-ionone	Inhibition of cell proliferation	Human gastric adenocarcinoma/ SGC-7901 cells	β-ionone had an inhibitory effect on the growth of SGC-7901 cells. Eight days after the treatment with β-ionone at concentrations of 25, 50, 100 and 200 μmol/L, the inhibition rates were 25.9%, 28.2%, 74.4% and 90.1%, respectively.	Liu, Yang et al. (2004b)
β-ionone	Induction of apoptosis/ Inhibition of PI3K and p-AKT	Human gastric adenocarcinoma/ SGC-7901 cells	β-ionone-induced apoptosis in a dose-dependent manner. β-ionone was also shown to induce the expression of cleaved-caspase 3 and inhibit Bcl-2 expression in SGC-7901 cells. The significantly decreased levels of p-PI3K and p-AKT expression were observed in SGC-7901 cells after β-ionone treatments in a time- and dose-dependent manner (P < 0.01, *t-test*).	Liu et al. (2013)

(continued)

TABLE 8.1 (CONTINUED)
Anticarcinogenic effects of apocarotenoids on different cancers

Apocarotenoid	Mechanism	Cancer/model	Findings	References
Bixin and norbixin	Inhibition of cell viability/ Induction of apoptosis	Melanoma/B16 murine melanoma cells	The apocarotenoids inhibited the cell viability in a time- and concentration-dependent manner. Also, activated caspase 9 and caspase 3 activities and It decreased and increased the gene expression of transcription factor, antiapoptotic Bcl-2 and proapoptotic Bax, respectively.	Anantharaman et al. (2016)
Bixin	Inhibition of growth/Induction of apoptosis and reduced cell migration	Melanoma/human melanoma cells (A2058)	Bixin promoted growth inhibition, reduced cell migration, induced apoptosis and cell cycle arrest in the G2/M phase. When associated with dacarbazine, bixin restored the sensitivity of A2058 cells to chemotherapy, enhancing its antiproliferative, anti-migratory and proapoptotic effects.	de Oliveira Júnior et al. (2019)
Crocin	Inhibition of cell proliferation/ Induction of apoptosis	Leukaemia/HL-60 cells	Crocin (0.625–5 mg/mL) inhibited HL-60 cell proliferation and induced apoptosis and cell cycle arrest at G0/G1 phase, in a concentration-and-time-dependent manner. In addition, crocin (6.25, 25 mg/kg) inhibited the tumour weight and size of HL-60 xenografts in nude mice, inhibited Bcl-2 expression, and increased Bax expression in xenografts.	Sun et al. (2013)
Crocin	Induction of apoptosis/ Inhibition of growth	Carcinoma of bladder/T24 cells	T24 cells were treated with crocin and the growth was remarkably inhibited. Flow cytometric profiles revealed that crocin led to the increase of the cells in G0/G1 phase, the percentage of cell apoptosis was also increased. Crocin could inhibit the growth of BALB/c xenograft tumour. Bcl-2, Cyclin D1 and survivin expressions determined by immunohistochemical staining were downregulated after treatment with Bax expression upregulated.	Zhao et al. (2008)
Crocin	Induction of apoptosis/ Decreasing cell growth	Breast cancer/ MCF-7 cells	Crocin decreased the growth of MCF-7 cells in a dose-and-time-dependent manner. The apoptosis rate was significantly higher when combined with Doxorubicin, as a chemotherapy drug, and X-ray or gamma radiation. The caspase-7 and p53 expression was increased.	Fanayi et al. (2016)

(*continued*)

TABLE 8.1 (CONTINUED)
Anticarcinogenic effects of apocarotenoids on different cancers

Apocarotenoid	Mechanism	Cancer/model	Findings	References
Crocin	Induction of apoptosis	Lung adenocarcinoma/ A549 and SPC-A1 cells	Crocin inhibited cell proliferation and induced apoptosis in A549 and SPC-A1 cells in a concentration-dependent manner, accompanied with an increase of G0/G1 arrest. Crocin significantly increased the mRNA levels of both p53 and Bax, while decreasing Bcl-2 mRNA expressions.	Chen, Zhao et al. (2015b)
Safranal	Induction of apoptosis/ Inhibition of cell growth	Neuroblastoma/ Mouse neuroblastoma N2A cell line	Safranal inhibited the growth of malignant cells in a dose-and-time-dependent manner. The IC (50) values against the neuroblastoma cell line were determined as 11.1 and 23.3 μg/mL after 24 and 48 h, respectively. Safranal induced a sub-G1 peak in the flow cytometry histogram of treated cells compared to control cells indicating that apoptotic cell death is involved in safranal toxicity.	Samarghandian et al. (2014)
Safranal	Induction of apoptosis/ Suppression of proliferation	Colon cancer/ colo-205 cells	Safranal suppressed the proliferation of colo-205 cancer cells with an IC50 of 20 μM. Furthermore, the anticancer effects of safranal were found to be due to accretion of ROS and decrease in the MMP, ultimately leading to apoptosis of colo-205 cancer cells. In addition, safranal caused increase in the expression of Bax in parallel with concomitant reduction in the expression of Bcl-2. Safranal also triggered G2/M cell cycle arrest and inhibition of PI3K/AKT/mTOR signalling pathway.	Zhang, Zhao et al. (2018b)
Crocetin	Induction of apoptosis/ Inhibition of proliferation	Leukaemia/ Primary APL cells	Crocetin (100 μM), significantly inhibited proliferation and induced apoptosis in primary APL cells. The effect was associated with the decreased expressions of prosurvival genes Akt and Bcl2, the multidrug resistance (MDR) proteins, ABCB1 and ABCC1 and the inhibition of tyrosyl-DNA phosphodiesterase 1 (TDP1), while the expressions of proapoptotic genes CASP3, CASP9, and BAX/BCL2 ratio were significantly increased.	Moradzadeh et al. (2018)

(*continued*)

TABLE 8.1 (CONTINUED)
Anticarcinogenic effects of apocarotenoids on different cancers

Apocarotenoid	Mechanism	Cancer/model	Findings	References
Crocetin	Induction of apoptosis/ Inhibition of proliferation	Pancreatic cancer/ MIA-PaCa-2 cells/Athymic nude mice	Crocetin inhibited proliferation of MIA-PaCa-2. Both the *in vitro* pancreatic cancer cells and *in vivo* athymic nude mice tumour, apoptosis was significantly stimulated as indicated by Bax/Bcl-2 ratio.	Dhar et al. (2009b)
ATRA	Enhanced GJC	Embryonal carcinoma/ Mouse F9 teratocarcinoma cells	ATRA at 1 μM increased GJC above the control, parallelly the Cx43 protein also increased. Cx43 mRNA levels didn't change.	Clairmont et al. (1996)
ATRA	Enhanced GJC	Glial tumour/Rat C6 glioma cells	ATRA at 1, 10, 100 μM increased the GJC function. ATRA didn't induce the transcription of Cx43 gene.	Zhang et al. (2002)
ATRA	Enhanced GJC	Oral cancer/Two cell lines (SCC9 and Tca8113) of oral squamous cell carcinoma (OSCC)	ATRA at 0.1–100 μM for 48 h significantly ($P > 0.05$, ANOVA) induced cell growth inhibition of SCC9 and Tca8113 cells in a dose-and-time-dependent manner. ATRA also enhanced GJC 2.69-fold in SCC9 cells ($P = 0.019$, ANOVA) and 2.06-fold in Tca8113 cells ($P = 0.034$, ANOVA). Enhanced the GJC was correlated with the induction of transcription Cx32, Cx43 and Cx45 genes by ATRA. Levels of Cx32 and Cx43 proteins also were increased by ATRA.	Wang et al. (2013)
Lycopene derivate, acyclo-retinoic acid (ARA)	Enhanced GJC	Human fetal skin fibroblast cells (HFFF2)	ARA at a concentration between 1 and 50 μM enhanced GJC in HFFF2 cells. ARA is less active than ATRA concerning GJC.	Stahl et al. (2000)
Canthaxanthin derivates, all-trans- and 13-cis-4-oxoretinoic	Enhanced GJC	Mouse embryo fibroblast cells C3H/10T1/2	Apocarotenoids derivates from canthaxanthin stimulated GJC function at 10 and 0.1 μM in C3H/10T1/2 cells. Apocarotenoids also increased mRNA of the Cx43 gene (4.7-fold) compared with control.	Hanusch et al. (1995)
Lycopene derivate, dialdehyde 2,7,11-trimethyl-tetradecahexaene-1,14-dial	Enhanced GJC	Rat liver epithelial cells (WB-F344)	The compound stimulates GJC in WB-F344, the GJC was significantly ($P < 0.001$) increased up to 140% at 0.1 μM.	Aust et al. (2003)

(continued)

TABLE 8.1 (CONTINUED)
Anticarcinogenic effects of apocarotenoids on different cancers

Apocarotenoid	Mechanism	Cancer/model	Findings	References
Crocin	Inhibition of HIF-1	Human gastric cancer/AGS and HGC-27 cell lines	Crocin at 2 or 3 μM suppresses the migration, invasion, and epithelial-mesenchymal transition. Crocin at 2 or 3 μM significantly ($P < 0.05$, *t*-test) reduced the gene expression of HIF-1 and KLF5 (transactivator of HIF-1). The reduction of HIF-1 and KLF5 mRNA levels was associated with the upregulation of miR320; this microRNA recognized KLF5.	Zhou et al. (2019)
10,10′ and 6,14′	Inhibition of the oestrogenic activity	Breast cancer/ T47D cells	Apocarotenoids at 2–4.5 μM significantly ($P < 0.05$, ANOVA) repressed oestrogenic activity. The reduction correlated to low OR protein levels in the nucleus of cells treated with oestradiol and 10,10′ and 6,14′.	Veprik et al. (2012)
Picrocrocin	Inhibition of the STAT-5 phosphorylation/ ROS/apoptosis	Melanoma/ SK-MEL-2 cells	Picrocrocin inhibited the cellular growth of SK-MEL-2 melanoma cells in a dose-dependent manner. Picrocrocin at 20 μM induce apoptosis through the increased protein Bax/ Bcl-2 ratio. Picrocrocin at 40 μM increased ROS levels up to 205%. Picrocrocin at 20 and 40 μM avoid the phosphorylation of STAT-5.	Yu et al. (2018)
Crocin	Inhibition of the STAT-3 phosphorylation /Apoptosis	Hepatocellular carcinoma/HepG2 and Hep3B cells	Crocin at 20 μM inhibited IL-6-induced phosphorylation of STAT3 in several cancer cells (HepG2, Hep3B, HCT116, MDA-MB-231, BxPC3, A549 and A2780). Crocin inhibited the DNA binding of STAT-3 in HepG2 and Hep3B cells. Also, crocin inhibited the JAK1, JAK2 and Src kinase phosphorylation. Crocin at 20 μM induces the protein levels of BAX and decreased antiapoptotic Bcl-2 (Bcl-2, survivin), proliferative (cyclin D1), invasive (CXCR4) and angiogenic (VEGF) proteins. Therefore Caspase 3 and 9 were activated.	Kim and Park (2018)

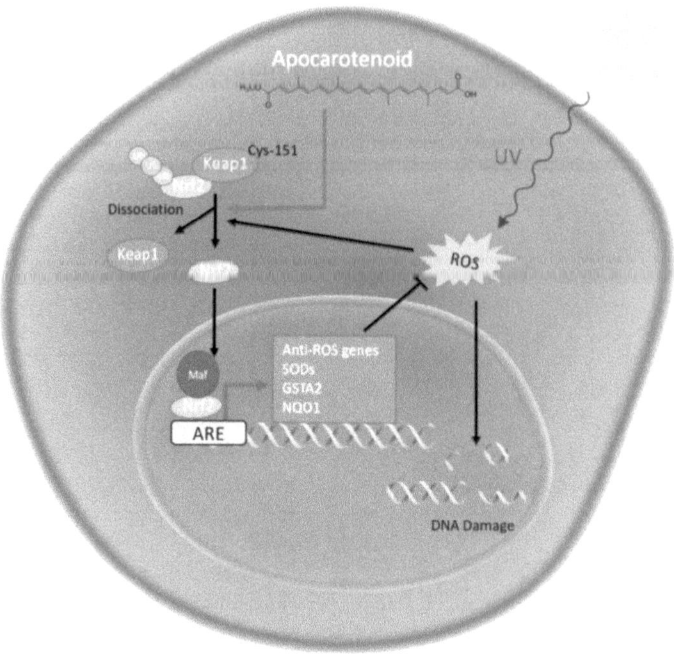

FIGURE 8.1 Activation of the Nrf2-Keap1 system by apocarotenoids.

Apocarotenoids can prevent carcinogenesis, promoting the Nrf2 activation in healthy cells (Table 8.1), however, the accumulation of Nrf2 in cancer cells promotes their survival against oxidative stress and cancer treatments (Wang et al., 2008; Rojo de la Vega et al., 2018a). According to this idea, pancreatic neoplasia was less progressive in Nrf2 knockout mice (DeNicola et al., 2011). Thus, therapeutic strategies against cancer include blocking the Nrf2 system (Panieri and Saso, 2019). Interestingly, ATRA has shown the ability to reduce the expression of genes regulated by Nrf2 in MCF-7 cells (breast cancer cell line) (Wang et al., 2007). ATRA did not block the Nrf2 accumulation, but promoted a complex formation of Nrf2:RARα, which reduces the binding between Nrf2 and AREs (Wang et al., 2007).

MODULATION OF NF-κB PATHWAY BY APOCAROTENOIDS

NF-κB is a transcription factor family (p50/p105, p52/p100, p65/RelA, RelB, c-Rel, Dosal, Dif, and Relish), which form homo- or hetero-dimeric complexes (Albensi 2019). NF-κB modulates genes related to several biological signals, such as inflammation, apoptosis, proliferation, innate immune response and tumourigenesis (Serasanambati and Chilakapati, 2016; Albensi, 2019). NF-κB also regulates the transcription of antioxidant and pro-oxidant genes (Morgan and Liu, 2011; Lingappan, 2018).

The NF-κB transcription system operates similarly to the Nrf2 system. In a summarized way, in an inactive state, NF-κB is located in the cytosol due to the association with cytoplasmic inhibitory κB (IκB). IkB contains six ANKYRIN repeats that are necessary for interaction with the Rel homology domain (RHD) of NF-κB. A nuclear-localization sequence (NLS) is located in RHD, therefore IkB masks the NLS of NF-κB, which prevents the entry of NF-κB into the nucleus. In response to the tumour necrosis factor-alpha (TNF-α), interleukin 1 (IL-1) or stress stimuli (free radicals, bacterial antigens, viral antigens, etc.), through several intermediary steps, I-κB proteins are phosphorylated by IκB kinases (IKK), promoting degradation of I-κB via ubiquitination. Therefore the cytoplasmic NF-κB:IκB complex is disrupted, which allows NF-κB to translocate and accumulate

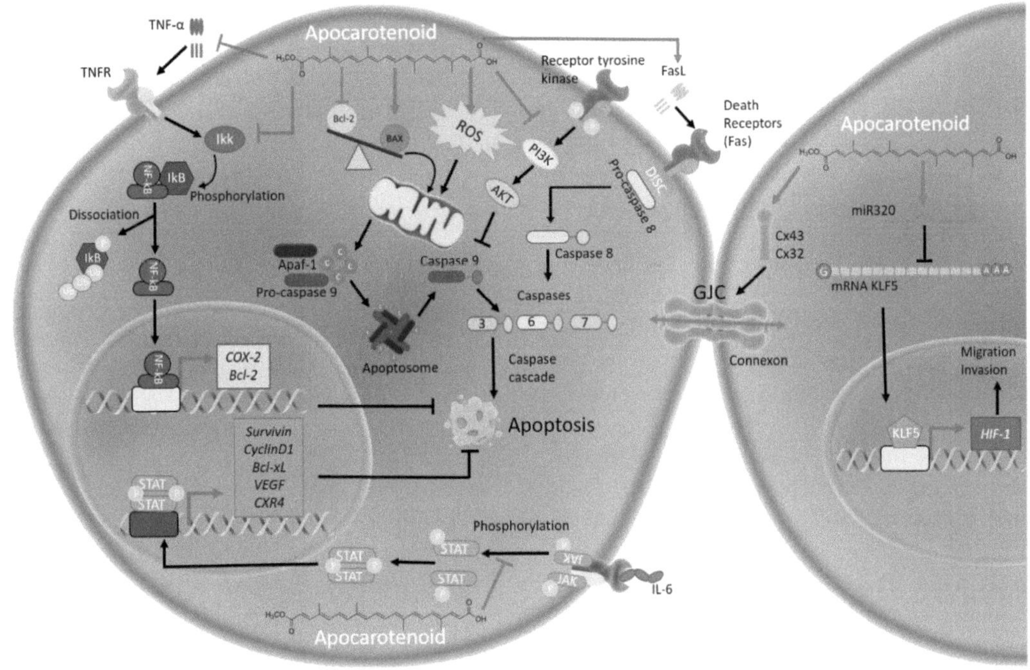

FIGURE 8.2 Model depicting the anticarcinogenic effects of apocarotenoids on diverse signalling pathways (see text for details).

in the nucleus where it binds to the promoter leading the activation of genes (Figure 8.2) (reviewed from Albensi, 2019; Karin et al., 2004).

Alteration in the NF-κB system could derive in cancer; for example, constitutive NF-κB activation could induce chronic inflammation, processes linked to cancer development (Karin, 2009; Xia et al., 2014). The suppression of NF-κB leads to tumour regression in some tumour cells (Xia et al., 2014). Thus, molecules with the ability to inactivate the NF-κB could be used as therapeutic drugs. Several lines of evidence suggest that apocarotenoids can suppress the NF-κB pathway (Table 8.1).

Bixin is able to block NF-κB pathway, but this ability has not been related to anticancer activity (Xu and Kong, 2017; Xue et al., 2018). However, other lycopene derivatives with the ability to modulate NF-κB pathways have shown anticancer activity (Linnewiel-Hermoni et al., 2014; Gong et al., 2016). Electrophilic lycopene derivatives 10,10′-diapocarotene-10,10′-dial (10,10′) and 6,14′-diapocarotene-6,14′-dial (6,14′) inhibited the induction of NF-κB via TNF-α (Linnewiel-Hermoni et al., 2014). The NF-κB activation was reduced by 52% in T47D cells (breast cancer cell line) using 10,10′, whereas the reduction caused by 6,14′ was higher (90%); probably apocarotenoids interfered with thiol groups of both IKK and NF-κB (p65) (Linnewiel-Hermoni et al., 2014).

On the other hand, apocarotenoids from saffron, also inhibitors of NF-κB, have been associated with anticancer properties (Amin et al., 2011; Kawabata et al., 2012; Chen et al., 2015a; Hashemi and Hosseinzadeh, 2019). For example, crocin protects mice against induced colorectal cancer (CRC) (Kawabata et al., 2012). Mice treated with azoxymethane (AOM) exposed to the inflammatory agent dextran sodium sulphate (DSS) dramatically develop adenocarcinomas through inflammatory processes. Diets containing crocin at 50 ppm, 100 ppm, and 200 ppm, supplied for one month to ICR mice, reduce the development of colonic adenocarcinomas induced by AOM/DSS. The protection provided for crocin was related to the modulation of the NF-κB and Nrf2 pathways. Crocin at 200 ppm suppressed the NF-κB protein and enhanced the expression of Nrf2 at 200 ppm crocin in the adenocarcinomas developed in the colons of the mice. Also, the expression of Nrf2 and NF-κB genes were modulated at 200 ppm crocin, the expression of Nrf2 was upregulated, whereas NF-κB was downregulated. Downregulation of NFk-B probably leads the downregulation of inflammatory enzymes, such as TNF-α, IL-1β, IL-6, interferon γ (IFN-γ), NF-κB, cyclooxygen ase 2 (COX-2) and inducible nitric oxide synthase (iNOS) (Kawabata et al., 2012). These examples confirm the anticancer activity of apocarotenoids via modulation of NF-κB.

APOPTOSIS INDUCTION IN CANCER CELLS BY APOCAROTENOIDS

Apoptosis is a self directed process that involves elegant molecular mechanisms leading to programmed cell death in order to maintain a stable internal environment in organisms (Nagata, 2018). Apoptosis is mediated by specific proteins named 'caspases' that belong to cysteine proteases family. They act in cascades in two main pathways: the extrinsic and the intrinsic pathway. Caspases are classified as initiator caspases, such as 2, 8, 9, and 10; and effector caspases such as 3, 6, and 7. It is well known that caspases 8 and 10 are important in the extrinsic pathway, and caspase 9 is vital in the intrinsic pathway (Chen et al., 2018; Nagata, 2018). The extrinsic pathway is triggered through the death receptor ligands closely related to the tumour necrosis factor (TNF) located on the cell surface, that oligomerizes with death receptors and combines with procaspases 8 and 10 to form the death-inducing signalling complex (DISC); which initiates the transducing downstream apoptosis signal (Pan et al., 2016). The intrinsic pathway of apoptosis, also known as the mitochondrial pathway, is regulated by Bcl-2 family, which includes members of proapoptotic effector molecules (Bax and Bak) and antiapoptotic proteins (Bcl-2 and Bcl-XL). The intrinsic pathway is triggered when mitochondria release cytochrome C into the cytosol through mitochondrial apoptosis-induced channel (MAC). Then the complex apoptosome is formed through the interaction of cytochrome C with the apoptotic protease-activating factor (Apaf-1). This complex activates procaspase 9, which in turn activates the downstream caspases (Chen et al., 2018; Nagata, 2018).

Noticeably, apocarotenoids can have apoptotic effects in cancer cells through the interaction at different levels of apoptosis intrinsic and extrinsic pathways (Table 8.1 and Figure 8.2). For example, β-ionone, an apocarotenoid enriched in many fruits, vegetables and grains, inhibit cell growth, proliferation, and induces apoptosis of SGC-7901, a human gastric adenocarcinoma cell line (Liu et al., 2004a; Liu et al., 2004b). The mechanism by which β-ionone initiates apoptosis *in vitro* is through the induction of the expression of cleaved-caspase 3 and inhibition of Bcl-2 expression. Additionally, authors suggest that β-ionone can regulate apoptosis through a PI3K-AKT pathway (Liu et al., 2013). Another example of anti-cancerogenic effect mediated by induction of apoptosis was reported for bixin and norbixin, apocarotenoids that are food colourants from *Bixa orellana* seeds. Aggressive B16 murine melanoma cells were treated with the apocarotenoids and showed work as activators of caspase 3 and 9 activities, and were also shown be able to interact hydrophobicly with the antiapoptotic proteins Bcl-2 and Bcl-XL, with resultant inhibition of these proteins. Bixin and norbixin decreased, and increased the gene expression of antiapoptotic Bcl-2 and proapoptotic Bax respectively (Anantharaman et al., 2016). These results are supported by other studies made in human melanoma cells (A2058) where bixin, in combination with dacarbazine treatment, an anticancer agent, proved to be more effective to induce the increase of caspase 3 activity (de Oliveira Júnior et al., 2019). These findings suggest that bixin and norbixin induce apoptosis by both intrinsic and extrinsic apoptosis pathways.

Safranal, crocetin and crocin are other important apocarotenoids derived from dried stigmas *of Crocus sativus* L. (saffron), widely used by their several biological activities. For instance, anticancer effects have been shown for safranal, since it inhibits the proliferation of malignant neuroblastoma N2A cell lines through the involvement of apoptosis (Samarghandian et al., 2014). Likewise, treatment of human colon cancer cells colo-205 with safranal triggered an increase in the expression of Bax, associated with an important reduction in the expression of Bcl-2 in a concentration-dependent manner. Additionally, safranal inhibited the PI3K/AKT/mTOR signalling pathway. These findings indicate that safranal-induced apoptosis follows the mitochondrial pathway. On the other hand, the effects of crocetin have been studied; for example, it was reported that the treatment with crocetin inhibited proliferation and induced apoptosis in primary acute promyelocytic leukaemia (APL) cells. These effects were concomitant with the reduction of the expression of Akt and Bcl-2 prosurvival genes (Moradzadeh et al., 2018). In other model of the human pancreatic adenocarcinoma cell line, MIA-PaCa-2, treatment with crocetin provoked an augmented Bax expression and an important decrease in the expression of Bcl-2 protein, which resulted in the 4–5-fold enhancement of the ratio of Bax/Bcl-2 (Dhar et al., 2009a).

Finally, crocin, which is considered to be one of the most effective cancer therapeutic constituents of saffron stigma, exhibiting apoptotic effects. The mechanism underlying the induction of apoptosis have been partially established through various studies that have been carried out to confirm the anticancer properties using different malignant cell lines such as leukaemia, colon carcinoma, bladder cancer, breast cancer, and lung adenocarcinoma (Zhao et al., 2008; Sun et al., 2013; Chen et al., 2015b; Ben Salem et al., 2016; Fanayi et al., 2016).

MODULATION OF RETINOID RECEPTORS BY APOCAROTENOIDS

Nuclear receptors (NRs) are ligand-activated transcription factors that are involved in complex gene regulatory networks, including cell differentiation, proliferation, metabolism, etc. Members of the NR superfamily include the family of retinoic acid receptors (RARs) and the retinoid X receptors (RXRs). Each family has three subtypes (α, β and γ) and it is well known that the RAR family heterodimerizes with the retinoid X receptor. Receptor heterodimers are attached to the retinoic acid response element (RARE) in the promoter region of RAR target genes, as part of complexes with histone deacetylases (HDACs) and other corepressors, forming a condensed chromatin structure (Figure 8.3a). The ligand binding, which in turn is either all-trans-retinoic acid (ATRA) or 9-cis-RA for the RARs and only 9-cis-RA for the RXRs (Heyman et al., 1992), results in corepressor releasing, chromatin derepression, and recruitment of coactivators such as histone acetyltransferases to form the activator complex that induces the transcriptional activation (Figure 8.3b) (Glass and Rosenfeld, 2000).

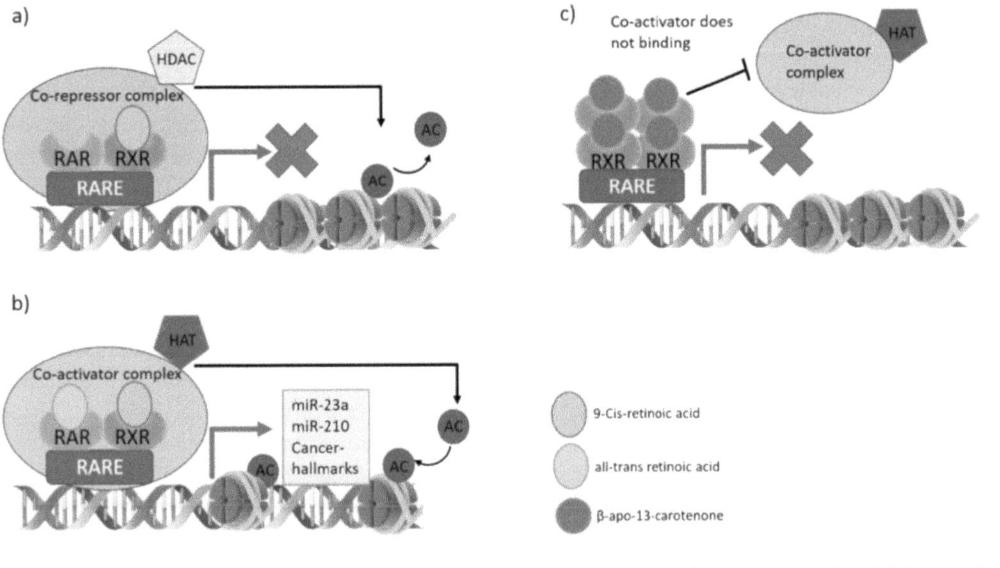

FIGURE 8.3 Schematic representation of modulation of retinoid receptors by apocarotenoids. (a) Repression gene transcription through chromatin condensation leader by the co-repressor complex and HDAC. (b) Transcriptional activation of target genes through chromatin decondensation leader by ATRA. (c) Blocking the binding of the activator complex as a result of the tetramerization of the RXRα induced by β-Apo-13-carotenone.

It is well known that RXR and RAR receptors are regulators of gene networks that control differentiation, cell growth, and apoptosis (Cunningham and Duester, 2015). Accumulating evidence suggests the involvement of RAR receptors in carcinogenesis, for instance, RARβ gene is found epigenetically silenced, and its expression is reduced in breast, cervical, and prostate cancer (Widschwendter et al., 2000; Ivanova et al., 2002; Dou et al., 2018); and some studies have reported that RARγ is downregulated in prostate cancer compared with other cancers (Cancer Genome Atlas Research Network, 2015). Similarly, when the *in situ* expression of nuclear retinoid receptors was analyzed in premalignant and malignant gastric tissues, the results showed a decrease in RARα expression tissues, suggesting an important role of RARα in the cancer gastric development (Jiang et al., 1999). In agreement with this, it was reported that the restoration of RARβ2 in a RARβ-negative lung cancer cell line inhibits tumorigenicity in nude mice (Houle et al., 1993).

It is therefore not surprizing that RAR and RXR receptors, as well as the ligands for these transcription factors and their agonists, have been used in experimental cancer therapies. Epidemiological studies indicate that a deficiency of vitamin A apparently increase the susceptibility to chemical carcinogens (Sun and Lotan, 2002). It has also been shown that vitamin A is a good chemopreventive, since the incidence of chemically induced experimental tumours decreased when treated with pharmacological doses of vitamin A (Niles, 2004). The most documented application of retinoids is in the first-line treatment of APL, where ATRA can overcome the block to cell differentiation (Cicconi and Lo-Coco, 2016). Studies showed that more than 90% of the patients observed a complete remission; therefore, these findings showed the effectiveness of ATRA in the treatment of APL. The acting mechanism of ATRA is not well understood; however, studies carried out in different cell lines of APL treated with ATRA showed an increased expression of the microRNAs miR-23a and miR-210. Further experiments performed in 293T cells have shown that RARα/RXRα heterodimers bind miR-210 and miR-23a/24-2 promoters (Saumet et al., 2009).

Apocarotenoids possess anticancer ability, which is exerted through the activation of several pathways, such as inhibition of cancer cell proliferation, activation of apoptosis programme, and the modulation activity of RAR and RXR receptors. For example, Apo-10′-lycopenoic acid, a derivative of lycopene, increased the mRNA levels of RARβ in NHBE, BEAS-2B and A549 cells, and transactivated the promoter of the RARβ gene inducing the transcriptional activity of this receptor. The authors suggest that the growth inhibitory effect on these cell lines was mediated by retinoid signalling (Lian et al., 2007). According to this mechanism, a study reported that healthy human bronchial epithelial cells treated with β-apo-14′-carotenoic acid (14′CA) showed increased levels of RARβ. The expression of this subtype receptor was downregulated by BP (benzo[a]pyrene) treatment, a primary lung carcinogen from cigarette smoke. Interestingly, treatment with 14′CA prevented the reduction of RARβ levels seen with BP insult (Prakash et al., 2004). The cell growth inhibition of β-apo-carotenoic acid is probably due to its conversion to RA and concomitant upregulation of RARβ; since the presence of 3-mercaptopropionic acid, an inhibitor of fatty acid oxidation, decreased both RA formation and transactivation activity from 14′CA.

In contrast, RXR overexpression is found in ~70% of ductal breast carcinomas (reviewed by Rodrigues Almeida, 2019) and is reported that RXRα is upregulated in a high-risk lesion for invasive breast cancer, ductal carcinoma *in situ* (Almeida and Conda-Sheridan, 2019). Increased expression of RXRα gene was observed in oesophageal carcinoma (EC) tissues, compared with normal oesophageal tissues. EC patients that presented RXRα high expression had significantly worse disease-free survival (Hong-Yue et al., 2018). Thus, therapeutic strategies may include RXRα as a target in cancer. In this sense, analysis carried out in Cos-1 cells showed that the apocarotenoid β-apo-13-carotenone was a potent antagonist of β-apo-13-carotenone in the activation of RXRα. Computational docking studies determined that β-apo-13-carotenone interacts with the ligand binding site of RXRα, acting as antagonist. Additional studies determined that β-Apo-13-carotenone induces the tetramerization of the RXRα, which blocks the binding of the activator complex, resulting in a 'transcriptionally silent' RXRα receptor (Figure 8.3c) (Eroglu et al., 2010; Sun et al., 2014). The antagonism of this apocarotenoid may have implications as a modulator of cancer.

APOCAROTENOIDS ENHANCED GAP JUNCTIONAL COMMUNICATION

Gap junction (GJ) refers to intercellular channels formed by hexameric assemblies of membrane proteins called connexins (Cx). The hexameric assembly of Cx leaves a narrow ~2 nm extracellular 'gap', which allows communication between neighbouring cells; a physiological phenomenon called gap junction communication (GJC). GJC permits the diffusion of metabolites, small hydrophilic molecules, and ions less than 1–2 kDa in size (Goodenough and Paul, 2009). GJC is considered crucial in the control of homeostasis, cell growth, and differentiation, so alteration in this intercellular communication has been associated with cancer (Aasen et al., 2016; Aasen et al., 2019). Several studies have demonstrated that the downregulation of Cx expression and loss of GJC are associated with human cancer (Cronier et al., 2009; Aasen et al., 2019). For instance, knockout mice for Cx32 or Cx43 are susceptible to developing lung and hepatic cancers (Moennikes et al., 2000; Evert et al., 2002; King and Lampe, 2004). High expression of Cx correlated with better prognosis in several cancers (Aasen et al., 2016) and restoration of Cx levels has therapeutic advantages in the early stages of cancer (Cronier et al., 2009).

The anticarcinogenic attribute of apocarotenoids also has been associated with the capacity to enhanced GJC (Table 8.1). Apocarotenoids enhanced the GJC via increased Cx protein levels (Clairmont et al., 1996; Wang et al., 2013). Cx43 and Cx32 have a vital role in cancer, and aberrant expression of these genes has been associated with cancer progression and metastasis (Moennikes et al., 2000; Evert et al., 2002; King and Lampe, 2004; Bonacquisti and Nguyen, 2019). ATRA at 1 μM enhanced the GJC in embryonal carcinoma cell line F9. Enhancement of the GJC was correlated with the increase in the amount of Cx43 protein (3-fold) (Clairmont et al., 1996). ATRA also significantly enhanced GJC in C6 glioma cells at 1, 10, and 100 μM (Zhang et al., 2002). Interestingly, the increased levels of Cx43 protein induced by ATRA did not correlate with an increase in Cx43 mRNA levels. It has been suggested that the elevation of Cx43 protein levels is related to a post-transcriptional effect; ATRA probably confers stability to the Cx43 mRNA (Clairmont and Sies, 1997). However, a more recent study showed that ATRA enhances GJC through increased protein and mRNA levels of Cx43 and Cx32 (Wang et al., 2013). Wang's study (2013) showed that ATRA is able to inhibit the growth of oral squamous cell carcinoma. This study suggests that the anti-tumour effect of ATRA is via restoring the GJC function. ATRA increased GJC 2.69-fold in SCC9 cells ($P = 0.019$, ANOVA; 15 μM) and 2.06-fold in Tca8113 cells ($P = 0.034$, ANOVA;10 μM). The enhancement of GCJ correlated with the induction of transcription of Cx32 and Cx43 by ATRA, which was reflected in increased levels of Cx32 and Cx43 proteins.

In addition to apocarotenoids derived from β-carotene, compounds resulting from lycopene and canthaxanthin also improved the GJC. From lycopene, the apocarotenoids acyclo-retinoic acid and dialdehyde 2,7,11-trimethyl-tetradecahexaene-1,14-dial were able to increase GJC in human fetal skin fibroblast cells (Stahl et al., 2000) and rat liver epithelial cells (Aust et al., 2003) respectively. From canthaxanthin, the apocarotenoids all-trans- and 13-cis-4-oxoretinoic acid stimulated GJC in mouse embryo fibroblast cells (Hanusch et al., 1995).

OTHER ANTICANCER MECHANISMS MODULATED BY APOCAROTENOIDS

In addition to anticancer activities of apocarotenoids described above, apocarotenoids can also fight cancer through the inhibition of the transcription hypoxia-inducible factor-1 (Zhou et al., 2019), oestrogen receptors (Veprik et al., 2012), and signal transducer and activator of transcription 5 (Table 8.1).

HIF-1 is a transcription factor that plays an integral role in the cell response to low oxygen concentrations; or hypoxia. HIF-1 activates over 100 genes involved in maintaining proper oxygen haemostasis. Solid tumours have inadequate O_2 supply since they compress and obstruct blood vessels. Then, tumour cells activate several survivors' pathways, such as Hif-1 pathway, which is the most critical pathway to overcome hypoxia. High levels of HIF-1 strongly correlate with tumour metastasis, resistance to therapy, and poor prognosis (reviewed in Masoud and Li 2015). Thus, HIF-1 is recognized as an essential cancer drug target. Recently it was shown that crocin could inhibit the HIF 1 pathway in an indirect manner in gastric cancer cells (Figure 8.2) (Zhou et al., 2019). Human gastric cancer cell lines AGS and HGC-27 treated with crocin at 3 μM induced the expression of microRNA-320 (miR-320). This microRNA is a tumour suppressor in gastric cancer, which reduces the mRNA and protein levels of Krüppel-like factor 5 (KLF5) (Zhou et al., 2019). KLF5 is identified as a transactivator of HIF-1 (Lee et al., 2013). High expression of miR-320 and low levels of KLF5 mRNA and protein led by crocin caused the inhibition of migration, invasion, and metastasis mechanism in gastric cancer cells (Zhou et al., 2019).

Oestrogen receptor (OR) is a transcription factors that is activated by the hormone oestrogen. This transcription factor is involved in several mechanisms, such as mammary gland development. When oestrogen is binding to OR, the complex OR-oestrogen translocate to the nucleus and binds to promoters of genes with oestrogen response element sequences (Saha Roy and Vadlamudi, 2012). OR is expressed in approximately 70% of breast cancer patients, hence, disrupts the function of OR is an effective therapy against breast cancer (Waks and Winer, 2019). In a human mammary cancer cell line (T47D cells), the apocarotenoids 10,10′ and 6,14′ significantly repressed the activation of the oestrogen response element (ORE) via reduction of OR protein levels in the nucleus of cells treated with oestradiol and apocarotenoids (Veprik et al., 2012).

Signal transducers and activators of transcription (STAT) are transcription factors involved in several pathways, such as apoptosis, cell differentiation and proliferation. Inactive STATs are located in the cytosol under basal condition; external signals as interleukin-2 or interleukin-6 induce the phosphorylation of Janus-activated kinases (JAKs), which recruits and activates via phosphorylation the STATs factors. Phosphorylation of STATs promotes the formation of dimers or tetramers, which are translocated to the nucleus to activate the transcription of target genes (Figure 8.2) (reviewed in: Mitchell and John, 2005; Able et al., 2017). Constitutive activation of STATs has been observed in a variety of cancer (Rani and Murphy, 2016). Picrocrocin, found in the spice saffron, showed that it could inhibit the cell proliferation of SK-MEL-2 melanoma cells in a dose-dependent manner. The inhibition was attributed to the induction of high ROS levels in the cell that trigger apoptosis. Additionally, the anticancer activity of picrocrocin was accredited to the inhibition of the STAT-5 phosphorylation (Yu et al., 2018).

Another saffron apocarotenoid, crocin, also was able to inhibit JAK/STAT pathway in hepatocellular carcinoma HepG2 and Hep3B cells (Kim and Park, 2018). Crocin at 20 μM inhibited the phosphorylation of JAK1 and 2; therefore, the activation of STAT3 was repressed, as well as its translocation to the nucleus. Crocin also enhanced the expression of SH2-containing tyrosine phosphatase (SHP-1), an enzyme with the ability to dephosphorylate STAT3. The inhibition of JAK/STAT pathway by crocin leads to an increased Bax/Bcl-2 ratio, and activation of caspase 9 and 3 (Kim and Park, 2018). Inhibition of STAT3 phosphorylation by crocin also was successful in other cancer cells (colon carcinoma (HCT116); breast cancer (MDA-MB-231); pancreatic cancer (BxPC3); lung adenocarcinoma (A549); ovarian cancer (A2780)) (Kim and Park, 2018).

CONCLUSION

Apocarotenoids comprise a group of natural and synthetic molecules that regulate multiple biological pathways. There is strong evidence from both epidemiologic studies, and animal and cell culture studies, that these metabolites may be used in the prevention or treatment of cancer. In general, the mechanisms underlying cancer chemopreventive activity of apocarotenoids include modulation of Nrf2, NF-κB and RAR/RXR pathways, enhancement of GJC, induction of apoptosis, inhibition of cell proliferation, inhibition of HIF-1, OR and JAK/STAT pathways, and regulation of cell growth and cell cycle progression. Although each cancer type seems to have an exclusive pattern of altered signalling pathways, the studies summarized here certainly encourage the use of apocarotenoids as cancer therapy agents. However, further detailed investigation is required to clarify the mechanism of actions of these metabolites, focusing on their interactions with genetic and epigenetic factors, as well as clinical validation, in order to gain insight into the roles and accurate applications of apocarotenoids in cancer.

ACKNOWLEDGEMENTS

This work was financially supported through grants from the Consejo Nacional de Ciencia y Tecnología (CONACYT) with Grant APN-2015-01-741 to YCC.

REFERENCES

Aasen T, Leithe E, Graham SV, Kameritsch P, Mayán MD, Mesnil M, Pogoda K, Tabernero A. 2019. Connexins in cancer: Bridging the gap to the clinic. *Oncogene.* 38:4429–4451.

Aasen T, Mesnil M, Naus CC, Lampe PD, Laird DW. 2016. Gap junctions and cancer: Communicating for 50 years. *Nat Rev Cancer.* 16:775–788.

Able AA, Burrell JA, Stephens JM. 2017. STAT5-interacting proteins: A synopsis of proteins that regulate STAT5 activity. *Biology (Basel).* 6:1–16.

Albensi BC. 2019. What is nuclear factor Kappa B (NF-κB) doing in and to the mitochondrion? *Front Cell Dev Biol.* 7:154.

Almeida NR, Conda-Sheridan M. 2019. A review of the molecular design and biological activities of RXR agonists. *Med Res Rev.* 39:1372–1397.

Amin A, Hamza AA, Bajbouj K, Ashraf SS, Daoud S. 2011. Saffron: A potential candidate for a novel anticancer drug against hepatocellular carcinoma. *Hepatology.* 54:857–867.

Anantharaman A, Hemachandran H, Mohan S, Manikoth Ayyathan D, Kumar DT, George Priya Doss C, Siva R. 2016. Induction of apoptosis by apocarotenoids in B16 melanoma cells through ROS-mediated mitochondrial-dependent pathway. *J Funct Foods.* 20:346–357.

Aust O, Ale-Agha N, Zhang L, Wollersen H, Sies H, Stahl W. 2003. Lycopene oxidation product enhances gap junctional communication. *Food Chem Toxicol.* 41:1399–1407.

Ben-Dor A, Steiner M, Gheber L, Danilenko M, Dubi N, Linnewiel K, Zick A, Sharoni Y, Levy J. 2005. Carotenoids activate the antioxidant response element transcription system. *Mol Cancer Ther.* 4:177–186.

Ben Salem I, Boussabbeh M, Kantaoui H, Bacha H, Abid-Essefi S. 2016. Crocin, the main active saffron constituent, mitigates dichlorvos-induced oxidative stress and apoptosis in HCT-116 cells. *Biomed Pharmacother.* 82:65–71.

Blomhoff R, Blomhoff HK. 2006. Overview of retinoid metabolism and function. *J Neurobiol.* 66:606–630.

Bonacquisti EE, Nguyen J. 2019. Connexin 43 (Cx43) in cancer: Implications for therapeutic approaches via gap junctions. *Cancer Lett.* 442:439–444.

Cancer Genome Atlas Research Network TCGAR. 2015. The molecular taxonomy of primary prostate cancer. *Cell.* 163:1011–1025.

Chen B, Hou Z-H, Dong Z, Li C-D. 2015a. Crocetin downregulates the proinflammatory cytokines in methylcholanthrene-induced rodent tumor model and inhibits COX-2 expression in cervical cancer cells. *Biomed Res Int.* 2015:829513.

Chen L, Zeng Y, Zhou S-F. 2018. Role of apoptosis in cncer resistance to chemotherapy. In: *Current Understanding of Apoptosis—Program Cell Death.* [place unknown]: InTech.

Chen S, Zhao S, Wang X, Zhang L, Jiang E, Gu Y, Shangguan AJ, Zhao H, Lv T, Yu Z. 2015b. Crocin inhibits cell proliferation and enhances cisplatin and pemetrexed chemosensitivity in lung cancer cells. *Transl Lung Cancer Res.* 4:775–783.

Cicconi L, Lo-Coco F. 2016. Current management of newly diagnosed acute promyelocytic leukemia. *Ann Oncol.* 27:1474–1481.

Clairmont A, Sies H. 1997. Evidence for a posttranscriptional effect of retinoic acid on connexin43 gene expression via the 3′-untranslated region. *FEBS Lett.* 419:268–270.

Clairmont A, Tessmann D, Sies H. 1996. Analysis of *connexin43* gene expression induced by retinoic acid in F9 teratocarcinoma cells. *FEBS Lett.* 397:22–24.

Cronier L, Crespin S, Strale P-O, Defamie N, Mesnil M. 2009. Gap junctions and cancer: New functions for an old story. *Antioxid Redox Signal.* 11:323–338.

Cunningham TJ, Duester G. 2015. Mechanisms of retinoic acid signalling and its roles in organ and limb development. *Nat Rev Mol Cell Biol.* 16:110–123.

Degos L, Wang ZY. 2001. All trans retinoic acid in acute promyelocytic leukemia. *Oncogene.* 20:7140–7145.

DeNicola GM, Karreth FA, Humpton TJ, Gopinathan A, Wei C, Frese K, Mangal D, Yu KH, Yeo CJ, Calhoun ES, et al. 2011. Oncogene-induced Nrf2 transcription promotes ROS detoxification and tumorigenesis. *Nature*. 475:106–109.

Dhar A, Mehta S, Dhar G, Dhar K, Banerjee S, Van Veldhuizen P, Campbell DR, Banerjee SK. 2009a. Crocetin inhibits pancreatic cancer cell proliferation and tumor progression in a xenograft mouse model. *Mol Cancer Ther.* 8:315–323.

Dhar A, Mehta S, Dhar G, Dhar K, Banerjee S, Van Veldhuizen P, Campbell DR, Banerjee SK. 2009b. Crocetin inhibits pancreatic cancer cell proliferation and tumor progression in a xenograft mouse model. *Mol Cancer Ther.* 8:315–323.

Dou M, Zhou X, Fan Z, Ding X, Li L, Wang S, Xue W, Wang H, Suo Z, Deng X. 2018. Clinical significance of retinoic acid receptor β promoter methylation in prostate cancer: A meta-analysis. *Cell Physiol Biochem.* 45:2497–2505.

Eroglu A, Hruszkewycz DP, Curley RW, Harrison EH, Harrison EH. 2010. The eccentric cleavage product of β-carotene, β-apo-13-carotenone, functions as an antagonist of RXRα. *Arch Biochem Biophys.* 504:11–16.

Evert M, Ott T, Temme A, Willecke K, Dombrowski F. 2002. Morphology and morphometric investigation of hepatocellular preneoplastic lesions and neoplasms in connexin32-deficient mice. *Carcinogenesis.* 23:697–703.

Fanayi AR, Changizi V, Safa M. 2016. Effect of crocin and doxorubicin/radiation on the breast cancer cell line, Michigan Cancer Foundation-7. *Biosci Biotechnol Res Commun.* 9:428–434.

Farmer EE, Davoine C. 2007. Reactive electrophile species. *Curr Opin Plant Biol.* 10:380–386.

Glass CK, Rosenfeld MG. 2000. The coregulator exchange in transcriptional functions of nuclear receptors. *Genes Dev.* 14:121–141.

Gong X, Marisiddaiah R, Zaripheh S, Wiener D, Rubin LP. 2016. Mitochondrial β-Carotene 9',10' oxygenase modulates prostate cancer growth via NF-κB inhibition: A lycopene-independent function. *Mol Cancer Res.* 14:966–975.

Goodenough DA, Paul DL. 2009. Gap junctions. *Cold Spring Harb Perspect Biol.* 1:a002576.

Hanusch M, Stahl W, Schulz WA, Sies H. 1995. Induction of gap junctional communication by 4-oxoretinoic acid generated from its precursor canthaxanthin. *Arch Biochem Biophys.* 317:423–428.

Harrison EH, Quadro L. 2018. Apocarotenoids: Emerging roles in mammals. *Annu Rev Nutr.* 38:153–172.

Harrison PJ, Bugg TDH. 2014. Enzymology of the carotenoid cleavage dioxygenases: Reaction mechanisms, inhibition and biochemical roles. *Arch Biochem Biophys.* 544:105–111.

Hashemi M, Hosseinzadeh H. 2019. A comprehensive review on biological activities and toxicology of crocetin. *Food Chem Toxicol.* 130:44–60.

Havaux M. 2014. Carotenoid oxidation products as stress signals in plants. *Plant J.* 79:597–606.

Heyman RA, Mangelsdorf DJ, Dyck JA, Stein RB, Eichele G, Evans RM, Thaller C. 1992. 9-cis retinoic acid is a high affinity ligand for the retinoid X receptor. *Cell.* 68:397–406.

Hong-Yue R, Yi-He G, Li-Min Y, Tai-Jian D, Ming-Zhu L, Le Y, Qi-Gan Z, Jia-Rong M. 2018. Expression and clinical significance of retinoid X receptor α in esophageal carcinoma. *Ann Diagn Pathol.* 34:110–115.

Hou X, Rivers J, León P, McQuinn RP, Pogson BJ. 2016. Synthesis and function of apocarotenoid signals in plants. *Trends Plant Sci.* 21:792–803.

Houle B, Rochette-Egly C, Bradley WE. 1993. Tumor-suppressive effect of the retinoic acid receptor beta in human epidermoid lung cancer cells. *Proc Natl Acad Sci U S A.* 90:985–989.

Ivanova T, Petrenko A, Gritsko T, Vinokourova S, Eshilev E, Kobzeva V, Kisseljov F, Kisseljova N. 2002. Methylation and silencing of the retinoic acid receptor-β2 gene in cervical cancer. *BMC Cancer.* 2:4.

Jaramillo M, Zhang D. 2013. The emerging role of the Nrf2–Keap1 signaling pathway in cancer. *Genes Dev.* 27:2179–2191.

Jiang SY, Shen SR, Shyu RY, Yu JC, Harn HJ, Yeh MY, Lee MM, Chang YC. 1999. Expression of nuclear retinoid receptors in normal, premalignant and malignant gastric tissues determined by in situ hybridization. *Br J Cancer.* 80:206–214.

Karin M. 2009. NF-kappaB as a critical link between inflammation and cancer. *Cold Spring Harb Perspect Biol.* 1:a000141.

Karin M, Yamamoto Y, Wang QM. 2004. The IKK NF-κB system: A treasure trove for drug development. *Nat Rev Drug Discov.* 3:17–26.

Kawabata K, Tung NH, Shoyama Y, Sugie S, Mori T, Tanaka T. 2012. Dietary crocin inhibits colitis and colitis-associated colorectal carcinogenesis in male ICR mice. *Evid Based Complement Alternat Med.* 2012:820415.

Kim B, Park B. 2018. Saffron carotenoids inhibit STAT3 activation and promote apoptotic progression in IL-6-stimulated liver cancer cells. *Oncol Rep.* 39:1883–1891.

King TJ, Lampe PD. 2004. The gap junction protein Connexin32 is a mouse lung tumor suppressor. *Cancer Res.* 64:7191–7196.

Knatko E V, Ibbotson SH, Zhang Y, Higgins M, Fahey JW, Talalay P, Dawe RS, Ferguson J, Huang JT-J, Clarke R, et al. 2015. Nrf2 activation protects against solar-simulated ultraviolet radiation in mice and humans. *Cancer Prev Res (Phila).* 8:475–486.

Lee S-J, No YR, Dang DT, Dang LH, Yang VW, Shim H, Yun CC. 2013. Regulation of hypoxia-inducible factor 1α (HIF-1α) by lysophosphatidic acid is dependent on interplay between p53 and Krüppel-like factor 5. *J Biol Chem.* 288:25244–25253.

Levonen A-L, Landar A, Ramachandran A, Ceaser EK, Dickinson DA, Zanoni G, Morrow JD, Darley-Usmar VM. 2004. Cellular mechanisms of redox cell signalling: role of cysteine modification in controlling antioxidant defences in response to electrophilic lipid oxidation products. *Biochem J.* 378:373–382.

Lian F, Smith DE, Ernst H, Russell RM, Wang X-D. 2007. Apo-10′-lycopenoic acid inhibits lung cancer cell growth *in vitro*, and suppresses lung tumorigenesis in the A/J mouse model *in vivo*. *Carcinogenesis.* 28:1567–1574.

Lian F, Wang X. 2008. Enzymatic metabolites of lycopene induce Nrf2-mediated expression of phase II detoxifying/antioxidant enzymes in human bronchial epithelial cells. *Int J Cancer.* 123:1262–1268.

Lingappan K. 2018. NF-κB in oxidative stress. *Curr Opin Toxicol.* 7:81–86.

Linnewiel-Hermoni K, Motro Y, Miller Y, Levy J, Sharoni Y. 2014. Carotenoid derivatives inhibit nuclear factor kappa B activity in bone and cancer cells by targeting key thiol groups. *Free Radic Biol Med.* 75:105–120.

Linnewiel K, Ernst H, Caris-Veyrat C, Ben-Dor A, Kampf A, Salman H, Danilenko M, Levy J, Sharoni Y. 2009a. Structure activity relationship of carotenoid derivatives in activation of the electrophile/antioxidant response element transcription system. *Free Radic Biol Med.* 47:659–667.

Linnewiel K, Ernst H, Caris-Veyrat C, Ben-Dor A, Kampf A, Salman H, Danilenko M, Levy J, Sharoni Y. 2009b. Structure activity relationship of carotenoid derivatives in activation of the electrophile/antioxidant response element transcription system. *Free Radic Biol Med.* 47:659–667.

Liu J-R, Chen B-Q, Yang B-F, Dong H-W, Sun C-H, Wang Q, Song G, Song Y-Q. 2004a. Apoptosis of human gastric adenocarcinoma cells induced by beta-ionone. *World J Gastroenterol.* 10:348–351.

Liu J-R, Yang B-F, Chen B-Q, Yang Y-M, Dong H-W, Song Y-Q. 2004b. Inhibition of β-ionone on SGC-7901 cell proliferation and upregulation of metalloproteinases-1 and -2 expression. *World J Gastroenterol.* 10:167–171.

Liu Q, Dong H-W, Sun W-G, Liu M, Ibla JC, Liu L-X, Parry JW, Han X-H, Li M-S, Liu J-R. 2013. Apoptosis initiation of β-ionone in SGC-7901 gastric carcinoma cancer cells via a PI3K-AKT pathway. *Arch Toxicol.* 87:481–490.

Luo L, Fang K, Dan X, Gu M. 2019. Crocin ameliorates hepatic steatosis through activation of AMPK signaling in db/db mice. *Lipids Health Dis.* 18:11.

Masoud GN, Li W. 2015. HIF-1α pathway: Role, regulation and intervention for cancer therapy. *Acta Pharm Sin B.* 5:378–389.

McQuate GT, Peck SL. 2001. Enhancement of Attraction of Alpha-Ionol to Male *Bactrocera latifrons* (Diptera: Tephritidae) by Addition of a Synergist, Cade Oil. *J Econ Entomol.* 94:39–46.

Mitchell TJ, John S. 2005. Signal transducer and activator of transcription (STAT) signalling and T-cell lymphomas. *Immunology.* 114:301–312.

Moennikes O, Buchmann A, Romualdi A, Ott T, Werringloer J, Willecke K, Schwarz M. 2000. Lack of phenobarbital-mediated promotion of hepatocarcinogenesis in connexin32-null mice. *Cancer Res.* 60:5087–5091.

Moradzadeh M, Ghorbani A, Erfanian S, Mohaddes ST, Rahimi H, Karimiani EG, Mashkani B, Chiang SC, El-Khamisy SF, Tabarraei A, Sadeghnia HR. 2018. Study of the mechanisms of crocetin-induced differentiation and apoptosis in human acute promyelocytic leukemia cells. *J Cell Biochem.* 120:1943–1957.

Morgan MJ, Liu Z. 2011. Crosstalk of reactive oxygen species and NF-κB signaling. *Cell Res.* 21:103–115.

Nagata S. 2018. Apoptosis and clearance of apoptotic cells. *Annu Rev Immunol.* 36:489–517.

Nguyen T, Nioi P, Pickett CB. 2009. The Nrf2-antioxidant response element signaling pathway and its activation by oxidative stress. *J Biol Chem.* 284:13291–13295.

Niles RM. 2004. Signaling pathways in retinoid chemoprevention and treatment of cancer. *Mutat Res Mol Mech Mutagen.* 555:97–105.

de Oliveira Júnior RG, Bonnet A, Braconnier E, Groult H, Prunier G, Beaugeard L, Grougnet R, da Silva Almeida JRG, Ferraz CAA, Picot L. 2019. Bixin, an apocarotenoid isolated from *Bixa orellana* L., sensitizes human melanoma cells to dacarbazine-induced apoptosis through ROS-mediated cytotoxicity. *Food Chem Toxicol.* 125:549–561.

Pan S-T, Li Z-L, He Z-X, Qiu J-X, Zhou S-F. 2016. Molecular mechanisms for tumour resistance to chemotherapy. *Clin Exp Pharmacol Physiol.* 43:723–737.

Panieri E, Saso L. 2019. Potential applications of Nrf2 inhibitors in cancer therapy. *Oxid Med Cell Longev.* 2019:1–34.

Prakash P, Liu C, Hu K-Q, Krinsky NI, Russell RM, Wang X-D. 2004. β-Carotene and β-Apo-14′-carotenoic acid prevent the reduction of retinoic acid receptor β in Benzo[a]pyrene-treated normal human bronchial epithelial cells. *J Nutr.* 134:667–673.

Rani A, Murphy JJ. 2016. STAT5 in cancer and immunity. *J Interf Cytokine Res.* 36:226–237.

Rojo de la Vega M, Chapman E, Zhang DD. 2018a. Nrf2 and the hallmarks of cancer. *Cancer Cell.* 34:21–43.

Rojo de la Vega M, Zhang DD, Wondrak GT. 2018b. Topical bixin confers Nrf2-dependent protection against photodamage and hair graying in mouse skin. *Front Pharmacol.* 9:287.

Saha Roy S, Vadlamudi RK. 2012. Role of estrogen receptor signaling in breast cancer metastasis. *Int J Breast Cancer.* 2012:654698.

Samarghandian S, Shoshtari ME, Sargolzaei J, Hossinimoghadam H, Farahzad JA. 2014. Anti-tumor activity of safranal against neuroblastoma cells. *Pharmacogn Mag.* 10:S419-S424.

Saumet A, Vetter G, Bouttier M, Portales-Casamar E, Wasserman WW, Maurin T, Mari B, Barbry P, Vallar L, Friederich E, et al. 2009. Transcriptional repression of microRNA genes by PML-RARA increases expression of key cancer proteins in acute promyelocytic leukemia. *Blood.* 113:412–421.

Serasanambati M, Chilakapati SR. 2016. Function of nuclear factor Kappa B (NF-kB) in human diseases: A review. *South Indian J Biol Sci.* 2:368.

Sharoni Y, Linnewiel-Hermoni K, Khanin M, Salman H, Veprik A, Danilenko M, Levy J. 2012. Carotenoids and apocarotenoids in cellular signaling related to cancer: A review. *Mol Nutr Food Res.* 56:259–269.

Siddikuzzaman, Guruvayoorappan C, Berlin Grace VM. 2011. All trans retinoic acid and cancer. *Immunopharmacol Immunotoxicol.* 33:241–249.

Stahl W, von Laar J, Martin H-D, Emmerich T, Sies H. 2000. Stimulation of gap junctional communication: Comparison of acyclo-retinoic acid and lycopene. *Arch Biochem Biophys.* 373:271–274.

Sun J, Narayanasamy S, Curley RW, Harrison EH, Harrison EH. 2014. β-Apo-13-carotenone regulates retinoid X receptor transcriptional activity through tetramerization of the receptor. *J Biol Chem.* 289:33118–33124.

Sun S-Y, Lotan R. 2002. Retinoids and their receptors in cancer development and chemoprevention. *Crit Rev Oncol Hematol.* 41:41–55.

Sun Y, Xu H-J, Zhao Y-X, Wang L-Z, Sun L-R, Wang Z, Sun X-F. 2013. Crocin exhibits antitumor effects on human leukemia HL-60 cells *in vitro* and *in vivo*. *Evid Based Complement Alternat Med.* 2013:690164.

Tao S, Park SL, Rojo de la Vega M, Zhang DD, Wondrak GT. 2015. Systemic administration of the apocarotenoid bixin protects skin against solar UV-induced damage through activation of Nrf2. *Free Radic Biol Med.* 89:690–700.

Veprik A, Khanin M, Linnewiel-Hermoni K, Danilenko M, Levy J, Sharoni Y. 2012. Polyphenols, isothiocyanates, and carotenoid derivatives enhance estrogenic activity in bone cells but inhibit it in breast cancer cells. *Am J Physiol Metab.* 303:E815–E824.

Waks AG, Winer EP. 2019. Breast cancer yreatment. *JAMA.* 321:288.

Walter MH, Strack D. 2011. Carotenoids and their cleavage products: Biosynthesis and functions. *Nat Prod Rep.* 28:663–692.

Wang J, Dai Y, Huang Y, Chen X, Wang H, Hong Y, Xia J, Cheng B. 2013. All-trans retinoic acid restores gap junctional intercellular communication between oral cancer cells with upregulation of Cx32 and Cx43 expressions *in vitro*. *Med Oral Patol Oral Cir Bucal.* 18:e569-e577.

Wang X-J, Sun Z, Villeneuve NF, Zhang S, Zhao F, Li Y, Chen W, Yi X, Zheng W, Wondrak GT, et al. 2008. Nrf2 enhances resistance of cancer cells to chemotherapeutic drugs, the dark side of Nrf2. *Carcinogenesis.* 29:1235–1243.

Wang XJ, Hayes JD, Henderson CJ, Wolf CR. 2007. Identification of retinoic acid as an inhibitor of transcription factor Nrf2 through activation of retinoic acid receptor alpha. *Proc Natl Acad Sci U S A.* 104:19589

Widschwendter M, Berger J, Hermann M, Muller HM, Amberger A, Zeschnigk M, Widschwendter A, Abendstein B, Zeimet AG, Daxenbichler G, Marth C. 2000. Methylation and silencing of the retinoic acid receptor-β2 gene in breast cancer. *JNCI J Natl Cancer Inst.* 92:826–832.

Wyss A, Wirtz G, Woggon W-D, Brugger R, Wyss M, Friedlein A, Bachmann H, Hunziker W. 2000. Cloning and expression of β,β-carotene 15,15′-dioxygenase. *Biochem Biophys Res Commun.* 271:334–336.

Xia Y, Shen S, Verma IM. 2014. NF-κB, an active player in human cancers. *Cancer Immunol Res.* 2:823–830.

Xu Z, Kong X-Q. 2017. Bixin ameliorates high fat diet-induced cardiac injury in mice through inflammation and oxidative stress suppression. *Biomed Pharmacother.* 89:991–1004.

Xue L, Zhang H, Zhang J, Li B, Zhang Z, Tao S. 2018. Bixin protects against particle-induced long-term lung injury in an Nrf2-dependent manner. *Toxicol Res (Camb).* 7:258–270.

Yu L, Li J, Xiao M. 2018. Picrocrocin exhibits growth inhibitory effects against SKMEL- 2 human malignant melanoma cells by targeting JAK/ STAT5 signaling pathway, cell cycle arrest and mitochondrial mediated apoptosis. *J Buon.* 23:1163–1168.

Zhang H, Xue L, Li B, Tian H, Zhang Z, Tao S. 2018a. Therapeutic potential of bixin in PM2.5 particles-induced lung injury in an Nrf2-dependent manner. *Free Radic Biol Med.* 126:166–176.

Zhang X, Ren Z, Zuo J, Su C, Wang R, Chang Y, Fang F. 2002. The effect of all-trans retinoic acid on gap junctional intercellular communication and connexin 43 gene expression in glioma cells. *Chinese Med Sci J = Chung-kuo i hsueh k'o hsueh tsa chih.* 17:22–26.

Zhang Y, Zhao Y, Guo J, Cui H, Liu S. 2018b. Anticancer activity of safranal against colon carcinoma is due to induction of apoptosis and G2/M cell cycle arrest mediated by suppression of mTOR/PI3K/Akt pathway. *J Buon.* 23:574–578.

Zhao P, Luo CL, Wu XH, Hu HB, Lv CF, Ji HY. 2008. Proliferation apoptotic influence of crocin on human bladder cancer T24 cell line. *Zhongguo Zhongyao Zazhi.* 33:1869–1873.

Zhou Y, Xu Q, Shang J, Lu L, Chen G. 2019. Crocin inhibits the migration, invasion, and epithelial-mesenchymal transition of gastric cancer cells via miR-320/KLF5/HIF-1α signaling. *J Cell Physiol.* 234:17876–17885.

9 An Overview of the Role of Abscisic Acid (ABA): From Plants to Humans

Rosa Us-Camas, Margarita Aguilar-Espinosa and Renata Rivera-Madrid

CONTENTS

INTRODUCTION

Carotenoids are naturally occurring pigments, essential for photosynthesis and plant photoprotection (Ruiz-Sola and Rodríguez-Concepción, 2012). Plant carotenoids are C_{40} isoprenoids with polyene chains that contain double conjugated bonds within their structure (up to 15 conjugated double bonds) (Hirschberg, 2001). Carotenoids are precursors of apocarotenoids, the end product of the carotenoid pathway. Apocarotenoids are compounds derived from the oxidative cleavage of carotenoids, in which the carbon skeleton has been shortened by the removal of fragments of one or both ends of the carotenoid structure (Rodriguez-Amaya, 2015; Beltran and Stange, 2016; Hou et al., 2016).

Apocarotenoids participate in various environmental and physiological processes of plants and are also essential for human health (Rivera-Madrid et al., 2016; Sah et al., 2016; Harrison and Quadro, 2018; Kim et al., 2019). For example, retinol (vitamin A) and bixin are apocarotenoids that have been related to cancer prevention (Rivera-Madrid et al., 2016; Harrison and Quadro, 2018; Kim et al., 2019). In plants, apocarotenoids participate as attractants of pollinators and seed dispersal, in the defence of pathogens and herbivores as well as signal molecules such as ABA and strigolactone (SL) that regulate several development processes (Beltran and Stange, 2016; Hou et al., 2016).

ABA is one of the most studied apocarotenoids. ABA is a sesquiterpenoid (C_{15}) phytohormone discovered in the early 1960s (Ohkuma et al., 1963). Originally named Abscisin II, it was first isolated from the mature fruit wall of cotton (Ohkuma et al., 1963). It was thought that it had an important role in accelerating the abscission of the fruit; however, various studies suggest that ABA does not directly control the abscission but promotes senescence and stress response, processes that precede abscission (Ohkuma et al., 1963; Finkelstein, 2013). ABA controls several biological processes in plants such as seed dormancy and germination (Finkelstein et al., 2008; Miransari and Smith, 2014), senescence (Xue-Xuan et al., 2010) embryo maturation (Gutmann et al., 1996; García-Martín et al., 2005), stomatal aperture (Kim et al., 2010; Saradadevi et al., 2017), modulation of root structure (Harris, 2015), floral transition (Shu et al., 2018), defence against pathogens (Cao et al., 2011; Alazem and Lin, 2017), abiotic stress tolerance (drought, salinity and cold stresses) (Xue-Xuan et al., 2010; Sah et al., 2016) and UV radiation protection (Berli et al., 2011).

The growth and productivity of plants ultimately depend on their ability to adapt and acclimatize to continually changing environmental conditions. Drought is one of the main abiotic stresses that negatively affect plant growth and productivity. Recently, damage to the ozone layer has placed UV radiation among the environmental factors that generate stress on plants and affect their productivity. Different studies have demonstrated the role of ABA as a signal molecule involved in the response of plants to drought and UV radiation stress. This chapter addresses general aspects of ABA in the response to drought stress and UV radiation in plants; mainly its role in counteracting drought stress and UV radiation and how its exogenous application in economically important crops can improve production, quality and yield, agronomic characteristics and maintain its commercialization even under these conditions of abiotic stress. Similarly, we address the findings that involve ABA with therapeutic potential against diseases such as atherogenesis, inflammatory bowel disease, ischaemic retinopathies, diabetes, and different types of cancer in humans.

ABA METABOLISM

ABA BIOSYNTHESIS

ABA is an isoprenoid (C_{15}), derived from a common five-carbon (C_5) precursor, isopentenyl diphosphate (IPP) (Nambara and Marion-Poll, 2005). Initially, it was thought to be a direct and indirect pathway for ABA biosynthesis. The direct pathway involved the C_{15} sesquiterpene precursor, farnesyl diphosphate (FDP), while the indirect pathway involved a precursor derived from carotenoids. In both cases, it was thought that the ultimate precursor is mevalonic acid (MVA) (Zeevaart and Creelman, 1988; Nambara and Marion-Poll, 2005). However, several pieces of evidence suggest that in plants the indirect pathway of ABA biosynthesis is the one that predominates over the direct pathway, in addition to the fact that the precursor C_{40} carotenoid comes from the MEP pathway and not from the mevalonic acid as previously thought (Zeevaart and Creelman, 1988; Cutler and Krochko, 1999; Nambara and Marion-Poll, 2005).

Lycopene represents a bifurcation point to obtain the cyclic carotenoids, α- and β-carotene. Hydroxylation of α-carotene produces lutein, a pigment that accumulates in high amounts in the chloroplast. Only β-carotene is metabolized to ABA through zeaxanthin. The conversion of β-carotene to zeaxanthin is catalysed by β-carotene hydroxylase (Ruiz-Sola and Rodríguez-Concepción, 2012; Finkelstein, 2013). ABA-deficient mutants have helped to elucidate the pathway of ABA biosynthesis. The *aba2* mutant of *Nicotiana plumbaginifolia* showed as impaired in the first step of the ABA biosynthesis pathway, the zeaxanthin epoxidation reaction. The results showed that *ABA2* encodes zeaxanthin epoxidase (ZEP), the enzyme that converts zeaxanthin to violaxanthin with antheraxanthin as intermediary (Marin et al., 1996; Liotenberg et al., 1999).

The next step of ABA biosynthesis is the conversion of violaxanthin to neoxanthin: this reaction is catalysed by the enzyme neoxanthin synthase (NSY). Neoxanthin together with violaxanthin can be used for the production of apocarotenoid ABA (Nambara and Marion-Poll, 2005; Ruiz-Sola and

Rodríguez-Concepción, 2012). The first committed step of ABA biosynthesis involves the cleavage of 9-cis-epoxy-carotenoids. The 9-cis-epoxycarotenoid dioxygenase (NCED) is the key enzyme in the biosynthesis of ABA, that is, they perform the cleavage of 9-cis-violaxanthin and 9-cis-neoxanthin to xanthoxin (C_{15}) and a C_{25}-epoxy-apo-aldehyde/C_{25}-allenic-apo-aldehyde (Schwartz et al., 2003). *VP14* (*viviparous 14*) was the first *NCED* gene reported (Tan et al., 1997). The *Vp14* (*viviparous 14*) mutant was found to be deficient in the double-bond cleavage reaction and biosynthesis of ABA (Tan et al., 1997). The biochemical characterization of the recombinant *VP14* showed that the protein catalyses the cleavage of 9-cis-epoxy-carotenoids to form C_{25} apo-aldehydes and xanthoxin, the precursor of ABA in higher plants (Schwartz et al., 1997). In *Arabidopsis*, nine genes encoding carotenoid cleavage dioxygenase have been identified, and five are *NCEDs* (*AtNCED*2, *AtNCED*3, *AtNCED*5, *AtNCED*6, and *AtNCED*9) involved in ABA biosynthesis (Iuchi et al., 2001; Tan et al., 2003).

The biologically active form of ABA is synthesized from xanthoxin through two oxidation processes. The first oxidation reaction is catalysed by a short-chain alcohol dehydrogenase/reductase (SDR) to produce ABA aldehyde. In the last reaction, the ABA aldehyde is converted to ABA by the activity of the ABA aldehyde oxidase (AAO) (Nambara and Marion-Poll, 2005; Cutler et al., 2010).

ABA CATABOLISM

ABA hydroxilation

ABA degradation is a mechanism that allows ABA levels to be regulated. *Arabidopsis* uses two reactions for ABA catabolism: hydroxylation and conjugation (Nambara and Marion-Poll, 2005; Schwartz and Zeevaart, 2010). The hydroxylation of ABA can occur at the C9′-methyl group, as well as at the C7′- and C8′-methyl groups of the ring structure (Zhou et al., 2004). The most common form of hydroxylation of ABA occurs at the C8′position (8′-OH-ABA) (Nambara and Marion-Poll, 2005; Schwartz and Zeevaart, 2010). The ABA 8′ hydroxylase catalyses the formation of 8′-OH-ABA, the first product of ABA degradation (Krochko et al., 1998). In *Arabidopsis*, the four members of the *CYP707A* (*CYP707A1-4*) gene family encode ABA 8′ hydroxylases, each of which exercises different functions in particular tissues and developmental stages (Kushiro et al., 2004; Saito et al., 2004; Matilla et al., 2015). The 8′-OH-ABA can be isomerized to phaseic acid (PA) which can be reduced by the enzyme phaseic acid reductase to form dihydrophaseic acid (DPA) (Cutler and Krochko, 1999). It has been observed that the three hydroxylated forms (7′-OH-ABA, 8′-OH-ABA, and 9′-OH-ABA) and PA exert significant ABA-like activity (Jitao et al., 1995; Zhou et al., 2004; Walker-Simmons et al., 2006). On the contrary, DPA does not exhibit ABA-like activity (Cutler and Krochko, 1999).

ABA conjugation

The carboxyl and hydroxyl group of ABA can be glycosylated to convert ABA into its inactive form (Nambara and Marion-Poll, 2005). The inactive conjugated form of ABA is predominantly found as ABA glucosyl ester (ABA-GE) (Boyer and Zeevaart, 1982; Nambara and Marion-Poll, 2005; Schwartz and Zeevaart, 2010). ABA-GE was isolated for the first time from the immature fruit of *Lupinus luteus* (Koshimizu et al., 1968). ABA uridine diphosphate transferases (UGTs) are responsible for converting ABA to ABA-GE. In *Arabidopsis*, *UGT71B6* and its two closely related homologs, *UGT71B7* and *UGT71B8*, encodes UGTs. These UGTs are localized in the cytosol and are shown to play crucial roles in ABA homeostasis and adaptation to dehydration, osmotic stress, and high-salinity stresses (Dong et al., 2014). In *Arabidopsis*, ABA-GE accumulates in the vacuole and also in the endoplasmic reticulum (Burla et al., 2013). ABA-GE contributes to the maintenance of ABA homeostasis.

Deconjugation of ABA-GE by the endoplasmic reticulum β-glucosidase homolog1 (BG1) and vacuolar β-glucosidase homolog2 (BG2), allows the rapid formation of free ABA in response to abiotic stress conditions such as dehydration and salt stress (Lee et al., 2006; Xu et al., 2012; Burla et al., 2013). BG2 overexpression results in enhanced resistance to dehydration and NaCl stress (Xu et al., 2012). On the other hand, the loss of AtBG1 function causes defective stomach movement and early germination, but plants with ectopic *AtBG1* accumulate higher levels of ABA and show greater tolerance to abiotic stress (Lee et al., 2006). It has been suggested that ABA is produced in multiple organelles by organelle-specific β-glucosidases in response to abiotic stress (Xu et al., 2012).

Other glycosylated conjugates such as 1′-O-ABA- β-D-glucoside, PA-β-D-glucosyl ester and 4′-O-DPA β-D-glucoside have been identified but ABA-GE appears to be the most widely distributed in plants (Zeevaart and Creelman, 1988; Cutler and Krochko, 1999; Nambara and Marion Poll, 2005).

ABA SIGNAL TRANSDUCTION

ABA exerts its action through the interaction of three core signalling components. In *Arabidopsis*, these three core components of ABA signalling have been identified:

1. The ABA receptors PYR/PYL/RCAR or PYLs (pyrabactin resistance/pyrabactin-like/regulatory part of ABA receptor) protein family;
2. The negative regulator type 2C protein phosphatases (PP2Cs); and
3. The positive regulator class III Sucrose non-fermenting-1-related protein kinase 2 (SnRK2) (Ma et al., 2009; Park et al., 2009; Raghavendra et al., 2010; Vilela and Riera 2015).

These three components are sufficient to carry out the ABA signalling pathway *in vitro* (Fujii et al., 2009).

The PYLs family can be divided into three evolutionary clades (Ma et al., 2009). The receptors belonging to clade I and II exist as monomers in solution, the members of clade III form dimers (Hao et al., 2011; Miyakawa et al., 2013). In *Arabidopsis*, 14 members of the PYLs receptor family have been reported, being classified as dimeric (PYR1 and PYL1-3) or monomeric (PYL4-13) receptors (Zhang et al., 2012; Miyakawa et al., 2013). Homologous proteins to ABA-PYLs receptors have been found in other species like soybean, cucumber, grape, citrus, rice, and strawberry (Kim et al., 2011; Li et al., 2011a; Boneh et al., 2012; Wang et al., 2012; Bai et al., 2013).

ABA signalling begins by the binding of ABA to PYL receptors. This leads to a conformational change in the receptors, generating a surface for its interaction with the active site of PP2C (Zhang et al., 2015). The formation of a ternary ABA-PYLs-PP2C complex, results in the inhibition of the phosphatase activity of PP2Cs, enabling autophosphorylation and the release of SnRKs inhibition required to activate the transcription factors that turn on the expression of ABA-responsive genes to convert ABA signals in appropriate cellular responses (Ma et al., 2009; Park et al., 2009; Raghavendra et al., 2010; Soon et al., 2012; Vilela and Riera 2015).

This indicates that the activity of phosphatases and kinases are an important mechanism in the regulation of the ABA signalling pathway. The type 2C protein phosphatases (PP2Cs) are negative regulators of ABA biosynthesis. In the absence of ABA, PP2Cs can bind to family members of SnRK2, inhibiting its kinase activity and therefore the transduction of ABA signalling. In plants, phosphatase proteins have been divided into two classes: serine/threonine (Ser/Thr) or tyrosine phosphatases, depending on the amino acid residue that dephosphorylates (Fuchs et al., 2013; Singh et al., 2015). PPCs are Ser/Thr phosphatases that require the metal ion for its activity (Fuchs et al., 2013; Singh et al., 2015). In model plants like *Arabidopsis* and rice, PP2C phosphatases have 80 and 90 members respectively, which are divided into 11 subclades (A-K) (Xue et al., 2008; Singh et al., 2010; Singh et al., 2015). Members of the clade A such as ABA1 (ABA insensitive 1) and ABA2 negatively regulate the response to ABA through its interaction with SnRK2s and ABA receptors (Leung et al., 1994; Leung et al., 1997; Xue et al., 2008; Singh et al., 2010; Fuchs et al., 2013;

Singh et al., 2015). ABA1 and ABA2 control abiotic stresses and development processes such as growth, and seed germination (Koornneef et al., 1984; Fuchs et al., 2013).

On the other hand, in *Arabidopsis*, three members of the subclass III of SnRK2s have been identified: SRK2I (also named SnRK2.3), SRK2D (also named SnRK2.2), and SRK2E (named SnRK2.6 or OPEN STOMATA 1 (OST1)), all differentially active the basic leucine zipper (bZIP) transcription factors (TFs) ABFs/AREBs. These TFs bind to ABA-responsive promoter elements (ABRE) to induce the gene expression of ABA-responsive genes during grown, reproduction and osmotic stress (Furihata et al., 2006; Fujii and Zhu 2009; Fujita et al., 2013).

ABA CHEMISTRY

After its discovery, the ABA structure was elucidated using spectroscopic methods and chemical synthesis (Milborrow, 1974). According to the Cahn, Ingold and Prelog rules, the carboxyl group at carbon 2 determine the *cis* and *trans* isomers, while the asymmetric carbon at 1′ in the ring, optically active, results in the S(+) and R(−) enantiomers (Milborrow, 1974; Berli and Bottini, 2013). The biologically active form of ABA is found as S-(+)-ABA ((+)-ABA) or 2-*cis*,-4-*trans* ABA (Milborrow 1974; Cutler et al., 2010). A minor inactive form, 2-*trans*, 4-*trans* ABA or R-(−)-ABA ((−)-ABA) was also detected (Nambara et al., 2002). Plants only synthesize (+)-ABA and (−)-ABA enantiomers and since its discovery, ABA enantiomers have been studied. The molecule of (+)-ABA is very similar in shape to (−)-ABA, it differs only in the disposition of the methyl groups on the ring (Figure 9.1). In some assays such as stomatal closure and seed germination, the (−)-ABA application showed similar activity to (+)-ABA isomer (Lin et al., 2005; Cutler et al., 2010). In *Arabidopsis*, the application of (−)-ABA present substantial similarities to (+)-ABA in gene regulation; about 80% of their significantly regulated genes were the same as in (+)-ABA(Huang et al., 2007).

It has been hypothesized that this response is because of the duality of the PYL receptors, members of the PYR/PYL/RCAR family of proteins that bind to the two stereoisomers (Nambara et al., 2002). The *cis-trans* isomerization of ABA occurs non-enzymatically and can occur under low light conditions or UV exposure (Milborrow, 1970; Cutler et al., 2010).

The physical properties of ABA have also been thoroughly studied by various researchers. The natural active form (+)-ABA has a molecular formula of $C_{15}H_{20}O_4$ and a molecular weight of 264.32 g/mol. The natural (+)-ABA melts at 160-161°C, whereas the synthetic racemic mixture (+)-ABA and (-)-ABA) melts at about 191°C. In acidic conditions, the maximum absorbance occurs at 262 nm, while in basic conditions the maximum absorption occurs at 245 nm (Zeevaart and Creelman, 1988).

ABA is a weak acid (pK 4.7) that dissociates depending on the pH of the different cell compartments. Its protonated form is permeable through the membrane, while its anionic form is

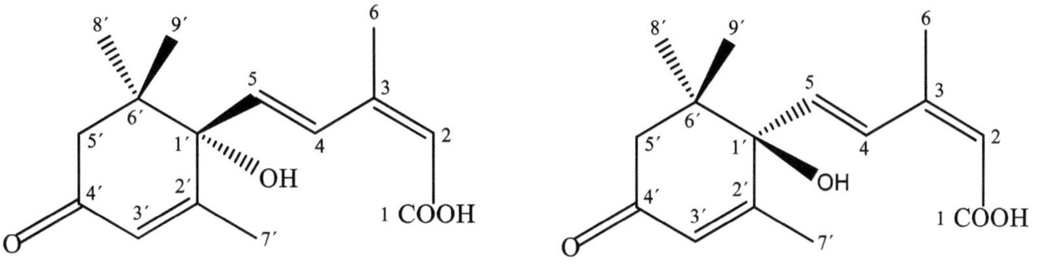

S-(+)-ABA **R-(-)-ABA**

FIGURE 9.1 Chemical structure of S-(+)-ABA and R-(−)-ABA enantiomers.

impermeable. This indicates that ABA will accumulate in greater quantity in an alkaline compartment (Zeevaart and Creelman, 1988). Based on the pH data of the free space, cytoplasm, chloroplasts, and vacuoles in leaf tissue, Cowan et al. (1982) calculated the distribution of ABA amongst these compartments. The results showed that 4.4% (8.2%), 68.4% (40.7%), 17.2% (32.4%), and 10% (18.7%) of the total cell ABA is be found in the apoplast, chloroplasts, cytoplasm, and vacuole, respectively, under light (dark) conditions (Cowan et al., 1982). Thus, the transition from light to darkness and vice versa causes a redistribution of ABA within cell compartments.

BIOTIC AND ABIOTIC STRESS IN PLANTS

In physical terms, stress is defined as a mechanical force per unit area applied to an object (Mahajan and Tuteja, 2005). In response to the applied stress, an object undergoes a change in dimension, which is also known as strain. Because plants are sensitive it is hard to measure the exact force exerted by stresses, and therefore it is difficult to exactly define stress in biological terms (Mahajan and Tuteja, 2005). The definition of a biological stress is then an adverse force (an environmental factor: biotic or abiotic) which inhibits the normal functioning of biological systems such as plants (Mahajan and Tuteja, 2005).

Throughout their life cycle, plants are exposed to a large number of stressful conditions or factors, which can be grouped into biotics and abiotic. Biotic stress is caused by the action of living beings, such as small or large animals, other plants, and pathogens (bacteria, fungi, viruses and viroids). Abiotic stress, depending on the nature of the causative agent, can be divided into physical and chemical. Among the physical abiotic stresses (or more accurately, physicochemical stresses), are the water deficiency, drought, salinity (the osmotic component), extreme temperatures (heat, cold, freezing), excessive or insufficient irradiation, excess of water or flooding, mechanical stress caused by wind or excessive soil compaction, and stress induced by wounds or injuries. On the other hand, chemical stress is caused by salinity (ionic or toxic component), by the lack of mineral elements, and by environmental pollutants, such as sulphur dioxide (SO2), nitrogen oxides (NOx), chlorofluorocarbon compounds (CFC), ozone (O_3) and metals (Mahajan and Tuteja, 2005; Azcón-Bieto and Talón, 2008).

Currently, drought is an abiotic stress that most affects the productivity of economically important crops, reducing quality and food security. Numerous studies have shown that an increase in temperature and drought could reduce crop productivity by up to 50% (Lamaoui et al., 2018). Similarly, the increase in UV radiation within the earth caused by damage to the ozone layer has placed UV radiation among the environmental factors that generate stress and therefore affect the productivity of plants (Bornman et al., 1997; Jansen et al., 1998; Berli et al., 2010; Berli et al., 2011). In this situation, the plants have developed dynamic responses at morphological, physiological and biochemical levels which allow them to adapt to these unfavourable environmental conditions.

Drought stress

Drought is one of the worst and most dynamic stresses that hinder plant growth and development. Drought stress has the potential to decrease crop yield and profitability, and induce adverse effects on the physiological and biochemical processes of plants. Drought stress imposes alterations in crucial plant growth and development processes such as germination, plant height, stem diameter, number of leaves, leaf size and area, flower and fruit production, and dry matter production (Anjum et al., 2017). Drought stress takes place when soil and atmospheric humidity is low and the ambient air temperature is high. This condition is the result of an imbalance between the evapotranspiration flux and water intake from the soil (Lipiec et al., 2013). As a consequence of drought stress, many changes occur in the cell. Removal of water from the membrane disrupts the normal bilayer structure and results in the membrane becoming porous when desiccated. Stress within the lipid bilayer may also result in displacement of membrane proteins, and this contributes to loss of membrane integrity, selectivity, disruption of cellular compartmentalization, and a loss of activity of enzymes, which are primarily membrane based. In addition, cytosolic and organelle proteins may exhibit reduced activity or may even undergo complete denaturation when dehydrated (Mahajan and Tuteja, 2005). There are also changes in the level of expression of the LEA/dehydrin-type genes, synthesis of molecular chaperones, and proteinases that function to remove denatured and damaged proteins (Cushman and Bohnert, 2000; Zhu, 2002). The over expression of barley group 3 LEA gene HVA1 in leaves and roots of rice and wheat lead to improved tolerance against osmotic stress as well as improved recovery after drought and salinity stress (Sivamani et al., 2000). Drought stress also leads to activation of enzymes involved in the production and removal of reactive oxygen species (ROS) (Cushman and Bohnert, 2000; Zhu, 2002).

Response of stomata to drought condition

The first response of virtually all plants to acute water deficit is the closure of their leaf stomata to prevent the transpirational water loss (Mansfield et al., 1990). Closure of stomata may result from direct evaporation of water from the guard cells with no metabolic involvement. This process of stomatal closure is referred to as hydropassive closure. Stomata closure may also be metabolically dependent and involve processes that result in ion fluxes. This process of stomatal closure, which requires ions and metabolites, is known as hydroactive closure (Blackman and Davies, 1985). This process seems to be ABA regulated. ABA promotes the efflux of K+ ions from the guard cells, which results in the loss of turgor pressure leading to stomata closure. Stomata closure also responds directly to soil desiccation even before there is any significant reduction in leaf mesophyll turgor pressure. Also, ABA can act as a long-distance communication signal due to water deficiency in roots, inducing the closure of stomata in leaves (Blackman and Davies, 1985; Wilkinson and Davies, 2002).

Under drought conditions, ABA induces the expression of a number of stress-related genes in plants, among which are positive and negative regulators of ABA signalling, TFs and genes encode enzymes that are involved in the synthesis of osmo-protectants. The 9-cis-epoxycarotenoid dioxygenase (NCED) encoding gene, one of the first enzymes of the ABA biosynthesis, is rapidly induced in response to the water deficit stress (Iuchi et al., 2001). The up-regulate of NCEDs expression has also been observed in *Arabidopsis* (Iuchi et al., 2001; Endo et al., 2008), maize (Schwartz et al., 1997; Tan et al., 1997), tomato (Sun et al., 2012), bean (Qin and Zeevaart, 1999) and rice (Huang et al., 2018).

Drought stress reduces plant growth and the production of cereal crops such as wheat (*Triticum aestivum* L.) and barley (*Hordeum vulgare* L.), which are among the most important cereal crops as a source of food for humans and animals (Turner and Asseng, 2005; Ghotbi-Ravandi et al., 2014; Naeem et al., 2015). The challenge of drought is even greater for crops such as rice because of its relatively high water needs (Haefele et al., 2009; Singh et al., 2012). Maize is more sensitive than other cereals to drought at flowering because anthers and silks are separated, and pollen and stigma are exposed to the environment, resulting in a lower than average yield (Quanqi et al., 2008; Efeoğlu et al., 2009). In orange trees, prolonged or excessive water stress can cause the fall of leaves, progressive drying of the branch tips, and drastic reduction in fruit yield due to flower and fruit abscission. In cases of very fast and severe stress, water stress can cause even more catastrophic events, such as xylem embolism and cell dehydration (Gomes et al., 2004). The predicted global warming and climate change will increase the frequency of drought, and hence decreases the agriculture crop productivity (Lipiec et al., 2013; Anjum et al., 2017; Lamaoui et al., 2018).

UV radiation stress

Ultraviolet radiation (UVR) is a type of solar radiation with wavelengths between 200 and 400 nm. The atmosphere absorbs all UV-C (< 280 nm), a significant part of UV-B (280–315 nm) but transmits most of the UV-A radiation (315–400 nm). UVR represents only about 7–9 % of total solar radiation reaching the biosphere, but unlike other types of solar radiation, UVR is highly energetic

(WHO, 1994; Lucas et al., 2008). This means that UVR can cause reactions among molecules and affect their structures and functions (WHO, 1994; Lucas et al., 2008). In plants, DNA damage will occur whenever the organism is exposed to UVR (Gill et al., 2015). The targets of UVR in any living cell are DNA, lipids, and proteins. Mutations are mostly negative, by inhibiting vital cellular processes (transcription and DNA replication), resulting in an altered cellular function, even causing cell death (Jansen et al., 1998; Frohnmeyer and Staiger, 2003; Jenkins, 2009; Gill et al., 2015). Researchers have shown that DNA damage due to UVR is mostly repaired by subsequent exposure to light in the blue or UV-A ranges of the spectrum (Hessen, 2008). Yet, DNA damage is not simply bad news: UVR has been an important evolutionary force, generating mutations, leading to acquisition of new traits, and driving the development of species diversity (Hessen, 2008).

One of the most consistent morphogenic responses of plants to solar UVR is synthesis and accumulation of UV absorbing compounds (Salama et al., 2011; Dhanya and Jos, 2017). UV-protective compounds in plants include mycosporine-like amino acids (MAAs), which are found in algae and in a variety of phenolic substances synthesized in vascular plants (Sinha et al., 2007; Bhatia et al., 2011; Carreto and Carignan, 2011). The concentration and type of these compounds generally depends on the group of organisms, and the level of UV-B radiation (Sinha et al., 2007; Bhatia et al., 2011). Another common protection mechanism development of plants is the synthesis of phenolic compounds in epidermal tissues (Bornman et al., 1997; Olsson et al., 1998). These compounds function as a barrier that reduces the penetration of UV-B radiation into the leaves (Burger and Edwards, 1996; Bornman et al., 1997). It also increases protection against oxidative stress because due to their chemical structures they are capable of scavenging free radicals, acting as antioxidants (Blokhina et al., 2003; Berli et al., 2010). Since UV absorbing compounds accumulate in the surface layers of plant tissue, they may significantly change optical properties of plant organs including fruits, flowers and leaves (Dhanya and Jos, 2017).

As mentioned previously, ABA regulates plant responses to various stressful abiotic and biotic factors, controlling many physiological and biochemical acclimation processes. Therefore, it is feasible to assume that ABA regulates plant responses to UV-B (Tossi et al., 2009; Berli et al., 2010; Berli et al., 2011; Berli and Bottini, 2013). In grapes and berries, significant alterations in the growth, composition of fruit, and taste and quality, are achieved by the elevated UV radiation (Tossi et al., 2009; Berli et al., 2010; Berli et al., 2011; Berli and Bottini, 2013). In a study conducted on grape leaves, a 2.7-fold increase in ABA levels was observed under high UV-B radiation compared to low UV-B radiation, suggesting that ABA is involved in plant protection against UV-B (Berli et al., 2010). Such responses include the increase in the activity of antioxidant enzymes and enhancement of membrane sterols that participate in structural defence (Berli et al., 2010). High UV-B levels may cause damage directly and/or indirectly, through overproduction of ROS, to a wide range of cellular constituents in different plant tissues. There are reports of impairment of nucleic acids, proteins, and lipids that inhibit photosynthetic reactions and different cell membrane processes because of ROS overproduction (Jansen et al., 1998; Frohnmeyer and Staiger, 2003; Jenkins, 2009; Gill et al., 2015). ABA stimulates the activity of NADPH oxidase and the generation of H_2O_2, which leads to increased NO production, suggesting that ABA is a key player in the activation of antioxidant enzymes (Tossi et al., 2009). These results suggest that the antioxidant defence system is initially activated by high UV-B levels and ABA acts downstream in the signalling pathways (Tossi et al., 2009).

EXOGENOUS APPLICATION OF ABA TO ECONOMICALLY IMPORTANT CROPS

Regulation of stomatal closing by ABA application

Shortly after its discovery, it was observed that the amount of ABA increased when the plant withered (Wright and Hiron, 1969), causing stomatic closure (Mittelheuser and Van Steveninck, 1969). This suggested an important role of ABA in water content regulation and stomatic closing during drought conditions. It is now known that drought stress induces the biosynthesis of the plant hormone ABA, activating complex signalling pathways in guard cells, which promotes stomatal closure and reduces transpirational water loss (Kim et al., 2010). The regulation of stomatal closing by ABA has been extensively studied (Kim et al., 2010; Merilo et al., 2013; McAdam et al., 2015; McAdam et al., 2016; Merilo et al., 2018). Plants regulate the use of water by the dynamic movement of stomata on the surface of the leaves (Kim et al., 2010). When air humidity decreases, guard cells shrink and stomata close in response to the increase in vapour pressure deficit (VPD) (the difference between the vapour pressures of leaf and atmosphere). In angiosperms, the stomatal responses to VPD appear to be regulated by the phytohormone ABA (Farquhar and Sharkey, 1982; Merilo et al., 2013; McAdam et al., 2015). When leaves of *A. thaliana*, *P. sativum*, and S. *lycopersicon* were exposed to a doubling in VPD for 20 minutes, the stomata closed rapidly, the foliar levels of ABA increased, and the expression of the NCED (9-cis-epoxycarotenoid dioxygenase) gene, specific to ABA foliar biosynthesis, increased significantly (McAdam et al., 2016). The guard cells respond to ABA through molecular signalling that involves the union of the ABA receptors PYR/PYL/RCAR protein family with members of the A subgroup of the PP2C family: this allows the release of *OST1* inhibition (Kim et al., 2010). OST1 is an ABA-activated protein kinase (AAPK), specifically involved in the regulation of the stomatal closure of ABA signalling (Mustilli et al., 2002). The *Arabidopsis* mutants, *ost1-1* and *ost1-2*, lost their ability to regulate their transpiration under drought conditions and are not responsive to ABA induction (Mustilli et al., 2002).

It is proposed that ABA promotes the closure of the stomata dually through its molecular and biochemical effects already known in the guard cells, and by a new indirect hydraulic effect through a decrease in water permeability within vascular tissues of the leaf (Pantin et al., 2013).

Stomatal closure reduces transpirational water loss but also can affect photosynthesis and thereby crop growth and productivity (Farquhar and Sharkey, 1982). Wheat (*Triticum aestivum* L.) is one of the most important food grains in the world. The global wheat production is approximately 766 million tonnes, of which 68% is used for food consumption (FAO, 2019). Drought stress is an environmental condition that reduces wheat production yields in the world's production areas (Saini and Aspinall, 1981). The cultivation of wheat in Mediterranean-type environments is exposed to 'terminal drought', which occurs after flowering and is characterized by a decrease in rainfall, increased temperatures, and VPD. The demand for evapotranspiration increases in spring and early summer, which often causes water deficiency when wheat crops enter the reproductive stage, causing pollen sterility and reducing grain yield (Turner and Asseng, 2005).

Several studies conducted on wheat show that the ABA concentration increases in the leaves under drought stress, inducing stomatal closure and inhibiting transpiration (Mittelheuser and Van Steveninck, 1969; Kim et al., 2010; Lim et al., 2015; Saradadevi et al., 2015; Saradadevi et al., 2017). It has been suggested that the application of ABA after anthesis can reduce stomatal conductance and transpiration, prolonging the availability of soil water and thus improving wheat yields in environments with water scarcity, as the terminal drought (Kirkegaard et al., 2007; Saradadevi et al., 2015). In a study conducted in the field, wheat plants at different phenological stages (at the beginning of the bud enlargement and repeated before) were sprayed with ABA (1 mM) for three consecutive winter-spring seasons. Plants treated with ABA showed an increase in the dry weight of the shoot, an increase in the content of photosynthetic pigments, carotene, and total chlorophylls. Also, the leaves of plants treated with ABA had a minimal opening of the ostiolus, which generated a higher conductance of gases and water vapour. ABA-treated plants also showed greater mobilization of total soluble carbohydrates towards developing grains relative to control plants. Besides, ABA-treated plants significantly increased the amount and weight of grains per spike and therefore crop yield per hectare (11%) (Travaglia et al., 2010). The results of this study suggest that the exogenous application of ABA could improve agricultural production in rainfed crops, with economic and environmental benefits, because wheat is valuable not only for the grain it produces but also for its use as stubble (Travaglia et al., 2010). In *Ilex paraguariensis* St. Hil. (yerba mate), sprayed leaves of with ABA alleviated diurnal water stress and promoted an increase in the leaf area, in the dry weight of the leaf (DW), in the length of the shoot, and the shoot DW (Sansberro et al., 2004). In another study, the exogenous application of ABA as a soil drench to two spring wheat (*Triticum aestivum* L.) cultivars released in different decades increased the concentration of ABA in the leaves, reduced stomatal conductance (gs), decreased the use rate of water, and increased antioxidant capacity and defence mechanisms by reducing the formation of ROS, which allows a reduction in the oxidative damage of the lipid membranes of both cultivars, also improved desiccation tolerance. Although no improvement in grain yield was found, it is essential to note that the ABA treatment induced a more significant adaptation to drought (Du et al., 2013).

The possibility of using the ABA response to facilitate rapid detection of a large number of wheat germplasm for drought tolerance has been explored (Ji et al., 2011; Du et al., 2013; Saradadevi et al., 2015; Saradadevi et al., 2017). Likewise, differential accumulation of ABA has been observed in different wheat genotypes under drought stress (Ji et al., 2011; Du et al., 2013; Saradadevi et al., 2015; Saradadevi et al., 2017). For example, a comparison of the ABA response between drought-sensitive and drought-tolerant wheat varieties showed that ABA levels significantly increase in the spike and leaf of the drought-sensitive wheat varieties Sundor and Cranbrook, compared to drought-tolerant wheat varieties Halberd and SYN604 (Ji et al., 2011).

ABA is synthesized at the roots and moves to the aerial parts of the plant through the xylem to regulate stomatic movement (Jiang and Hartung, 2008). ABA seems to be the signal that triggers this response; however, its role as the only molecule that regulates the communication between the root and shoot has not yet been confirmed (Munns and King, 1988; Zhang and Davies, 1990). In wheat, during the terminal drought, the upper layers of the soil dry, exposing the upper part of the wheat root to water deficit, while deep roots can still access the water (Saradadevi et al., 2015). The exposed roots of the upper soil layer regulate the stomatic response to drought, indicating that the soil distribution of the roots plays a vital role in the generation of the signal for stomatic closure (Saradadevi et al., 2015; Saradadevi et al., 2017). Saradadevi et al. (2015) found that wheat genotypes with more root biomass in the upper drying layer of the soil profile had a higher production of ABA in the leaf, and an early stomatic closure that allows them to regulate the water deficit under the terminal drought. The comparison between the wheat cultivar Drysdale and breeding line IGW-3262 showed that Drysdale started stomatal closure before IGW-3262, possibly due to a higher signal intensity generated due to a relatively higher proportion of roots in the upper drying segment. Similarly, the relationship between the amount of ABA in the leaf and the stomatal conductance was strong in Drysdale but weak in IGW-3262 (Saradadevi et al., 2015). A similar result was observed in two barley (*Hordeum vulgare*) genotypes: ABA concentrations were higher in plants with more roots in dry soil (Martin-Vertedor and Dodd, 2011). In maize (*Zea mays*) ABA concentrations in the xylem of seedlings with soil drying treatment are substantially increased, resulting in a restriction of stomatal conductance and a limitation in the growth of shoots (Zhang and Davies, 1990).

Improvement of the organoleptic characteristics of grapes by ABA

Grapes are one of the world's most commonly produced fruit crops, with approximately 75 million tonnes produced each year. Almost 50% of grapes are used to make wine, one-third is consumed as fresh fruit, and the rest is dried, consumed as grape juice, or stored in the form of grape musts (FAO-OIV, 2016). Table grapes (*Vitis* spp.) have become an important fresh product for human consumption around the world (FAO-OIV, 2016). A problem faced by table grape crops that grow in tropical and warm regions is high temperatures during ripening, which inhibit anthocyanin biosynthesis, so they don't achieve a good uniform colour (Spayd et al., 2002; Rybka et al., 2015). This decreases market acceptance of table grapes, and therefore decreases their economic value (Roberto et al., 2012). It has long been known that the application of ethephon, an ethylene release agent, improves the colour of table grapes when applied at the beginning of the ripening (veraison) (Jensen et al., 1975). Recently, it has been observed that the exogenous application of (S)-*cis*-abscisic acid (S-ABA) promotes the anthocyanin biosynthesis in the skins of grapes, improving the colour much better than ethephon (Ban et al., 2003; Peppi et al., 2006; Peppi et al., 2007; Koyama et al., 2018). The timing application of S-ABA had few effects on weight, juice composition and increased softness, firmness and anthocyanin concentrations in the 'Flame Seedless' grape skins (Peppi et al., 2006). Also, any ABA concentration between 75 and 300 mg.L^{-1} applied after veraison improved the colour of 'Flame Seedless' better that ethephon used at the same time (Peppi et al., 2006). Similar results were observed in 'Redglobe' grapes when approximately 300 mg. L^{-1} of ABA was applied at veraison (Peppi et al., 2007). In another study, the applications of 400 mg. L^{-1} of S-ABA at different times to the seedless hybrid grape (*V. vinifera* x *V. labrusca*), which presents a lack of red colour when grown in subtropical areas, resulted in a more significant accumulation of total anthocyanins in the berry skin and improved the colour attributes of the berries. The treatment of 400 mg. L^{-1} of S-ABA in two applications at 7 and 21 days after the veraison (DAV) resulted in a greater accumulation of total anthocyanins in the skin of the berries. The treatment with 400 mg. L^{-1} of S-ABA significantly increased the expression of the TFs *VvMYBA1* and *VvMYBA2* and the expression of the flavonoid biosynthetic genes *CHI*, *F3H*, *DFR* and *UFGT* in seedless grape (Koyama et al., 2018).

Effects of ABA and UV-B radiation on plants

Ultraviolet radiation (UVR) is part of the spectrum of electromagnetic radiation in the range 200–400 nm emitted by the sun. Depending on the wavelength, UV energy is divided into UV-A (315–400 nm), UV-B (280–315 nm) and UV-C (100–280 nm) (Lucas et al., 2008). The UV-C has the shortest and most energetic wavelengths; they are fully absorbed by the atmosphere with minimal penetration into the earth. UV-A crosses the atmosphere with few changes, while 90% or more of UV-B is absorbed by the atmospheric ozone layer (Roy et al., 1998). UV-B radiation is vital for the induction of vitamin D biosynthesis and, although UV-B radiation represents only a fraction of the total radiation, the action of the UV-B spectrum allows DNA to adsorb it and causes damage to nucleic acids (WHO 1994; Lucas et al., 2008). In humans, high exposure to UV-B radiation seems to be a critical factor in the beginning of the carcinogenic process in the skin (WHO, 1994; Lucas et al., 2008). In plants, exposure to high levels of UV-B radiation induces alterations in plant development, morphology, and physiology (Frohnmeyer and Staiger, 2003). UV-B has the potential to damage macromolecules, including DNA, lipids, proteins, cell membranes, generate ROS, and impair cellular processes which initiate diverse responses that affect metabolism, development, and viability of higher plants (Jansen et al., 1998; Frohnmeyer and Staiger, 2003; Jenkins, 2009). However, plants have evolved mechanisms to counteract the damage caused by UV-B radiation. One of the most common protection mechanisms is the synthesis of phenolic compounds in epidermal tissues (Bornman et al., 1997; Olsson et al., 1998). These compounds reduce the penetration of UV-B radiation into the leaves by transforming short-wave radiation with high energy, which is potentially harmful, into longer wavelengths and less destructive radiation (Burger and Edwards, 1996; Bornman et al., 1997). It also increases protection against oxidative stress due to its chemical structures being capable of scavenging free radicals, acting as antioxidants (Blokhina et al., 2003; Berli et al., 2010). Additionally, phenolic compounds are particularly important because of their wide distribution in plants and because they provide important organoleptic characteristics for wine quality, such as bitterness, structure, and colour (Cheynier, 2005).

It has been suggested that the defence response against UV-B radiation involved the transduction of the ABA signal that acts downstream of the pathway by triggering the biosynthesis of different phenols and improving the enzymatic and non-enzymatic response systems (Berli et al., 2010). In grapes, the application of ABA induces the biosynthesis of polyphenols such as anthocyanins and enhanced the mRNA accumulation of *VvmybA1*, a putative regulatory gene of anthocyanin biosynthesis of grape, as well as all the genes analysed in the anthocyanin biosynthesis pathway (Jeong et al., 2004). The combined effect of UV-B radiation and ABA application on *Vitis vinifera* L. cv. Malbec of one year old and cultivated in the field, increased the ABA content 12.5 times, the plants also showed a higher content of carotenoids, anthocyanins, flavonols (quercetin and kaempferol), hydroxycinnamic acids (caffeic acid and ferulic acid) and membrane sterols (β-sitosterol) that participate in structural defence. The combined effect of UV-B and ABA radiation promoted an increase in the activities of APX (ascorbate peroxidase) and POX (peroxidase), suggesting that ABA promotes the antioxidant defence by inducing the APX and POX activities in a UV-B dependent manner (Berli et al., 2010). In plants, ABA can increase ROS production as a second messenger and induce SOD, CAT and APX activities (Jiang and Zhang, 2001; Sakamoto et al., 2008). In another study conducted in *Vitis vinifera* L. cv. Malbec from a vineyard at an altitude of 1450 m, regarding the interaction between ABA and UV-B on berry growth, sugar accumulation, and phenol profiles across berry development and ripening, it was found that the combined treatment of UV B and ABA (1mM) increased the ABA levels of the berry skin, which also accelerated the onset of ripening by up to 20 days, and decreased towards the harvest, which implies a possible role of ABA in the control of ripening berries (Berli et al., 2011). The combined treatment also reduced berry growth and decreased berry sugar without affecting the concentration of sugar (°Brix) at harvest. Similarly, the level of phenols in the skin of the berries increased with UV-B treatment, with an additional increase when ABA was applied, inducing a change in the anthocyanin and non-anthocyanin profiles and an increase in the proportion of phenols with high antioxidant capacity. The levels of these phenols were maintained until harvest (Berli et al., 2011). The UV-B stimulus and ABA also induce hydrogen peroxide (H_2O_2) production and the nitric oxide (NO) synthesis pathway (Tossi et al., 2009). H_2O_2 plays a crucial role as a signalling molecule in several physiological processes, its biosynthesis increases during environmental stress and in the systematic response to pathogens (Slesak et al., 2007). The participation of NO has been reported against drought and saline stress and against cell damage induced by UV-B (Garcıa-Mata and Lamattina, 2002; Zhao et al., 2004; Shi et al., 2005). In maize (*Zea mays*), the ABA concentration increased by 100%, while H_2O_2 and NO concentrations increased by 70% and 100% respectively, in UV-B irradiated leaves. On the other hand, the *viviparous 14* (*vp14*) mutant, defective in ABA synthesis, was more sensitive to UV-B irradiation and presented more damage than wild-type (wt). The application of ABA (100 mM) attenuated UV-B induced damage in both wt and *vp14* and restored the production of H_2O_2 and NO in *vp14* mutants. The results suggest that UV-B irradiation increases the concentration of ABA, which activates the generation of NO and H_2O_2, indicating that ABA is necessary to increase NO and H_2O_2 production and thus maintain cell homeostasis and attenuate harmful cell damage derived from UV-B irradiation (Tossi et al., 2009). The interaction between UV-B and ABA has also been addressed in plants such as *Populus yunnanensis* (Duan et al., 2008) and *A. thaliana* (Rakitin et al., 2008).

ABA APPLICATION FOR POST-HARVEST MAINTENANCE OF ORNAMENTAL AND BEDDING PLANTS

The ornamental and floricultural sectors represent an important economic market in the world (Thomas, 2014). Before the plants go on sale, they are exposed to adverse environmental conditions such as high temperatures and lack of irrigation, which causes wilts in the plants, affecting their quality, and causing economic losses. Drought stress is a major cause of the postproduction decline of ornamental and bedding plants. However, the exogenous application of ABA or ABA analogues has received significant interest in counteracting the effects caused by drought stress in this type of plant and thus avoiding the losses caused by this condition (Sharma et al., 2005; Sharma et al., 2006; Blanchard et al., 2007). The foliar application of ABA at 200 or 400 $mg.L^{-1}$ to the flowering herbaceous bedding plant *Impatiens walleriana* caused fewer symptoms of moisture stress, and the ABA-treated plants presented better quality than the untreated plants. Similarly, ABA extended the marketability of other bedding plants such as *Catharanthus roseus, Impatiens hawkeri, Pelargonium x hortorum, Verbena x hibrida and Petunia* from 1.1 to 5.8 days (Blanchard et al., 2007). In another study, the application of 0 or 500 mg L^{-1} of s-ABA by spraying and drenching the bedding plants *Impatiens walleriana* (impatiens), *Pelargonium hortorum* (seed geranium), *Petunia hybrida* (petunia), *Tagetes patula* (marigold), *Salvia splendens* (savia) and *Viola wittrockiana* (pansy) delayed wilting in all crops (1.7 to 4.3 days) and allowed plants to survive for short periods of time of severe drought (Waterland et al., 2010). Conversely, in hydrangea (*Hydrangea macrophylla*), the ABA drenching treatment was more effective than spraying to reduce transpiration and extend the life of the plant. The effect of ABA on the available water of the plant was dependent on the ABA concentration. A higher concentration of ABA (1,000 ppm) correlated with a greater availability of water for the plant (Iersel et al., 2009). In another experiment, the root immersion treatment with ABA analogues, 8′-methylene ABA methyl ester (PBI 365) and 8′-acetylene ABA methyl ester (PBI 429) to tomato (*Lycopersicon esculetum*), snapdragon (*Antirrhinum majus*) and nasturtium (*Tropelaum majus*) reduced slow growth and allowed control of moisture loss in a dosage-dependent manner, with minimal adverse effects on the appearance of these three transplantable horticultural crops (Sharma et al., 2006). This suggests the potential use of this ABA analogue in controlling grown and extending the marketing period of seedlings during handling.

All these studies showed that ABA applications to ornamental and bedding plants before shipping and retailing would allow extending the post-harvest longevity of several plant species, and maintain the commercialization of ornamental and bedding plants even under severe drought stress conditions.

ABA AND ITS POTENTIAL USE IN HUMAN HEALTH

Since its discovery in plants, the ABA phytohormone has captured the attention of the scientific community. Now, the ABA mode of action, the regulation of its biosynthesis, the signal transduction and the different biological functions performed by ABA in plants are known in detail. The most surprising thing is that ABA is not only limited to plants but also has been found in mosses, sponges, protozoans, mammals, and humans, suggesting an ancient origin of ABA and its signal transduction mechanisms (Le Page-Degivry et al., 1986; Zocchi et al., 2001; Nagamune et al., 2008; Wasilewska et al., 2008). In mammals, ABA has been detected in the central nervous system of pigs and rats (Le Page-Degivry et al., 1986). In humans, ABA was detected in mesenchymal stem cells (Scarfì et al., 2008), granulocytes (Bruzzone et al., 2007), pancreatic islets (Bruzzone et al., 2008) and monocytes (Magnone et al., 2009). In recent years, ABA has received considerable attention in the medical field due to its potential for biomedical applications (Bassaganya-Riera et al., 2010b; Li et al., 2011b). Bruzzone et al. (2007) found that the release of ABA by human granulocytes has pro-inflammatory cytokine properties by stimulating the production of ROS, NO, phagocytosis, chemotaxis, and chemokinesis through an $[Ca^{2+}]_i$ increase. In plants, $[Ca^{2+}]_i$ functions as a second messenger in response to ABA for stomatic closure (Kim et al., 2010). In granulocytes, the signalling pathway involves the pertussis toxin (PTX)-sensitive receptor/ protein G complex, activation of cAMP overproduction, PKA (protein kinase A)-mediated ADPRC (ADP-ribosyl cyclase) activation and overproduction of cADPR (cyclic-ADP-ribose), which allows the increase of $[Ca^{2+}]_i$ (Bruzzone et al., 2007). Also, ABA biosynthesis in granulocytes was stimulated by the chemical application of 0.1 g/ml PMA (Phorbol 12-myristate 13-acetate), an activator of protein kinase C, or fever-like temperatures (39°C). Besides, the presence of ABA was reported freely and conjugated in human granulocytes. Approximately 30% of the ABA conjugate was found as an alkaline-hydrolysable conjugated ABA, representing a form of intracellular storage of the hormone (Bruzzone et al., 2007). Similar ABA signalling steps for the increase of $[Ca^2+]$ in granulocytes have been found in plants, suggesting a conservation of the ABA signalling path between plants and humans (Bruzzone et al., 2007).

In another study, the functional activation of monocytes and vascular cells was induced by ABA, through the signalling cascade that includes the release of $[Ca^{2+}]_i$ and an increase in the production of PGE2 (prostaglandin E2), MCP-1 (monocyte chemoattractant protein-1) and MMP-9 (metalloprotease-9), essential participants in the development of atherogenesis (Magnone et al., 2009). ABA acts as a monocyte chemoattractant to atherosclerotic plaque, stimulates the proliferation of vascular smooth muscle cells (VSMC), and stimulates the migration and proliferation of aortic smooth muscle cells (AoSMC), critical events in the development of atherogenesis (Magnone et al., 2009). Understanding the molecular mechanisms and signalling pathway of granulocyte activation by ABA would allow the development of new drugs with anti-inflammatory effects (Bruzzone et al., 2007). Similarly, ABA as a new signal molecule involved in the development of atherosclerosis could suggest a possible new target for antiatherosclerotic therapy (Magnone et al., 2009).

In metabolic syndrome and type II diabetes, the cascade signalling between pancreatic β-cells and inflammatory cells is recognized as a pathogenic mechanism of these diseases. In rat insulinoma cells (RIN-m and INS-1) and human and murine pancreatic β-cells, ABA stimulates insulin secretion. The ABA biosynthesis is also stimulated at high glucose concentrations, and an increase in nanomolar ABA concentration is sufficient to induce insulin secretion from RIN-m and INS-1 cells and murine and human pancreatic islets. The signalling cascade triggered by ABA in insulin-releasing cells is very similar to the signalling pathway found in granulocytes (Bruzzone et al., 2007) and monocytes (Magnone et al., 2009), suggesting an important role of ABA in the signalling network between inflammatory cells and pancreatic cells during the development of metabolic syndrome and diabetes (Bruzzone et al., 2008).

In a study conducted using annotated databases from humans (Remm et al., 2001; Bossi and Lehner, 2009) and *Arabidopsis* (Swarbreck et al., 2007; Peng et al., 2008), 1,677 human genes were identified that presented orthology with *Arabidopsis* genes related to ABA signalling. Analysis of the protein interaction network showed that 1,048 of the ABA-related human genes are highly connected (3,178 interactions between themselves) (Bassaganya-Riera et al., 2010b). Four genes in humans related to ABA signalling in plants were identified, which could be anchored in a peroxisome proliferator activator receptor γ (PPARγ) network. In mammals, PPARγ is a crucial metabolic regulator and a transcription factor that promotes adipogenesis and reduces the production of pro-inflammatory cytokines (Xu et al., 2003; Bassaganya-Riera et al., 2005). In mice fed high-fat diets, dietary supplementation with ABA (100 mg/kg was the lowest effective dose) for 36 days improves glucose tolerance and obesity-related inflammation through PPARγ transactivation in white adipose tissue (WAT). In addition, ABA-fed mice showed an increase in PPARγ expression and its responsive genes. During the onset of obesity, adipocytes become more hypertrophic and increase the production of pro-inflammatory cytokines. The drug class TZDs (thiazolidinedione) are used because they improve insulin sensitivity through its binding and activation of PPARγ in WAT (Xu et al., 2003). However, the use of TZDs is associated with side effects such as weight gain, hepatoxicity and heart failure that limit its use in thousands of patients with diabetes (Xu et al., 2003; Bassaganya-Riera et al., 2005). Based on the results found in this research, ABA has been proposed as an alternative to the class of TZD drugs to improve glucose homeostasis (Guri et al., 2007). Clinical findings provide evidence that ABA-induced improvements in severe colitis depend on PPARγ activity (Bassaganya-Riera et al., 2010a). Guri et al., (2007) evaluated the anti-inflammatory efficacy of ABA previously reported. The results showed that ABA significantly improved experimental inflammatory bowel disease (IBD), reducing colitis, leukocyte infiltration in the colon and inflammation. The use of ABA has also been suggested as a therapeutic alternative to prevent or mitigate angiogenesis involved in the development of ischaemic retinopathies (IRs) (Baliño et al., 2019). ABA is believed to be an essential part of human health. ABA is produced in the liver and is considered as a derivative of vitamin A. Therefore, the use of ABA and its analogues has been suggested as a dietary supplement in humans and as a food additive in animals and birds (Herrero and Shafer, 2012, 2015).

The potential use of ABA in cancer prevention has been addressed. The first report on the use of ABA in the treatment of cancer was a patent developed by Virginia Livingston in 1976. She found that the intervention of mice with ABA neutralize microbic Chorionic Gona- 65 dotropin, which was similar to the Human Chorionic Gona dotropin (hCG). hCG is a negatively charged glycoprotein that prevented the immune response by covering cancer cells. Their results also showed that after 14 days of ABA treatment there was an inhibition of the C1498 transplanted tumour in the C57Bl / 6J mouse. The C1498 tumour is a myeloid leukaemia that is lethal to the host in 10–15 days (Livingston, 1976). In another patent as a continuation of the previous one, it was found that the intervention with ABA counteracted and reversed the tissue damage syndrome caused in different types of cancer (colorectal cancer, breast cancer, gastric cancer, and all malignant neoplasms) by decreasing the metabolic movement of Na^+ and K^+ ions. It is known that there is a positive correlation between the increase in Na^+ in human tissue and the incidence of cancer (Romero, 2006). It is suggested that this depolarization in malignant tissue produces changes in the configuration of the cell, going from a damaged state to a normal one (Romero, 2006).

Previous reports indicate that ABA exerts its action through the signalling cascade that primarily involves the increase and release of $[Ca^{2+}]_i$, as well as the synthesis of ROS and NO (Bruzzone et al., 2007; Magnone et al., 2009). For mitochondrial dysfunction and apoptosis, an increase in the release of Ca^{2+} and the production of ROS and NO are necessary (Kruman et al., 1998; Parihar et al., 2008). Anti-cancer drugs such as tamoxifen and etoposide induce apoptosis of human breast cancer cell lines by increasing $[Ca^{2+}]$ and mitochondrial oxygen (Parihar et al., 2008). Due to its ability to modulate calcium, ROS, and NO signalling, ABA could be considered as a possible treatment against cancer. Zhou et al. (2016) found that ABA promotes apoptosis and cell differentiation of human glioma cells (U87-MG and A172 cell lines) through a pathway similar to retinoid acid (RA). Tumour tissues of patients with low-grade glioma (LGG) had ABA levels approximately twice as high as high-grade glioma (HGG) tissues. Likewise, a negative correlation was observed between ABA levels and the expression of the pathological biomarker of glioma KI-16 in a dose-dependent manner. AR signalling is involved in the pathogenesis of glioma. RA participates in cell differentiation by activating two receptors, the RA receptor (RAR) and PPAR. RA is transported to RAR by the cellular retinoic acid-binding protein 2 (CRABP2) and to PPAR by the fatty-acid-binding protein 5 (FABP5). The RA-CRABP2-RAR pathway leads to apoptosis and cell differentiation, playing an anti-cancer role, while the RA-FABP5-PPAR pathway promotes cell survival, tumour growth and proliferation (Schug et al., 2007). Interestingly, a positive correlation was observed between the increase in ABA and the expression and accumulation of CRABP2 proteins in glioma tissue. ABA treatment also induced a significant increase in the expression of *CRABP2* and *RXRγ* in these tissues. In contrast, a negative correlation was detected between ABA levels and the expression and accumulation of FABP5 and PPARγ proteins in glioma tissue. ABA also induces the expression of the pro-apoptotic proteins BAX and cleaved caspase-3 in U87MG and A172 cells, probably mediated by the CRABP2-RAR interaction. All these results suggest that ABA induces apoptosis and cell differentiation through the CRABP2-RAR interaction of the RA signal pathway (Zhou et al., 2016). Retinoic acid is used in cancer prevention and its therapeutic action has been tested in several types of human cancers (Soprano et al., 2004; Freemantle et al., 2006). However, its use may cause toxicity (Freemantle et al., 2006; Lo-Coco et al., 2013). The findings of this study may shed light on the development of new chemotherapeutic approaches against glioma based on ABA.

CONCLUSION

ABA is a phytohormone that controls various biological processes in plants, the response to drought stress being one of the most important. The knowledge of signal transduction and the mechanisms of stomatal movement mediated by ABA have made possible the efficient use of water and thus improve tolerance to drought stress of agronomic importance crops that grow in arid or dry environmental conditions. The evidence found in different studies presented in this chapter shows that the exogenous application of ABA can improve the production of crops grown on a large scale and under drought stress conditions. It can also enhance the organoleptic characteristics of fruits for consumption and extend the post-harvest life of various bedding plant species to maintain their commercialization. Although cost-benefit analyses still need to be done for exogenous application of ABA, the evidence provides the possibility of applying ABA as an option to improve the productivity of other agronomic interest crops that are sensitive to drought stress or that grow in adverse environmental conditions. Furthermore, ABA is a natural compound, so it does not pose any environmental threat.

ABA has recently been considered with the potential for biomedical applications in humans. Some human tissues produce and accumulate ABA; besides, it has been found that specific genes related to ABA signalling are conserved between humans and *Arabidopsis*, suggesting conservation of the ABA-mediated signal transduction mechanism in humans. Positive results have been found in the essays on the use of ABA and the induction of its signalling pathway to counteract diseases such as atherogenesis, inflammatory bowel disease, ischaemic retinopathies, diabetes, and different types of cancer in humans. The previous has allowed ABA to be considered as a therapeutic alternative to prevent and mitigate the development of these diseases in humans. However, it is necessary to carry out more tests on humans and to pass the experimental phase. A deeper understanding of ABA-mediated signal transduction mechanisms in humans will give us the ability to modulate best or manipulate this response in the benefit of human health.

ACKNOWLEDGEMENTS

The authors thank Consejo Nacional de Ciencia y Tecnología (CONACYT) for their support with the project: Fronteras de la Ciencia No. 2016-01-1716. RUC and MAE co-authors, postdoctoral position (FC No. 2016-01-1716) and Laboratory technical, respectively, were financially supported by CONACYT.

REFERENCES

Alazem M, Lin NS. 2017. Antiviral roles of abscisic acid in plants. *Front Plant Sci.* 8:1760–1760.

Anjum S, Ashraf U, Zohaib A, Tanveer M, Naeem M, Ali I, Tabassum T, Nazir U. 2017. Growth and developmental responses of crop plants under drought stress: a review. *Zemdirbyste-Agriculture.* 104(3):267–276.

Azcón-Bieto J, Talón M, editors. 2008. *Fundamentos de fisiología vegetal*, Second ed. McGraw-Hill Interamericana, España S. L.

Bai G, Yang DH, Zhao Y, Ha S, Yang F, Ma J, Gao XS, Wang ZM, Zhu JK. 2013. Interactions between soybean ABA receptors and type 2C protein phosphatases. *Plant Mol Biol.* 83(6):651–664.

Baliño P, Gómez-Cadenas A, López-Malo D, Romero FJ, Muriach M. 2019. Is there a role for abscisic acid, a proven anti-inflammatory agent, in the treatment of ischemic retinopathies?. *Antioxidants (Basel).* 8(4):104.

Ban T, Ishimaru M, Kobayashi S, Goto Yamamoto N, Horiuchi S. 2003. Abscisic acid and 2, 4-dichlorophenoxyacetic acid affect the expression of anthocyanin biosynthetic pathway genes in 'Kyoho' grape berries. *J Hortic Sci Biotechnol.* 78(4):586–589.

Bassaganya-Riera J, Guri A, Hontecillas R, inventors; Virginia Tech Intellectual Properties, Inc., Aug 26, 2010a. Method of Using Abscisic Acid to Treat and Prevent Diseases and Disorders. United States Patent. Patent number: US20100216883A1. Blacksburg, VA.

Bassaganya-Riera J, Guri A, King J, Hontecillas R. 2005. Peroxisome proliferator-activated receptors: the nutritionally controlled molecular networks that integrate inflammation, immunity and metabolism. *Curr Nutr Food Sci.* 1(2):179–187.

Bassaganya-Riera J, Skoneczka J, Kingston DGJ, Krishnan A, Misyak SA, Guri AJ, Pereira A, Carter AB, Minorsky P, Tumarkin R et al. 2010b. Mechanisms of action and medicinal applications of abscisic acid. *Curr Med Chem.* 17(5):467–478.

Beltran JCM, Stange C. 2016. Apocarotenoids: a new carotenoid-derived pathway. In: Stange C, editor. *Carotenoids in Nature: Biosynthesis, Regulation and Function.* Cham: Springer International Publishing; 239–272.

Berli F, Bottini R. 2013. UV-B and abscisic acid effects on grape berry maturation and quality. *J Berry Res.* 3(1):1–14.

Berli F, Fanzone M, Piccoli P, Bottini R. 2011. Solar UV-B and ABA are involved in phenol metabolism of *Vitis vinifera* L. increasing biosynthesis of berry skin polyphenols. *J Agric Food Chem.* 59(9):4874–4884.

Berli F, Moreno D, Piccoli P, Hespanhol-Viana L, Silva M, Bressan-Smith R, Cavagnaro JB, Bottini R. 2010. Abscisic acid is involved in the response of grape (*Vitis vinifera* L.) cv. Malbec leaf tissues to UV-B radiation by enhancing UV-absorbing compounds, antioxidant enzymes and membrane sterol. *Plant Cell Environ.* 33(1):1–10.

Bhatia S, Garg A, Sharma K, Kumar S, Sharma A, Purohit AP. 2011. Mycosporine and mycosporine-like amino acids: a paramount tool against ultra violet irradiation. *Pharmacogn Rev.* 5(10):138–146.

Blackman PG, Davies WJ. 1985. Root to shoot communication in maize plants of the effects of soil drying. *J Exp Bot.* 36(1):39–48.

Blanchard MG, Newton LA, Runkle ES, Woolard D, Campbell CA. 2007. Exogenous applications of abscisic acid improved the postharvest drought tolerance of several annual bedding plants. *Acta Hort.* 755:127–133.

Blokhina O, Virolainen E, Fagerstedt KV. 2003. Antioxidants, oxidative damage and oxygen deprivation stress: A review. *Ann Bot.* 91(2):179–194.

Boneh U, Biton I, Zheng C, Schwartz A, Ben Ari G. 2012. Characterization of potential ABA receptors in *Vitis vinifera* [journal article]. *Plant Cell Rep.* 31(2):311–321.

Bornman JF, Reuber S, Cen YP, Weissenböck G. 1997. Ultraviolet radiation as a stress factor and the role of protective pigments. In: Lumsden P, editor. *Plants and UV-B: Responses to Environmental Change.* Cambridge: Cambridge University Press; 157-168.

Bossi A, Lehner B. 2009. Tissue specificity and the human protein interaction network. *Mol Syst Biol.* 5(260): 1–7.

Boyer GL, Zeevaart JAD. 1982. Isolation and quantitation of β-D-glucopyranosyl abscisate from leaves of *Xanthium* and Spinach. *Plant Physiol.* 70(1):227–231.

Bruzzone S, Bodrato N, Usai C, Guida L, Moreschi I, Nano R, Antonioli B, Fruscione F, Magnone M, Scarfì S et al. 2008. Abscisic acid is an endogenous stimulator of insulin release from human pancreatic islets with cyclic ADP ribose as second messenger. *J Biol Chem.* 283(47):32188–32197.

Bruzzone S, Moreschi I, Usai C, Guida L, Damonte G, Salis A, Scarfì S, Millo E, De Flora A, Zocchi E. 2007. Abscisic acid is an endogenous cytokine in human granulocytes with cyclic ADP-ribose as second messenger. *Proc Natl Acad Sci U.S.A.* 104(14):5759–5764.

Burger J, Edwards GE. 1996. Photosynthetic efficiency, and photodamage by UV and visible radiation, in red versus green leaf coleus varieties. *Plant Cell Physiol.* 37(3):395–399.

Burla B, Pfrunder S, Nagy R, Francisco RM, Lee Y, Martinoia E. 2013. Vacuolar transport of abscisic acid glucosyl ester is mediated by ATP-binding cassette and proton-antiport mechanisms in *Arabidopsis. Plant Physiol.* 163(3):1446.

Cao FY, Yoshioka K, Desveaux D. 2011. The roles of ABA in plant–pathogen interactions. *J Plant Res.* 124(4):489–499.

Carreto JI, Carignan MO. 2011. Mycosporine-like amino acids: relevant secondary metabolites. Chemical and ecological aspects. *Mar Drugs.* 9(3):387–446.

Cowan IR, Raven JA, Hartung W, Farquhar GD. 1982. A possible ole for abscisic acid in coupling stomatal conductance and photosynthetic carbon metabolism in leaves. *Funct Plant Biol.* 9(4):489–498.

Cushman JC, Bohnert HJ. 2000. Genomic approaches to plant stress tolerance. *Curr Opin Plant Biol.* 3(2):117–124.

Cutler AJ, Krochko JE. 1999. Formation and breakdown of ABA. *Trends Plant Sci.* 4(12):472–478.

Cutler SR, Rodriguez P, Finkelstein R, Abrams S. 2010. Abscisic acid: Emergence of a core signaling network. *Annu Rev Plant Biol.* 61:651–679.

Cheynier V. 2005. Polyphenols in foods are more complex than often thought. *Am J Clin Nutr.* 81(1):223S–229S.

Dhanya TTT, Jos TP. 2017. UV radiation priming: a means of amplifying the inherent potential for abiotic stress tolerance in crop plants. *Environ Exp Bot.* 138:57–66.

Dong T, Xu ZY, Park Y, Kim DH, Lee Y, Hwang I. 2014. Abscisic acid uridine diphosphate glucosyltransferases play a crucial role in abscisic acid homeostasis in *Arabidopsis. Plant Physiol.* 165(1):277.

Du YL, Wang ZY, Fan JW, Turner NC, He J, Wang T, Li F-M. 2013. Exogenous abscisic acid reduces water loss and improves antioxidant defence, desiccation tolerance and transpiration efficiency in two spring wheat cultivars subjected to a soil water deficit. *Funct Plant Biol.* 40(5):494–506.

Duan B, Xuan Z, Zhang X, Korpelainen H, Li C. 2008. Interactions between drought, ABA application and supplemental UV-B in *Populus yunnanensis. Physiol Plant.* 134(2):257–269.

Efeoğlu B, Ekmekci Y, Cicek N. 2009. Physiological responses of three maize cultivars to drought stress and recovery. *S African J Bot.* 75(1):34–42.

Endo A, Sawada Y, Takahashi H, Okamoto M, Ikegami K, Koiwai H, Seo M, Toyomasu T, Mitsuhashi W, Shinozaki K. 2008. Drought induction of Arabidopsis 9-cis-epoxycarotenoid dioxygenase occurs in vascular parenchyma cells. *Plant Physiol.* 147(4):1984–1993.

FAO-OIV. 2016. Table and dried grapes. *Nonalcoholic Pproducts of the Vitivinicultural Sector Intended for Human Consumption.* Published by Food and Agriculture Organization of the United Nations and the International Organisation of Vine and Wine. ISBN 978-92-5-109708-3. Paris, France. 64 pp.

FAO. November 2019. *Food Outlook: Biannual Report on Global Food Markets.* Food and Agriculture Organization of the United Nations, Rome, Italy, 104 pp. ISBN: ISBN 978-92-5-131932-1.

Farquhar GD, Sharkey TD. 1982. Stomatal conductance and photosynthesis. *Annu Rev Plant Physiol.* 33(1):317–345.

Finkelstein R. 2013. Abscisic acid synthesis and response. *Arabidopsis Book.* 11:e0166–e0166.
Finkelstein R, Reeves W, Ariizumi T, Steber C. 2008. Molecular aspects of seed dormancy. *Annu Rev Plant Biol.* 59(1):387–415.
Freemantle SJ, Dragnev KH, Dmitrovsky E. 2006. The retinoic acid paradox in cancer chemoprevention. *J Natl Cancer Inst.* 98(7):426–427.
Frohnmeyer H, Staiger D. 2003. Ultraviolet-B radiation-mediated responses in plants. Balancing damage and protection. *Plant Physiol.* 133(4):1420–1428.
Fuchs S, Grill E, Meskiene I, Schweighofer A. 2013. Type 2C protein phosphatases in plants. *FEBS J.* 280(2):681–693.
Fujii H, Chinnusamy V, Rodrigues A, Rubio S, Antoni R, Park SY, Cutler SR, Sheen J, Rodriguez PL, Zhu JK. 2009. *In vitro* reconstitution of an abscisic acid signalling pathway. *Nature.* 462(7273):660–664.
Fujii H, Zhu JK. 2009. *Arabidopsis* mutant deficient in 3 abscisic acid-activated protein kinases reveals critical roles in growth, reproduction, and stress. *Proc Natl Acad Sci U.S.A.* 106(20):8380–8385.
Fujita Y, Yoshida T, Yamaguchi-Shinozaki K. 2013. Pivotal role of the AREB/ABF-SnRK2 pathway in ABRE-mediated transcription in response to osmotic stress in plants. *Physiol Plant.* 147(1):15–27.
Furihata T, Maruyama K, Fujita Y, Umezawa T, Yoshida R, Shinozaki K, Yamaguchi-Shinozaki K. 2006. Abscisic acid-dependent multisite phosphorylation regulates the activity of a transcription activator AREB1. *Proc Natl Acad Sci U.S.A.* 103(6):1988–1993.
García-Martín G, Manzanera JA, González-Benito ME. 2005. Effect of exogenous ABA on embryo maturation and quantification of endogenous levels of ABA and IAA in *Quercus suber* somatic embryos. *Plant Cell Tiss Org.* 80(2):171–177.
García-Mata C, Lamattina L. 2002. Nitric oxide and abscisic acid cross talk in guard cells. *Plant Physiol.* 128(3):790–792.
Ghotbi-Ravandi AA, Shahbazi M, Shariati M, Mulo P. 2014. Effects of mild and severe drought stress on photosynthetic efficiency in tolerant and susceptible barley (Hordeum vulgare L.) genotypes. *J Agron Crop Sci.* 200(6):403–415.
Gill SS, Anjum NA, Gill R, Jha M, Tuteja N. 2015. DNA damage and repair in plants under ultraviolet and ionizing radiations. *Sci World J.* 2015: 250158.
Gomes M, Lagôa A, Medina, Machado O, Machado M. 2004. Interactions between leaf water potential, stomatal conductance and abscisic acid content of orange trees submitted to drought stress. *Braz J Plant Physiol.* 16(3):155–161.
Guri AJ, Hontecillas R, Si H, Liu D, Bassaganya-Riera J. 2007. Dietary abscisic acid ameliorates glucose tolerance and obesity-related inflammation in db/db mice fed high-fat diets. *Clin Nutr.* 26(1): 107–116.
Gutmann M, von Aderkas P, Label P, Lelu M-A. 1996. Effects of abscisic acid on somatic embryo maturation of hybrid larch. *J Exp Bot.* 47(12):1905–1917.
Haefele SM, Siopongco JDLC, Boling AA, Bouman BAM, Tuong TP. 2009. Transpiration efficiency of rice (Oryza sativa L.). *Field Crops Res.* 111(1):1–10.
Hao Q, Yin P, Li W, Wang L, Yan C, Lin Z, Wu Jim Z, Wang J, Yan SF, Yan N. 2011. The molecular basis of ABA-independent inhibition of PP2Cs by a subclass of PYL proteins. *Mol Cell.* 42(5):662–672.
Harris JM. 2015. Abscisic acid: hidden architect of root system structure. *Plants (Basel).* 4(3):548–572.
Harrison EH, Quadro L. 2018. Apocarotenoids: emerging roles in mammals. *Annu Rev Nutr.* 38(1):153–172.
Herrero MP, Shafer WE, inventors; Herrero Maria Pilar, Shafer Warren E., assignee. March 22, 2012. Pharmaceutical and Nutraceutical Compositions of Abscisic Acid. United States Patent. Patent number: US20120071556A1. United States of America.
Herrero MP, Shafer WE, inventors; Valent biosciences corporation, assignee. September 10, 2015. *Animal Feed Compositions of Abscisic Acid United States Patent. Patent number: US20150250209A1.* United States of America.
Hessen D. 2008. Solar radiation and the evolution of life. In *Solar Radiation and Human Health.* The Norwegian Academy of Science and Letters, Oslo. 122–136.
Hirschberg J. 2001. Carotenoid biosynthesis in flowering plants. *Curr Opin Plant Biol.* 4(3):210–218.
Hou X, Rivers J, León P, McQuinn RP, Pogson BJ. 2016. Synthesis and function of apocarotenoid signals in plants. *Trends Plant Sci.* 21(9):792–803.
Huang D, Jaradat MR, Wu W, Ambrose SJ, Ross AR, Abrams SR, Cutler AJ. 2007. Structural analogs of ABA reveal novel features of ABA perception and signaling in *Arabidopsis. Plant J.* 50(3):414–428.
Huang Y, Guo Y, Liu Y, Zhang F, Wang Z, Wang H, Wang F, Li D, Mao D, Luan S et al. 2018. 9-cis-epoxycarotenoid dioxygenase 3 regulates plant growth and enhances multi-abiotic stress tolerance in rice. *Front Plant Sci.* 9:162.

Iersel MWV, Seader K, Dove S. 2009. Exogenous abscisic acid application effects on stomatal closure, water use, and shelf life of Hydrangea (*Hydrangea macrophylla*). *J Environ Hortic.* 27(4):234–238.

Iuchi S, Kobayashi M, Taji T, Naramoto M, Seki M, Kato T, Tabata S, Kakubari Y, Yamaguchi-Shinozaki K, Shinozaki K. 2001. Regulation of drought tolerance by gene manipulation of 9-cis-epoxycarotenoid dioxygenase, a key enzyme in abscisic acid biosynthesis in *Arabidopsis*. *Plant J.* 27(4):325–333.

Jansen MAK, Gaba V, Greenberg BM. 1998. Higher plants and UV-B radiation: balancing damage, repair and acclimation. *Trends Plant Sci.* 3(4):131–135.

Jenkins GI. 2009. Signal transduction in responses to UV-B radiation. *Annu Rev Plant Biol.* 60:407–431.

Jensen FL, Kissler JJ, Peacock WL, Leavitt GM. 1975. Effect of ethephon on color and fruit characteristics of 'Tokay' and 'Emperor' table grapes. *Am J Enol Vitic.* 26(2):79–81.

Jeong ST, Goto-Yamamoto N, Kobayashi S, Esaka M. 2004. Effects of plant hormones and shading on the accumulation of anthocyanins and the expression of anthocyanin biosynthetic genes in grape berry skins. *Plant Sci.* 167(2):247–252.

Ji X, Dong B, Shiran B, Talbot MJ, Edlington JE, Hughes T, White RG, Gubler F, Dolferus R. 2011. Control of abscisic acid catabolism and abscisic acid homeostasis is important for reproductive stage stress tolerance in cereals. *Plant Physiol.* 156(2):647–662.

Jiang F, Hartung W. 2008. Long-distance signalling of abscisic acid (ABA): the factors regulating the intensity of the ABA signal. *J Exp Bot.* 59:37–43.

Jiang M, Zhang J. 2001. Effect of abscisic acid on active oxygen species, antioxidative defence system and oxidative damage in leaves of maize seedlings. *Plant Cell Physiol.* 42(11):1265–1273.

Jitao Z, Abrams GD, Barton DL, Taylor D, Pomeroy MK, Abrams S. 1995. Induction of lipid and oleosin biosynthesis by (+)-abscisic acid and its metabolites in microspore-derived embryos of *Brassica napus* L. cv Reston. Biological responses in the presence of 8'-hydroxyabascisic acid. *Plant Physiol.* 108:563–571.

Kim H, Hwang H, Hong JW, Lee YN, Ahn IP, Sun Yoon I, Yoo SD, Lee S, Lee SC, Kim BG. 2011. A rice orthologue of the ABA receptor, OsPYL/RCAR5, is a positive regulator of the ABA signal transduction pathway in seed germination and early seedling growth. *J Exp Bot.* 63:1013–1024.

Kim J, Park MK, Li WQ, Qureshi AA, Cho E. 2019. Association of vitamin A intake with cutaneous squamous cell carcinoma risk in the united statesassociation of vitamin a intake with cutaneous squamous cell carcinoma riskassociation of vitamin a intake with cutaneous squamous cell carcinoma risk. *JAMA Dermatol.* 155(11):1260–1268.

Kim T-H, Böhmer M, Hu H, Nishimura N, Schroeder JI. 2010. Guard cell signal transduction network: advances in understanding abscisic acid, CO_2, and Ca_2+ signaling. *Annu Rev Plant Biol.* 61(1):561–591.

Kirkegaard J, Lilley J, Howe G, Graham J. 2007. Impact of subsoil water use on wheat yield. *Aust J Agric Res.* 58(4):303–315.

Koornneef M, Reuling G, Karssen CM. 1984. The isolation and characterization of abscisic acid-insensitive mutants of *Arabidopsis thaliana*. *Physiol Plant.* 61(3):377–383.

Koshimizu K, Inui M, Fukui H, Mitsui T. 1968. Isolation of (+)-abscisyl-β-D-glucopyranoside from immature fruit of Lupinus luteus. *Agric Biol Chem.* 32(6):789–791.

Koyama R, Roberto SR, de Souza RT, Borges WFS, Anderson M, Waterhouse AL, Cantu D, Fidelibus MW, Blanco-Ulate B. 2018. Exogenous abscisic acid promotes anthocyanin biosynthesis and increased expression of flavonoid synthesis genes in *Vitis vinifera* × *Vitis labrusca* table grapes in a subtropical region. *Front Plant Sci.* 9:323–323.

Krochko JE, Abrams GD, Loewen MK, Abrams SR, Cutler AJ. 1998. (+)-Abscisic acid 8′-hydroxylase is a cytochrome P450 monooxygenase. *Plant Physiol.* 118(3):849–860.

Kruman I, Guo Q, Mattson MP. 1998. Calcium and reactive oxygen species mediate staurosporine-induced mitochondrial dysfunction and apoptosis in PC12 cells. *J Neurosci Res.* 51(3):293–308.

Kushiro T, Okamoto M, Nakabayashi K, Yamagishi K, Kitamura S, Asami T, Hirai N, Koshiba T, Kamiya Y, Nambara E. 2004. The *Arabidopsis* cytochrome P450 CYP707A encodes ABA 8'-hydroxylases: key enzymes in ABA catabolism. *EMBO J.* 23(7):1647–1656.

Lamaoui M, Jemo M, Datla R, Bekkaoui F. 2018. Heat and drought stresses in crops and approaches for their mitigation. *Front Chem.* 6(26).

Le Page-Degivry MT, Bidard JN, Rouvier E, Bulard C, Lazdunski M. 1986. Presence of abscisic acid, a phytohormone, in the mammalian brain. *Proc Natl Acad Sci U.S.A.* 83(4):1155–1158.

Lee KH, Piao HL, Kim H-Y, Choi SM, Jiang F, Hartung W, Hwang I, Kwak JM, Lee I-J, Hwang I. 2006. Activation of glucosidase via stress-induced polymerization rapidly increases active pools of abscisic acid. *Cell.* 126(6):1109–1120.

Leung J, Bouvier-Durand M, Morris P-C, Guerrier D, Chefdor F, Giraudat J. 1994. *Arabidopsis* ABA response gene ABI1: features of a calcium-modulated protein phosphatase. *Science.* 264(5164):1448–1452.

Leung J, Merlot S, Giraudat J. 1997. The *Arabidopsis* ABSCISIC ACID-INSENSITIVE2 (ABI2) and ABI1 genes encode homologous protein phosphatases 2C involved in abscisic acid signal transduction. *Plant Cell.* 9(5):759–771.

Li C, Jia H, Chai Y, Shen Y. 2011a. Abscisic acid perception and signaling transduction in strawberry: a model for non-climacteric fruit ripening. *Plant Signal Behav.* 6(12):1950–1953.

Li H-H, Hao R-L, Wu S-S, Guo P-C, Chen C-J, Pan L-P, Ni H. 2011b. Occurrence, function and potential medicinal applications of the phytohormone abscisic acid in animals and humans. *Biochem Pharmacol.* 82(7):701–712.

Lim CW, Baek W, Jung J, Kim J-H, Lee SC. 2015. Function of ABA in stomatal defense against biotic and drought stresses. *Int J Mol Sci.* 16(7):15251–15270.

Lin BL, Wang HJ, Wang JS, Zaharia LI, Abrams SR. 2005. Abscisic acid regulation of heterophylly in *Marsilea quadrifolia* L.: effects of R-(−) and S-(+) isomers. *J Exp Bot.* 56(421):2935–2948.

Liotenberg S, North H, Marion-Poll A. 1999. Molecular biology and regulation of abscisic acid biosynthesis in plants. *Plant Physiol Biochem.* 37(5):341–350.

Lipiec J, Doussan C, Nosalewicz A, Kondracka K. 2013. Effect of drought and heat stresses on plant growth and yield: a review. *Int Agrophys.* 27(4):463–477.

Livingston VW-C, inventor; Livingston; Virginia W-C, assignee. May 18, 1976. Abscisic Acid Tablets and Process. United States Patent. Patent number: 3958025, United States of America.

Lo-Coco F, Avvisati G, Vignetti M, Thiede C, Orlando SM, Iacobelli S, Ferrara F, Fazi P, Cicconi L, Di Bona E. 2013. Retinoic acid and arsenic trioxide for acute promyelocytic leukemia. *N Engl J Med.* 369(2):111–121.

Lucas RM, McMichael AJ, Armstrong BK, Smith WT. 2008. Estimating the global disease burden due to ultraviolet radiation exposure. *Int J Epidemiol.* 37(3):654–667.

Ma Y, Szostkiewicz I, Korte A, Moes D, Yang Y, Christmann A, Grill E. 2009. Regulators of PP2C phosphatase activity function as abscisic acid sensors. *Science.* 324(5930):1064–1068.

Magnone M, Bruzzone S, Guida L, Damonte G, Millo E, Scarfì S, Usai C, Sturla L, Palombo D, De Flora A et al. 2009. Abscisic acid released by human monocytes activates monocytes and vascular smooth muscle cell responses involved in atherogenesis. *J Biol Chem.* 284(26):17808–17818.

Mahajan S, Tuteja N. 2005. Cold, salinity and drought stresses: an overview. *Arch Biochem Biophys.* 444(2):139–158.

Mansfield TA, Hetherington AM, Atkinson CJ. 1990. Some current aspects of stomatal physiology. *Annu Rev Plant Physiol Plant Mol Biol.* 41(1):55–75.

Marin E, Nussaume L, Quesada A, Gonneau M, Sotta B, Hugueney P, Frey A, Marion-Poll A. 1996. Molecular identification of zeaxanthin epoxidase of *Nicotiana plumbaginifolia*, a gene involved in abscisic acid biosynthesis and corresponding to the ABA locus of *Arabidopsis thaliana. EMBO J.* 15:2331–2342.

Martin-Vertedor AI, Dodd IC. 2011. Root-to-shoot signalling when soil moisture is heterogeneous: increasing the proportion of root biomass in drying soil inhibits leaf growth and increases leaf abscisic acid concentration. *Plant Cell Environ.* 34(7):1164–1175.

Matilla AJ, Carrillo-Barral N, Rodríguez-Gacio MdC. 2015. An update on the role of NCED and CYP707A ABA metabolism genes in seed dormancy induction and the response to after-ripening and nitrate. *J Plant Growth Regul.* 34(2):274–293.

McAdam SA, Sussmilch FC, Brodribb TJ. 2016. Stomatal responses to vapour pressure deficit are regulated by high speed gene expression in angiosperms. *Plant Cell Environ.* 39(3):485–491.

McAdam SA, Sussmilch FC, Brodribb TJ, Ross JJ. 2015. Molecular characterization of a mutation affecting abscisic acid biosynthesis and consequently stomatal responses to humidity in an agriculturally important species. *AoB Plants.* 7: plv091.

Merilo E, Laanemets K, Hu H, Xue S, Jakobson L, Tulva I, Gonzalez-Guzman M, Rodriguez PL, Schroeder JI, Broschè M. 2013. PYR/RCAR receptors contribute to ozone-, reduced air humidity-, darkness-, and CO_2-induced stomatal regulation. *Plant Physiol.* 162(3):1652–1668.

Merilo E, Yarmolinsky D, Jalakas P, Parik H, Tulva I, Rasulov B, Kilk K, Kollist H. 2018. Stomatal VPD response: There is more to the story than ABA. *Plant Physiol.* 176(1):851–864.

Milborrow B. 1970. The metabolism of abscisic acid. *J Exp Bot.* 21(1):17–29.

Milborrow B. 1974. The chemistry and physiology of abscisic acid. *Annu Rev Plant Physiol.* 25(1):259–307.

Miransari M, Smith DL. 2014. Plant hormones and seed germination. *Environ Exp Bot.* 99:110–121.

Mittelheuser CJ, Van Steveninck RFM. 1969. Stomatal closure and inhibition of transpiration induced by (RS)-abscisic acid. *Nature.* 221(5177):281–282.

Miyakawa T, Fujita Y, Yamaguchi-Shinozaki K, Tanokura M. 2013. Structure and function of abscisic acid receptors. *Trends Plant Sci.* 18(5):259–266.

Munns R, King RW. 1988. Abscisic acid is not the only stomatal inhibitor in the transpiration stream of wheat plants. *Plant Physiol.* 88(3):703–708.
Mustilli A-C, Merlot S, Vavasseur A, Fenzi F, Giraudat J. 2002. *Arabidopsis* OST1 protein kinase mediates the regulation of stomatal aperture by abscisic acid and acts upstream of reactive oxygen species production. *Plant Cell.* 14(12):3089–3099.
Naeem MK, Ahmad M, Kamran M, Shah MKN, Iqbal MS. 2015. Physiological responses of wheat (*Triticum aestivum* L.) to drought stress. *J Plant Nutr Soil Sci.* 6(1):1–9.
Nagamune K, Hicks LM, Fux B, Brossier F, Chini EN, Sibley LD. 2008. Abscisic acid controls calcium-dependent egress and development in *Toxoplasma gondii. Nature.* 451(7175):207.
Nambara E, Marion-Poll A. 2005. Abscisic acid biosynthesis and catabolism. *Annu. Rev. Plant Biol.* 56: 165–185.
Nambara E, Suzuki M, Abrams S, McCarty DR, Kamiya Y, McCourt P. 2002. A screen for genes that function in abscisic acid signaling in *Arabidopsis thaliana. Genetics.* 161(3):1247–1255.
Ohkuma K, Lyon JL, Addicott FT, Smith OE. 1963. Abscisin II, an abscission-accelerating substance from young cotton fruit. *Science.* 142(3599):1592–1593.
Olsson L, Veit M, Weissenböck G, Bornman J. 1998. Differential flavonoid response to enhanced UV-B radiation in *Brassica napus. Phytochemistry.* 49(4):1021–1028.
Pantin F, Monnet F, Jannaud D, Costa JM, Renaud J, Muller B, Simonneau T, Genty B. 2013. The dual effect of abscisic acid on stomata. *New Phytol.* 197(1):65–72.
Parihar A, Parihar MS, Ghafourifar P. 2008. Significance of mitochondrial calcium and nitric oxide for apoptosis of human breast cancer cells induced by tamoxifen and etoposide. In *J Mol Med.* 21(3):317–324.
Park S-Y, Fung P, Nishimura N, Jensen DR, Fujii H, Zhao Y, Lumba S, Santiago J, Rodrigues A, Tsz-fung FC. 2009. Abscisic acid inhibits type 2C protein phosphatases via the PYR/PYL family of START proteins. *Science.* 324(5930):1068–1071.
Peng ZY, Zhou X, Li L, Yu X, Li H, Jiang Z, Cao G, Bai M, Wang X, Jiang C. 2008. *Arabidopsis* hormone database: A comprehensive genetic and phenotypic information database for plant hormone research in *Arabidopsis. Nucleic Acids Res.* 37(suppl_1):D975–D982.
Peppi MC, Fidelibus MW, Dokoozlian N. 2006. Abscisic acid application timing and concentration affect firmness, pigmentation, and color of 'Flame Seedless' grapes. *HortScience.* 41(6):1440–1445.
Peppi MC, Fidelibus M, Dokoozlian N. 2007. Application timing and concentration of abscisic acid affect the quality of 'Redglobe'grapes. *J Hortic Sci Biotechnol.* 82(2):304–310.
Qin X, Zeevaart JAD. 1999. The 9-cis-epoxycarotenoid cleavage reaction is the key regulatory step of abscisic acid biosynthesis in water-stressed bean. *Proc Natl Acad Sci U.S.A.* 96(26):15354–15361.
Quanqi L, Yuhai C, Mengyu L, Xunbo Z, Baodi D, Songlie Y. 2008. Water potential characteristics and yield of summer maize in different planting patterns. *Plant Soil Environ.* 54(1):14–19.
Raghavendra AS, Gonugunta VK, Christmann A, Grill E. 2010. ABA perception and signalling. *Trends Plant Sci.* 15(7):395–401.
Rakitin VY, Karyagin VV, Rakitina TY, Prudnikova ON, Vlasov PV. 2008. UV-B stress-induced ABA production in *Arabidopsis thaliana* mutants defective in ethylene signal transduction pathway. *Russ J Plant Physiol.* 55(6):854–856.
Remm M, Storm CE, Sonnhammer EL. 2001. Automatic clustering of orthologs and in-paralogs from pairwise species comparisons. *J Mol Biol.* 314(5):1041–1052.
Rivera-Madrid R, Aguilar-Espinosa M, Cárdenas-Conejo Y, Garza-Caligaris LE. 2016. Carotenoid derivates in Achiote (*Bixa orellana*) seeds: synthesis and health promoting properties. *Front Plant Sci.* 7(1406): 1–7
Roberto SR, de Assis AM, Yamamoto LY, Miotto LCV, Sato AJ, Koyama R, Genta W. 2012. Application timing and concentration of abscisic acid improve color of 'Benitaka'table grape. *Sci Hortic.* 142:44–48.
Rodriguez-Amaya DB. 2015. *Nomenclature, Structures, and Physical and Chemical Properties. Food Carotenoids: Chemistry, Biology, and Technology*; 1–23.
Romero GM, inventor; Romero, Gonzalo M., assignee. December 28, 2006. Abscisic Acid Against Cancer. United States Patent. Patent number: US20060292215A1. United States of America.
Roy CR, Gies HP, Lugg DJ, Toomey S, Tomlinson DW. 1998. The measurement of solar ultraviolet radiation. *Mutat Res.* 422(1):7–14.
Ruiz-Sola MÁ, Rodríguez-Concepción M. 2012. *Carotenoid biosynthesis in Arabidopsis: a colorful pathway.* The Arabidopsis book/American Society of Plant Biologists.; 10: e0158.
Rybka ACP, de Freitas ST, Netto AF, Biasoto ACT. 2015. Central composite rotatable design approach to optimize 'Italia'raisin drying conditions. *Com Sci.* 6(4):454–462.
Sah SK, Reddy KR, Li J. 2016. Abscisic acid and abiotic stress tolerance in crop plants. *Front Plant Sci.* 7(571): 1–26.

Saini HS, Aspinall D. 1981. Effect of water deficit on sporogenesis in wheat (*Triticum aestivum* L.). *Ann Bot.* 48(5):623–633.

Saito S, Hirai N, Matsumoto C, Ohigashi H, Ohta D, Sakata K, Mizutani M. 2004. Arabidopsis CYP707As encode (+)-abscisic acid 8′-hydroxylase, a key enzyme in the oxidative catabolism of abscisic acid. *Plant Physiol.* 134:1439–1449.

Sakamoto H, Matsuda O, Iba K. 2008. ITN1, a novel gene encoding an ankyrin-repeat protein that affects the ABA-mediated production of reactive oxygen species and is involved in salt-stress tolerance in *Arabidopsis thaliana. Plant J.* 56(3):411–422.

Salama HMH, Al Watban AA, Al-Fughom AT. 2011. Effect of ultraviolet radiation on chlorophyll, carotenoid, protein and proline contents of some annual desert plants. *Saudi J Biol Sci.* 18(1):79–86

Sansberro PA, Mroginski LA, Bottini R. 2004. Foliar sprays with ABA promote growth of Ilex paraguariensis by alleviating diurnal water stress. *Plant Growth Regul.* 42(2):105–111.

Saradadevi R, Bramley H, Palta J, Edwards E, Siddique K. 2015. Root biomass in the upper layer of the soil profile is related to the stomatal response of wheat as the soil dries. *Funct Plant Biol.* 43(1):62–74.

Saradadevi R, Palta JA, Siddique KHM. 2017. ABA-mediated stomatal response in regulating water use during the development of terminal drought in wheat. *Front Plant Sci.* 8(1251):1–14.

Scarfì S, Ferraris C, Fruscione F, Fresia C, Guida L, Bruzzone S, Usai C, Parodi A, Millo E, Salis A et al. 2008. Cyclic ADP-ribose-mediated expansion and stimulation of human mesenchymal stem cells by the plant hormone abscisic acid. *Stem Cells.* 26(11):2855–2864.

Schug TT, Berry DC, Shaw NS, Travis SN, Noy N. 2007. Opposing effects of retinoic acid on cell growth result from alternate activation of two different nuclear receptors. *Cell.* 129(4):723–733.

Schwartz SH, Qin X, Zeevaart J. 2003. Elucidation of the indirect pathway of abscisic acid biosynthesis by mutants, genes, and enzymes. *Plant Physiol.* 131(4):1591–1601.

Schwartz SH, Tan BC, Gage DA, Zeevaart JA, McCarty DR. 1997. Specific oxidative cleavage of carotenoids by VP14 of maize. *Science.* 276(5320):1872–1874.

Schwartz SH, Zeevaart JA. 2010. Abscisic acid biosynthesis and metabolism. *Plant Hormones* Davies P.J. (editor). Springer, Dordrecht, 137–155.

Sharma N, Abrams S, Waterer DR. 2005. Uptake, movement, activity, and persistence of an abscisic acid analog (8′ acetylene ABA methyl ester) in Marigold and Tomato. *J Plant Growth Regul.* 24:28–35.

Sharma N, Waterer DR, Abrams SR. 2006. Evaluation of abscisic acid analogs as holding agents for bedding plant seedlings. *HortTechnology.* 16(1):71–77.

Shi S, Wang G, Wang Y, Zhang L, Zhang L. 2005. Protective effect of nitric oxide against oxidative stress under ultraviolet-B radiation. *Nitric Oxide.* 13(1):1–9.

Shu K, Luo X, Meng Y, Yang W. 2018. Toward a molecular understanding of abscisic acid actions in floral transition. *Plant Cell Physiol.* 59(2):215-221.

Singh A, Giri J, Kapoor S, Tyagi AK, Pandey GK. 2010. Protein phosphatase complement in rice: genome-wide identification and transcriptional analysis under abiotic stress conditions and reproductive development. *BMC Genomics.* 11(1):435.

Singh A, Pandey A, Srivastava A, Tran L-S, Pandey G. 2015. Plant protein phosphatases 2C: From genomic diversity to functional multiplicity and importance in stress management. *Crit Rev Biotechnol.* 36(6):1023–1035.

Singh C, Kumar B, Mehandi S, Chandra K. 2012. Effect of drought stress in rice: a review on morphological and physiological characteristics. *TIBS.* 5(4):261–265.

Sinha RP, Singh SP, Häder D-P. 2007. Database on mycosporines and mycosporine-like amino acids (MAAs) in fungi, cyanobacteria, macroalgae, phytoplankton and animals. *J Photochem Photobiol B.* 89(1):29–35.

Sivamani E, Bahieldin A, Wraith JM, Al-Niemi T, Dyer WE, Ho T-HD, Qu R. 2000. Improved biomass productivity and water use efficiency under water deficit conditions in transgenic wheat constitutively expressing the barley HVA 1 gene. *Plant Sci.* 155(1):1–9.

Slesak I, Libik M, Karpinska B, Karpinski S, Miszalski Z. 2007. The role of hydrogen peroxide in regulation of plant metabolism and cellular signalling in response to environmental stresses. *Acta Biochim Pol.* 54(1):39–50.

Soon FF, Ng LM, Zhou XE, West GM, Kovach A, Tan ME, Suino Powell KM, He Y, Xu Y, Chalmers MJ. 2012. Molecular mimicry regulates ABA signaling by SnRK2 kinases and PP2C phosphatases. *Science.* 335(6064):85–88.

Soprano DR, Qin P, Soprano KJ. 2004. Retinoic acid receptors and cancers. *Annu. Rev. Nutr.* 24:201–221.

Spayd SE, Tarara JM, Mee DL, Ferguson JC. 2002. Separation of sunlight and temperature effects on the composition of *Vitis vinifera* cv. Merlot Berries. *Am J Enol Vitic.* 53(3):171–182.

Sun L, Sun Y, Zhang M, Wang L, Ren J, Cui M, Wang Y, Ji K, Li P, Li Q et al. 2012. Suppression of 9-cis-epoxycarotenoid dioxygenase, which encodes a key enzyme in abscisic acid biosynthesis, alters fruit texture in transgenic tomato. *Plant Physiol.* 158(1):283–298.

Swarbreck D, Wilks C, Lamesch P, Berardini TZ, Garcia-Hernandez M, Foerster H, Li D, Meyer T, Muller R, Ploetz L. 2007. The Arabidopsis Information Resource (TAIR): gene structure and function annotation. *Nucleic Acids Res.* 36(suppl_1):D1009–D1014.

Tan B-C, Joseph LM, Deng WT, Liu L, Li QB, Cline K, McCarty DR. 2003. Molecular characterization of the *Arabidopsis* 9-cis epoxycarotenoid dioxygenase gene family. *Plant J.* 35(1):44–56.

Tan B-C, Schwartz SH, Zeevaart JA, McCarty DR. 1997. Genetic control of abscisic acid biosynthesis in maize. *Proc Natl Acad Sci U.S.A.* 94(22):12235–12240.

Thomas G. 2014. Growing greener cities in Latin America and the Caribbean: An FAO report on urban and peri-urban horticulture in the region. Graeme Thomas (editor). FAO, Rome (Italy).

Tossi V, Lamattina L, Cassia R. 2009. An increase in the concentration of abscisic acid is critical for nitric oxide-mediated plant adaptive responses to UV-B irradiation. *New Phytol.* 181:871–879.

Travaglia C, Reinoso H, Cohen A, Luna C, Tommasino E, Castillo C, Bottini R. 2010. Exogenous ABA increases yield in field-grown wheat with moderate water restriction. *J Plant Growth Regul.* 29(3):366–374.

Turner NC, Asseng S. 2005. Productivity, sustainability, and rainfall-use efficiency in Australian rainfed Mediterranean agricultural systems. *Aust J Agric Res.* 56(11):1123–1136.

Vilela B, Riera M. 2015. Emerging roles of protein kinase CK2 in abscisic acid signaling. *Front Plant Sci.* 6:966.

Walker-Simmons M, Holappa L, Abrams G, Abrams S. 2006. ABA metabolites induce group 3 LEA mRNA and inhibit germination in wheat. *Physiol Plant.* 100:474–480.

Wang Y, Wu Y, Duan C, Chen P, Li Q, Dai S, Sun L, Ji K, Sun Y, Xu W et al. 2012. The expression profiling of the CsPYL, CsPP2C and CsSnRK2 gene families during fruit development and drought stress in cucumber. *J Plant Physiol.* 169(18):1874–1882.

Wasilewska A, Vlad F, Sirichandra C, Redko Y, Jammes F, Valon C, Frey NFD, Leung J. 2008. An update on abscisic acid signaling in plants and more... *Mol Plant.* 1(2):198–217.

Waterland NL, Campbell CA, Finer JJ, Jones ML. 2010. Abscisic acid application enhances drought stress tolerance in bedding plants. *HortScience.* 45(3):409–413.

WHO. 1994. *Ultraviolet Radiation.* Geneva, Switzerland: World Health Organization. (Environmental Health Criteria; 160).

Wilkinson S, Davies WJ. 2002. ABA-based chemical signalling: the co-ordination of responses to stress in plants. *Plant Cell Environ.* 25(2):195–210.

Wright STC, Hiron RWP. 1969. (+)-Abscisic Acid, the growth inhibitor induced in detached wheat leaves by a period of wilting. *Nature.* 224(5220):719–720.

Xu H, Barnes GT, Yang Q, Tan G, Yang D, Chou CJ, Sole J, Nichols A, Ross JS, Tartaglia LA. 2003. Chronic inflammation in fat plays a crucial role in the development of obesity-related insulin resistance. *J Clin Invest.* 112(12):1821–1830.

Xu Z-Y, Lee KH, Dong T, Jeong JC, Jin JB, Kanno Y, Kim DH, Kim SY, Seo M, Bressan RA et al. 2012. A vacuolar β-glucosidase homolog that possesses glucose-conjugated abscisic acid hydrolyzing activity plays an important role in osmotic stress responses in *Arabidopsis*. *Plant Cell.* 24(5):2184–2199.

Xue-Xuan X, Hong-Bo S, Yuan-Yuan M, Gang X, Jun-Na S, Dong-Gang G, Cheng-Jiang R. 2010. Biotechnological implications from abscisic acid (ABA) roles in cold stress and leaf senescence as an important signal for improving plant sustainable survival under abiotic-stressed conditions. *Crit Rev Biotechnol.* 30(3):222–230.

Xue T, Wang D, Zhang S, Ehlting J, Ni F, Jakab S, Zheng C, Zhong Y. 2008. Genome-wide and expression analysis of protein phosphatase 2C in rice and *Arabidopsis*. *BMC Genomics.* 9(1):550.

Zeevaart J, and, Creelman R. 1988. Metabolism and physiology of abscisic acid. *Annu Rev Plant Physiol Plant Mol Biol.* 39(1):439–473.

Zhang J, Davies WJ. 1990. Changes in the concentration of ABA in xylem sap as a function of changing soil water status can account for changes in leaf conductance and growth. *Plant Cell Environ.* 13(3): 277–285.

Zhang X, Zhang Q, Xin Q, Yu L, Wang Z, Wu W, Jiang L, Wang G, Tian W, Deng Z et al. 2012. Complex structures of the abscisic acid receptor PYL3/RCAR13 reveal a unique regulatory mechanism. *Structure.* 20(5):780–790.

Zhang XL, Jiang L, Xin Q, Liu Y, Tan JX, Chen ZZ. 2015. Structural basis and functions of abscisic acid receptors PYLs. *Front Plant Sci.* 6:88.

Zhao L, Zhang F, Guo J, Yang Y, Li B, Zhang L. 2004. Nitric oxide functions as a signal in salt resistance in the calluses from two ecotypes of reed. *Plant Physiol.* 134(2):849–857.
Zhou N, Yao Y, Ye H, Zhu W, Chen L, Mao Y. 2016. Abscisic-acid-induced cellular apoptosis and differentiation in glioma via the retinoid acid signaling pathway. *Int J Cancer.* 138(8):1947–1958.
Zhou R, Cutler AJ, Ambrose SJ, Galka MM, Nelson KM, Squires TM, Loewen MK, Jadhav AS, Ross AR, Taylor DC. 2004. A new abscisic acid catabolic pathway. *Plant Physiol.* 134(1):361–369.
Zhu J-K. 2002. Salt and drought stress signal transduction in plants. *Annu Rev Plant Biol.* 53(1):247–273.
Zocchi E, Carpaneto A, Cerrano C, Bavestrello G, Giovine M, Bruzzone S, Guida L, Franco L, Usai C. 2001. The temperature-signaling cascade in sponges involves a heat-gated cation channel, abscisic acid, and cyclic ADP-ribose. *Proc Natl Acad Sci U.S.A.* 98(26):14859–14864.

10 C_{13} α-Ionol (Blumenol) Glycosides and C_{14} Mycorradicin: Apocarotenoids Accumulating in Roots during the Arbuscular Mycorrhizal Symbiosis

Michael H. Walter

CONTENTS

INTRODUCTION

Plants need to interact with their abiotic and biotic environment and adapt their growth and development accordingly. One important determinant for plant growth is the availability of mineral nutrients from the soil and their mode of acquisition into roots. This acquisition can be accomplished through direct uptake mechanisms or by indirect acquisition with the help of microbial partners living in the rhizosphere next to roots (Bucher, 2007; Martin et al., 2017). Such an assistance by beneficial microbes is often decisive for plant survival in nutrient-poor soils and can increase whole-plant performance in regular soils with less or no fertilizer input. Potentially beneficial microbes in the huge variety of rhizosphere organisms must be recognized and supported by the plant to enter into a stable long-term relationship with reciprocal benefits for both partners called mutualistic symbiosis (Plett and Martin, 2018).

The arbuscular mycorrhizal (AM) symbiosis between roots and a special clade of fungi is arguably the most prevalent, the most ancestral, and the most relevant mutualistic root symbiosis on earth, helping plants in mineral nutrient acquisition (Parniske, 2008; Harrison, 2012; Recorbet et al., 2013; Martin et al., 2017). In nutrient-poor soils, the symbiosis can develop between the roots of most terrestrial plants and a small group of specialized fungi belonging to the phylum Glomeromycota. It has been estimated that the AM symbiosis can be formed by more than 80% of land plants and that the oldest fossil traces of arbuscules date back to more than 400 million years ago (Remy et al., 1994; Martin et al., 2017). The AM symbiosis was instrumental in the evolution of land plants coevolving with early roots for establishing terrestrial ecosystems (Brundrett, 2002). The fungi are obligate biotrophs depending on their plant hosts to complete their life cycle. The name-giving fungal organs, which develop inside roots cells, are arbuscules, which literally means 'little trees' but these 'trees' are tiny: only 20–80 μm in size, and can only be seen with a microscope. Similar to real trees with the need to capture a maximum of light, a large interface area for exchanging nutrients between the fungus and the root cell determines the highly branched tree-like morphology of arbuscules (Luginbuehl and Oldroyd, 2017). These organs develop intracellularly from fungal hyphae, which colonize the root extra- and intra-cellularly (Gutjahr and Parniske, 2013). Both the root colonization and arbuscule development and decay require tight regulation in order to get the most benefit with the least cost out of the symbiosis. The major mineral nutrient obtained via assistance by AM fungi is phosphate, which is often in very limited supply in soils (Bucher, 2007). Special plant phosphate transporter proteins for phosphate uptake are located in the periarbuscular membrane, which constitutes the interface between plant and fungi and the site of nutrient exchange (Harrison et al., 2002). Given the paramount importance of phosphate concentrations for the symbiosis, it is not surprising that phosphate seems to be not only a nutrient but also a decisive regulatory molecule, whose availability can determine symbiotic structures, in particular the development and the decay of arbuscules (Javot et al., 2007; Carbonnel and Gutjahr, 2014; Luginbuehl and Oldroyd, 2017; Lanfranco et al., 2018). The cost paid by the plant to the microbial partner typically consists of sugars exported to the rhizosphere. Recently, lipids have also been identified as a plant currency to pay for phosphate delivery (Keymer et al., 2017). The mutualistic nature of the interaction must be controlled in order not to waste resources on a partner which is ineffective in delivering the expected benefits (Kiers et al., 2011).

AM fungi are but one type of microorganism among many other beneficial or harmful organisms in the rhizosphere. Communication between rhizosphere friends or foes occurs through hormones and other small molecules requiring very low concentrations, which result in the activation of many activities at the plant tissue, cell, or metabolic level (MacLean et al., 2017). In addition, secondary (specialized) metabolites produced by one or both interactors usually in higher concentrations are known to often act as antimicrobial chemical defences against harmful microbes (Massalha et al., 2017). However, such compounds might also be produced in a beneficial relationship to control one partner by the other to keep the interaction in a balanced and mutualistic state.

At least three major structural classes of apocarotenoids are currently known to occur and potentially play a role in the AM symbiosis: (i) the strigolactones having 19 C-atoms, (ii) linear mycorradicin derivatives with 14 C-atoms and (iii) various derivatives of ring-containing α-ionols with 13 C-atoms (also called blumenols or cyclohexenones), which are usually glycosylated (Klingner et al., 1995a, b; Maier et al., 1995, 2000; Walter et al., 2010, 2015, Fiorilli et al., 2019). Abscisic acid (ABA, C_{15}) and zaxinones (C_{18}) are additional apocarotenoids involved in the AM symbiosis, which should be mentioned here, but their role is less well studied (Felemban et al., 2019). Further apocarotenoid signal molecules have been reviewed (Hou et al., 2016). The strigolactones act in minute amounts and can be classified among the hormones and regulatory molecules thought to affect primarily the development of the symbiosis (Al-Babili and Bouwmeester, 2015; López-Ráez et al., 2017). Since strigolactones have many other functions beyond their regulatory role in the AM symbiosis, and since they are molecules that have been studied intensely in the recent past, strigolactones will be dealt with in a separate chapter, which also details their role in the AM symbiosis (see Chapter 11). This chapter will focus on the C_{14} mycorradicin and C_{13} α-ionol derivatives, which accumulate preferentially in later stages of the symbiosis and occur at least sometimes in higher concentrations pointing to a role beyond simple signalling in the AM symbiosis.

DISCOVERY AND ANALYSIS OF C_{14} MYCORRADICIN AND ACCOMPANYING C_{13} α-IONOL APOCAROTENOIDS

The phenomenon of an occasional yellow or yellowish colouration visible by eye of plant roots upon colonization by AM fungi is long known to many mycorrhiza researchers. It was first described as early as 1924 for a legume plant (Jones, 1924). Unusual plastids, identified as globular chromoplasts, have later been observed in *Ornithogalum umbellatum* probably describing the same phenomenon (Scannerini and Bonfante-Fasolo, 1977). Cereal roots can exhibit a particularly pronounced yellow colouration when strongly colonized by AM fungi (Figure 10.1A). It took more than 70 years from the first description of such a 'yellow pigment' in mycorrhizal roots to the elucidation of the chemical structure of the chromophore. It was first isolated and characterized from maize roots as a linear carotenoid cleavage product with 14 C-atoms by the group of H. Bothe (Klingner et al., 1995a). This C_{14} all-*E*-4,9-dimethyldodeca-2,4,6,8,10-pentaenedioic acid (10,10′-diapocarotene-10,10′-dioic acid) compound was termed 'mycorradicin' (Figure 10.1B). The name given already implies

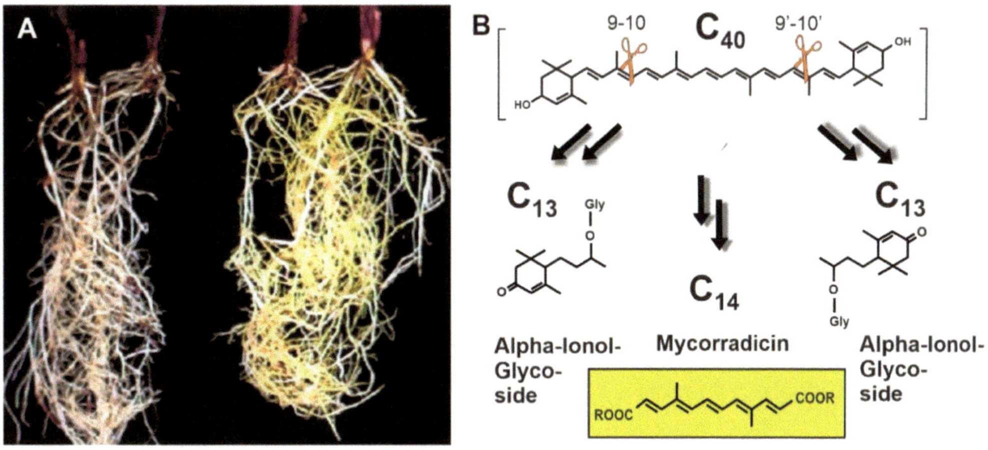

C

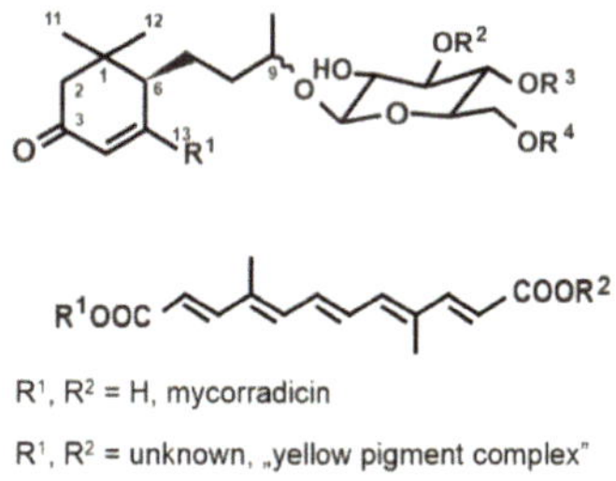

	R^1	R^2	R^3	R^4
1	CH2O-Glc	H	H	H
2	CH2OH	H	H	H
3	CH2OH	-COCH3-C(OH)(CH3)-CH2COOH	H	H
6	CH3	H	H	Ara
7	CH3	H	Glc	H
8	CH2OH	-COCH3-C(OH)(CH3)-CH2COOH	H	-COCH2COOH
9	CH3	-COCH3-C(OH)(CH3)-CH2COOH	H	H
10	CH3	H	H	-COCH2COOH

D

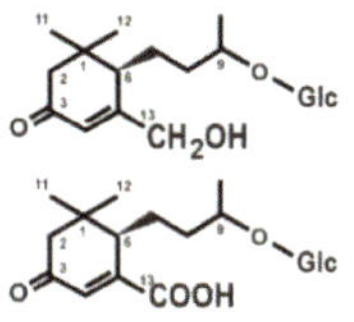

Compound 1	CH2OH	H	H	H
Compound 2	COOH	H	H	H

FIGURE 10.1 Occurrence, biogenesis from carotenoids and structural diversity of C_{13} α-ionol glycosides and C_{14} mycorradicin derivatives accumulating in mycorrhizal roots. (A) Yellow colouration due to mycorradicin accumulation in mycorrhizal roots of maize after six weeks of root colonization by AM fungi (roots on the right). Nonmycorrhizal roots (left) do not display this coloration. (B) Simple scheme for biogenesis of C_{13} α-ionol glycosides and C_{14} mycorradicin derivatives from a common C_{40} carotenoid or xanthophyll precursor, presumably zeaxanthin, via carotenoid cleavage at the 9,10 and 9′,10′ double bonds. (C) Structural diversity of C_{13} α-ionol glycosides both at the aglycon and the glycoside moieties. Adapted from work on *Allium porrum* as described by Schliemann et al. (2008b). Additional variants have been found in other plant species (see text). C_{14} mycorradicin is linked in its native state to unknown moieties probably by ester bonds within the 'yellow pigment complex.' (D) Structural variants of C_{13} α-ionol glycosides (blumenol glycosides, compounds 1 and 2) detected in both roots and leaves of *Nicotiana attenuata* (Wang et al., 2018). Their accumulation in leaves correlates with the mycorrhization status of roots enabling their use as a metabolic shoot marker of mycorrhization.

an arbuscular mycorrhiza-specific accumulation. Indeed, no reports of yellow colouration or mycorradicin accumulation exist for ectomycorrhizal or nodulated roots. Moreover, pathogen infection or abiotic stresses of cereal roots also did not result in mycorradicin accumulation (Maier et al., 1997). Similarly, phytohormones or supply of mineral nutrients did not induce mycorradicin accumulation in mycorrhizal maize roots (Fester et al., 2002). Small amounts of mycorradicin in non-mycorrhizal roots of a few plant species were found, but in all these cases, mycorradicin content of the corresponding mycorrhizal samples were markedly higher (Fester et al., 2002; Schliemann et al., 2006). This result also strongly argues for a plant and not a fungal origin of mycorradicin, which has been questioned in some of the early papers. These results also support earlier conclusions that, apart from minimal basal levels in rare cases, mycorradicin accumulation is an AM-specific phenomenon. Interestingly, whereas mycorradicin was accumulating strongly in mycorrhizal *Ornithogalum* roots (Schliemann et al., 2006), it is absent in some plant species like *Tagetes erecta* despite strong root colonization by AM fungi (Bothe et al., 1994; Klingner et al., 1995b). It can thus not be used as a universal marker of AM fungal root colonization. In a species where it accumulates abundantly (maize), the lack of its occurrence has been used to screen for mutants defective in the AM symbiosis (Paszkowski et al., 2006). Mycorradicin has also been chemically synthesized in a quest for novel yellow compounds to be utilized as food colourants (Giger, 2002) but its instability makes it a poor choice for such an application.

In 1995, another paper reported an AM-mediated accumulation of a so-called cyclohexenone sesquiterpenoid glycoside (blumenin), which initially appeared to be unrelated to the yellow root phenomenon (Maier et al., 1995). The aglycons have 13 C-atoms and their structure is similar to norisoprenoids, ionols, and ionones, known to occur in a variety of plant organs other than roots (Figure 10.1B, C; Walter et al., 2010). The AM-specific nature of the accumulation of this and other structurally related sesquiterpenoid-like glycosides was demonstrated by treatments of roots with heat, cold, drought, heavy metals and pathogens, none of which resulted in increased levels of these compounds (Maier et al., 1995, 1997; Fester et al., 1999). The first arguments for an apocarotenoid nature of these compounds came from NMR analyses showing a biosynthetic origin from the plastidial MEP pathway, which is known to produce precursors for carotenoids (Maier et al., 1998). By the year 2000 a tight correlation of accumulation of C_{13} cyclohexenone glycosides and C_{14} mycorradicin was shown and the biosynthesis of both compounds from a common C_{40} carotenoid precursor was first proposed (Walter et al., 2000). Two molecules of C_{13} and one molecule of C_{14} nicely fit into a conceptual cleavage scheme of a common C_{40} carotenoid or xanthophyll precursor (Figure 10.1B). Based on structural similarities of the C_{13} compounds to ABA the proposed precursor was a xanthophyll, as is the case in ABA biogenesis, but this has not been verified yet. In the meantime, we have referred to the aglycons of these compounds as C_{13} α-ionol derivatives, which is the correct classification within the (apo)carotenoid nomenclature (Walter et al., 2015; Stauder et al., 2018). They had also been called blumenols earlier (Maier et al., 1995; Kato-Noguchi et al., 2012; Matsunami et al., 2010) and also recently (Wang et al., 2018), based on the aglycon of blumenin. As mentioned above, structurally related ionones and ionols in both glycosylated and nonglycosylated forms are found frequently in leaves, flowers, and fruits, where they can be scent and aroma compounds (Walter et al., 2010; Serra, 2015).

The structural diversity of the AM-specific C_{13} α-ionol glycoside derivatives (correct name adopted from here onwards, except when referring to the old literature) in mycorrhizal roots is high, and has been studied in quite some detail, particularly in roots of *Medicago truncatula*, *Allium porrum*, and *O. umbellatum* (Figure 10.1C). Variation is found both in the C_{13} aglycon and in the sugar moieties of the glycosides, but interestingly, the aglycon is always of the α-ionol type and never of the β-ionol type (Figure 10.1C) (Schliemann et al., 2006; Strack and Fester, 2006; Schliemann et al., 2008a). Additional variants have been found for *M. truncatula* later in a metabolomics study at late stages of the AM symbiosis (Laparre et al., 2014). Mono- and di-glycosides of 13-hydroxyblumenol C and blumenol C acylated with 3-hydroxy-3-methylglutaric and/or malonic acid have been described (Schliemann et al., 2008a,b). *O. umbellatum* contains the highest diversity experienced so far, exhibiting mono-, di-, and branched tri-glycosides of blumenol C, 13-hydroxyblumenol C, and 13-nor-5-carboxyblumenol C, some of which contain terminal rhamnose as sugar moiety (Schliemann et al., 2006). This plant also has strongly yellow mycorrhizal roots and high amounts of mycorradicin derivatives. Native root extracts contained up to 60 nmol/g root fresh weight of native 'yellow pigment' and many different compounds with the typical mycorradicin UV spectrum, which could not be identified in detail but very little free mycorradicin was found. Only upon alkaline hydrolysis mycorradicin derivatives are liberated from its unknown conjugates (Schliemann et al., 2006). Thus the native 'yellow pigment' is rather a 'yellow pigment complex' apparently composed of a highly complex mixture of various glycosylated and esterified apocarotenoids conjugated to each other and additional unknown components (Schliemann et al., 2006; Walter, 2013). It consists of insoluble hydrophobic droplets accumulating in vacuoles and sometimes also in the apoplast around arbusculated cells (Fester et al., 2002). Recently, the structural diversity of C_{13} α-ionol glycosides (blumenols) has been further extended by studies in *Nicotiana attenuata* and other plants (Wang et al., 2018). These new studies, which revealed an application potential of these compounds as shoot indicators of mycorrhizal colonization will be further discussed below.

BIOSYNTHESIS

Apocarotenoid biosynthesis in general can be subdivided into three parts: (i) synthesis of the central intermediates of isoprenoid biosynthesis, isopentenyl diphosphate (IPP, C_5) and dimethylallyl diphosphate (DMAPP); (ii) the subsequent assembly of C_5 units and their modification to tetraterpene C_{40} carotenoids and xanthophylls; and (iii) the carotenoid cleavage steps and subsequent modifications of the primary carotenoid cleavage products (Figure 10.2). Since the biosynthesis of C_{13} α-ionol derivatives and C_{14} mycorradicin overlaps with other apocarotenoid pathways, two other pathways to ABA or strigolactones are also shown, since they might compete for carotenoid precursors (Figure 10.2). These other pathways are covered in a recent review (Felemban et al., 2019). The strigolactone pathway is described in detail in Chapter 11.

At the start of molecular genetic analyses of AM-induced apocarotenoids in the late 1990s, very little was known on the regulation of carotenoid biosynthetic genes by AM fungal colonization, let alone having any knowledge on the late steps. We focused our interest on the early steps to IPP/DMAPP, since a new non-mevalonate pathway for IPP formation in bacteria and plants had just been described (Rohmer et al., 1993). It is located in plastids and it is now known as the methylerythritol phosphate (MEP) pathway, providing building blocks to carotenoid biosynthesis. Initial experiments on the biosynthetic route to mycorrhizal C_{13} α-ionol structures by ^{13}C-labelling and NMR analysis indicated an involvement of the MEP pathway, and delivered early arguments for an apocarotenoid nature of these compounds (Maier et al., 1998).

For the initial molecular studies on the MEP pathway, heterologous probes from rice for the first biosynthetic genes of this pathway were used, encoding 1-deoxy-D-xylulose 5-phosphate synthase (DXS), and 1-deoxy-D-xylulose 5-phosphate reductoisomerase (DXR). Northern-type RNA blot analyses of wheat RNA samples clearly demonstrated strongly elevated transcript levels of both *DXS* and *DXR* genes in mycorrhizal vs. nonmycorrhizal root samples (Walter et al., 2000). This increase in steady-state levels of transcripts nicely correlated with the accumulation of both C_{13} α-ionol glycosides and C_{14} mycorradicin derivatives as well as with the development of mycorrhizal structures (hyphae, arbuscules, vesicles and spores). Subsequently, we obtained strong evidence for the existence of various differentially regulated *DXS* isogenes in *Medicago truncatula* and other plant species, of which only *DXS2* was strongly regulated by mycorrhizal colonization of roots (Figure 10.2, Walter et al., 2002). Silencing the expression of *DXS2* by RNAi techniques led to strong reduction of both C_{13} α-ionol and C_{14} mycorradicin derivatives in mycorrhizal roots (Floss et al., 2008a). The marked alterations in mycorrhizal structures correlated with the reduction of

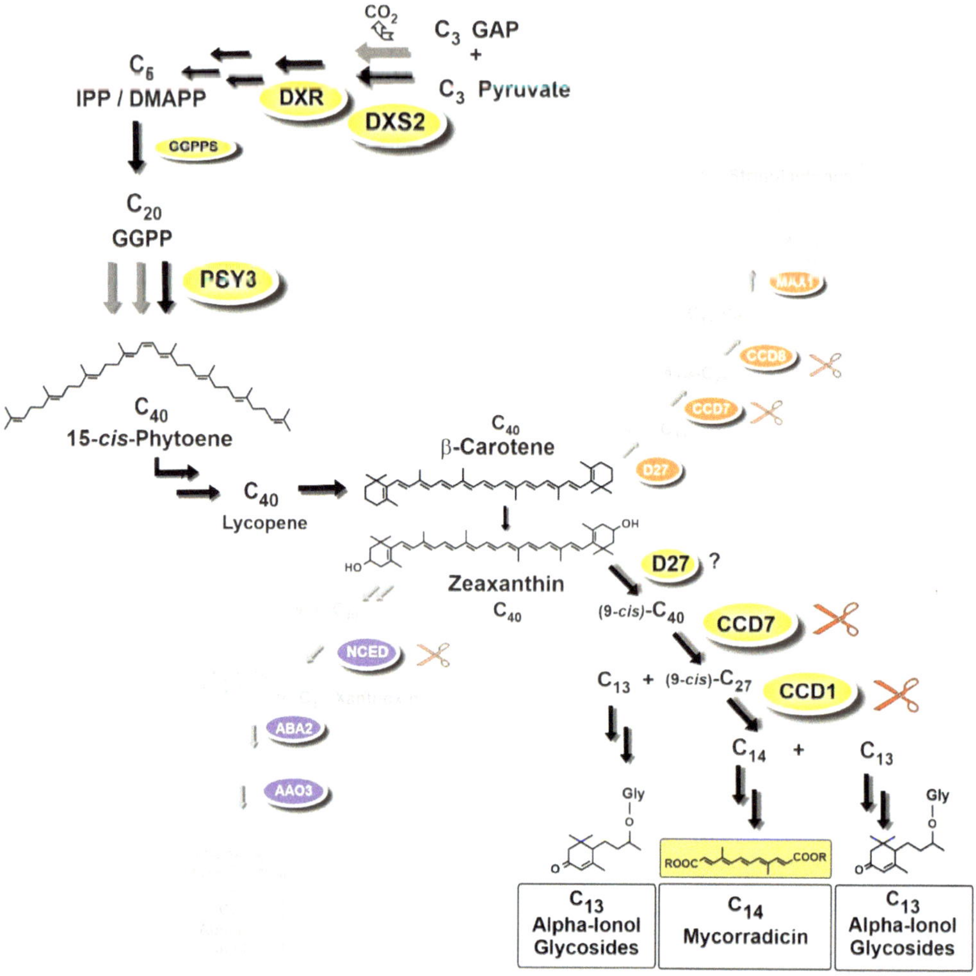

FIGURE 10.2 Apocarotenoid biosynthetic pathways in roots and their C_{40} carotenoid and C_5 IPP precursor supply. The pathway to C_{13} α-ionol glycosides and C_{14} mycorradicin derivatives is shown in full print while potentially competing side pathways to strigolactones and ABA are faded out. The simplified scheme indicates key enzymatic steps (some with numbered specialized isoforms) of the MEP pathway (horizontal to the left, top), of carotenoid biosynthesis (vertical, centre) and of carotenoid cleavage steps (below, inclined) marked in yellow. These steps are catalysed by 1-deoxy-D-xylulose 5-phosphate synthase (DXS), 1-deoxy-D-xylulose 5-phosphate reductoisomerase (DXR), GGPP synthase, (GGPPS) and phytoene synthase (PSY). Further steps lead to β-carotene and zeaxanthin apocarotenoid precursors. Cleavage by CCDs or 9-*cis*-epoxy-carotenoid cleavage dioxygenases (NCEDs) typically starts from 9-*cis* isomerized substrates (brought about by D27 *cis–trans* isomerase) and is carried out by a single or by sequential double cleavage reactions followed by modification steps leading to various apocarotenoid products (boxed). A potential involvement of D27 in C_{13}/C_{14} apocarotenoid biosynthesis is still unresolved indicated by a question mark.

these compounds will be discussed in a separate subchapter below, since they provided the first hints on the potential function of these compounds in mycorrhizal roots.

Other studies to locate a biosynthetic enzyme (DXR) at the subcellular level were conducted on mycorrhizal maize roots. Immunolocalization using affinity-purified antibodies raised from maize DXR recombinant protein revealed a DXR localization specifically in plastids around arbuscule branches. DXR protein accumulation was tightly correlated with arbuscule development. The highest level of DXR protein was reached around maturity and at initial senescence of arbuscules but not at early stages (Hans et al., 2004).

Turning next to the analysis of a carotenoid cleavage step provided an unexpected but very clear result. RNAi-silencing of carotenoid cleavage dioxygenase 1 (CCD1) in mycorrhizal roots of *M. truncatula* expected to catalyse two concomitant cleavage steps at the 9,10 and 9′,10′ positions (Figure 10.1B) did not result in an equal reduction of both types of mycorrhizal apocarotenoids. Only C_{14} mycorradicins were strongly reduced, while the C_{13} α-ionol derivatives were only reduced to about 50%. Concomitantly, a new yellow-orange colour appeared in RNAi roots. Based on ultraviolet light spectra and mass spectrometry analyses, the new compounds produced are C_{27} apocarotenoic acid derivatives, which never appeared in roots transformed with an empty vector control construct or in wild-type plants (Floss et al., 2008b). The unexpected outcome of the RNAi approach suggested C_{27} apocarotenoids as the major substrates of CCD1 in mycorrhizal root cells. A revised scheme of plant carotenoid cleavage in two consecutive steps was proposed, in which CCD1 catalyses only the second step in the cytosol ($C_{27} \rightarrow C_{14} + C_{13}$), while the first step ($C_{40} \rightarrow C_{27} + C_{13}$) is catalysed by another carotenoid cleavage dioxygenase (CCD) within plastids (Figure 10.2, Floss et al., 2008b; Walter et al., 2010). CCD7 was later shown to be regulated by mycorrhization and to be the missing enzyme catalysing the first cleavage step in mycorradicin/α-ionol biosynthesis (Figure 10.2, Vogel et al., 2010). CCD7 also catalyses the first cleavage step in strigolactone biosynthesis where it requires *cis*-substrates (Alder et al., 2012).

The most recent study on the biosynthesis of the mycorrhizal C_{13}/C_{14} apocarotenoids was focused on the key enzyme of carotenoid biosynthesis, phytoene synthase (PSY). We identified novel *PSY3* isogenes from *M. truncatula* (*MtPSY3*) and *Solanum lycopersicum* (*SlPSY3*) strongly expressed exclusively upon root interaction with AM fungi, and moderately in response to phosphate starvation (Stauder et al., 2018). The PSY proteins encoded belong to a widespread clade of conserved PSYs restricted to dicots (dPSY3), distinct from the Poaceae-PSY3s involved in ABA formation. An ancient origin of dPSY3s and a potential co-evolution of this PSY class with the AM symbiosis was discussed in the context of PSY evolution. Knockdown of *MtPSY3* in hairy roots of *M. truncatula* strongly reduced bot C_{13} α-ionol glycosides and mycorradicin levels and also strigolactones (Stauder et al., 2018). d*PSY3* isogenes from tomato were coregulated with upstream isogenes and downstream carotenoid cleavage steps towards strigolactones (*DXS2*, *GGPPS1*, *D27*, *CCD7*) suggesting a combined carotenoid/apocarotenoid pathway, which provides a demand-oriented timely delivery of precursors for apocarotenoid formation.

Taken together, all the evidence available indicates that mycorrhizal C_{13} and C_{14} apocarotenoids are not synthesized from existing carotenoid pools in roots (which are very low or not-existing anyway) but are rather synthesized *de novo* from the C_3 building blocks pyruvate and glyceraldehyde 3-phosphate in the MEP pathway. This reaction chain involves specific isoforms or enzyme variants (DXS2, GGPPS1, PSY3, CCD7, CCD1) in several but not all steps (Figure 10.2, Stauder et al., 2018). Isogene availability and/or expression committed to production of mycorrhizal apocarotenoids can also be species- or plant family-dependent. For example, the step catalysed by DXR seems to have a single-copy gene in cereals, but two differentially regulated isogenes in legumes. A dPSY3-type enzyme regulated by symbionts in roots has now also been identified in a monocot (*Crocus sativus*, Ahrazem et al., 2019). Since this class is absent in cereals and the cereal-type PSY3 is not regulated by mycorrhization, the job of synthesizing C_{13}/C_{14} apocarotenoids in cereals (which are abundant also in this plant family) appears to be done by PSY1 and PSY2 isoforms (Stauder et al., 2018).

Multi-enzyme complexes also called metabolons, in which intermediates are directly passed on to the next enzymic step without being released to the cytosol is a concept described for several pathways and also for carotenogenesis (Nisar et al., 2015). Our results suggest that such still largely hypothetical metabolons may exist not only for carotenogenesis, but also for production of mycorrhizal C_{13}/C_{14} apocarotenoids encompassing early steps of building block formation up to carotenoid cleavage. Having a 'just in time' delivery of carotenoid precursors for apocarotenoid biogenesis would keep the carotenoid concentrations low, and intermediates would show up in elevated concentrations only upon interrupting metabolite flux. Indeed, only interrupting delivery of intermediates by a norflurazon treatment (inhibitor of phytoene desaturase) revealed a level of phytoene elevated up to 240-fold in mycorrhizal roots relative to negative controls (Stauder et al., 2018). Similarly, only upon blocking the CCD1-catalysed step, the C_{27} apocarotenoid intermediates could be identified (Floss et al., 2008b).

What still remains to be investigated is a potential involvement of a β-carotene isomerase (D27) providing *cis*-substrates for CCD7 as has been described for strigolactone biosynthesis (Figure 10.2, Alder et al., 2012). AtD27 from Arabidopsis appears to be a D27, which is focused on β-carotene isomerization preferentially if not exclusively. It has very little activity with all-*trans*-β,β-cryptoxanthin (having one hydroxy-group at the ionone ring) and none with lutein (a xanthophyll with two hydroxy-groups) (Abuauf et al., 2018). However, Arabidopsis is a non-host plant to AM fungi, and therefore other D27 enzymes from host plants should be investigated. As for the actual cleavage step, recombinant CCD7 enzymes from both Arabidopsis and pea accepted the xanthophyll zeaxanthin when provided with 9-*cis*-zeaxanthin yielding 9-*cis*-3-OH-β-apo-10-carotenal (C_{27}) (Bruno et al., 2014). The structure of this product is compatible with a still-conceptual structure of an intermediate in C_{13} α-ionol and C_{14} mycorradicin biosynthesis as shown earlier (Vogel et al., 2010), and also with the findings after silencing the second cleavage step by CCD1 (Floss et al., 2008b). Indeed, there is a symbiosis-regulated *D27* sequence in *M. truncatula* databases (Stauder et al., 2018), but whether it is involved in the pathway is still an open question.

What is also unresolved is the nature of the carotenoid or xanthophyll intermediate in C_{13}/C_{14} apocarotenoid formation. The 3-oxo-keto groups at the ionone ring in all the mycorrhizal α-ionol derivatives suggests the xanthophyll zeaxanthin but this has never been proven. By association with ABA biosynthesis, which has a similar ionone ring (cyclohexenone) configuration and does have a zeaxanthin intermediate, we searched for genes with sequence similarity to the *ABA2* gene of ABA biosynthesis bringing about the α-ionol ring configuration. There is an AM-regulated candidate gene, which we call *ABA2*-like, but there is not yet sufficient information to claim that the enzyme encoded is indeed part of the reaction chain to C_{13}/C_{14} apocarotenoids (R. Stauder, personal communication; see Figure 10.2 for position of ABA2).

PROSPECTIVE ROLES IN THE AM SYMBIOSIS

The AM symbiosis is a highly complex interaction between two symbiotic partners, which appears to be mutualistic in most, if not all, cases (Gutjahr and Parniske, 2017). That means there is always a cost to having the fungal partner entering the root and delivering nutrients, which is paid in plant sugars or lipids exported to the fungus (Harrison, 2012; Gutjahr and Parniske, 2013; Recorbet et al., 2013; Keymer et al., 2017). AM fungi live and thrive only when plant compounds are obtained in return for mineral nutrient delivery. Thus, two important parameters that need to be controlled by the plant are: (i) the extent of colonization of roots, that is, the magnitude of growth of the fungal hyphal network outside the roots including fungal entry into roots; and (ii) the formation, maintenance and degradation of arbuscules arising from internal hyphae in the inner root cortex as these are the fungal organs delivering mineral nutrients. Different classes of apocarotenoids appear to be involved in these two different tasks: strigolactones appear to support the symbiosis by affecting colonization and entry into roots (see Chapter 11); on the other hand, C_{13} and C_{14} apocarotenoids appear to be involved in the regulation of the second control point, namely a control of the arbuscular life cycle and the lifetime of individual arbuscules. But how did this current model of involvement of C_{13}/C_{14} apocarotenoids in arbuscule lifetime regulation arise, and how does it fit into other current models of AM symbiosis functioning?

In recent years, the arbuscular life cycle and its regulation has been studied extensively (Luginbuehl and Oldroyd, 2017). Arbuscules are ephemeral structures and highly dynamic organs. Typically, there is a very heterogeneous population of individual arbuscules, which apparently act cell-autonomously. Their composition from young, early developing, mature, and degrading individuals, can vary considerably. One important lesson learned during these studies was that there is not only a transcriptional programme for arbuscule inception and development, but also one for active arbuscule degradation (not simple decay). A highly complex interaction of signals with MYB transcription factors and DELLA proteins regulates a proper degradation programme for arbuscules, if they are no longer needed, or are not functioning properly (Floss et al., 2017). Intriguing results from cellular and subcellular studies on arbuscule degradation are also available (Kobae and Hata, 2010). The first hints towards such an active degradation mechanism came from the observation that loss of function by mutation of the plant phosphate transporter PT4 residing in the periarbuscular membrane and enabling phosphate transport from the fungus to the plant leads to accelerated and premature degeneration of arbuscules (Harrison et al., 2002; Javot et al., 2007). Such an active degradation programme needs regulators, but also components being 'executers' that is, they have a biochemical activity enabling or supporting one or several cellular processes towards the desired modification directly. One example for such 'executers' may be hydrolytic enzymes such as chitinases, lipases, and proteases, to help in digestion and disassembly of the arbuscule and its periarbuscular membrane (Gutjahr and Parniske, 2017). One prominent example for an 'executer' recently investigated is indeed a hydrolase, which is regulated during active arbuscule degradation (Floss et al., 2017).

Since the C_{13} and C_{14} apocarotenoids are synthesized only at distinct arbuscule stages, namely mature or early degenerating, but not in young stages as judged from the localization of the DXR biosynthetic enzyme (Hans et al., 2004), it was tempting to speculate that they are not only accompanying arbuscule degradation, but may also be actively involved in the process. Both C_{13} and yellow C_{14} apocarotenoids or their glycosides or esters, respectively, are deposited in or next to arbusculated cells at rather high concentrations (Figure 10.3), which is reminiscent to phytoalexin deposition and executive antimicrobial action at sites of local pathogen attack. Interestingly, among the diterpenes, there are regulatory compounds (the gibberellins) and executors (phytoalexins such as the momilactones of rice) that originate from the same pathway, and this might also be the case for apocarotenoids.

One question asked early on was whether both types of apocarotenoids accumulating in mycorrhizal roots might be involved in such processes, or only one of them, and if so, which one of the two is it (Walter et al., 2007). There is now a considerable body of arguments that it is the C_{13} α-ionols, and not the C_{14} mycorradicin:

(i) Mycorradicin does not occur in all mycorrhizal roots despite strong root colonization and arbuscule formation. It is low in *Allium porrum* (Schliemann et al., 2008b) and completely absent in *Tagetes erecta* (Klingner et al., 1995b). At the same time there are abundant α-ionol glycosides accumulating in mycorrhizal roots of *A. porrum* (Schliemann et al., 2008b). The complete lack of mycorradicin from mycorrhizal roots of *T. erecta* being strongly colonized, exhibiting very high numbers of arbuscules and accumulating many C_{13} α-ionol glycosides has recently been confirmed in my laboratory (M.H. Walter, unpublished results).
(ii) Silencing of *DX2* expression in mycorrhizal roots (strong reduction of both C_{13} and C_{14}) did strongly affect mycorrhizal performance (Floss et al., 2008a), whereas *CCD1* silencing (strong reduction of C_{14} but not C_{13}) did not (Floss et al., 2008b).
(iii) No biochemical activity has yet been demonstrated for C_{14} mycorradicin (Schröder et al., 2001), whereas C_{13} ionol derivatives are known to have some antimicrobial activity (Park et al., 2004).

Taken together, the evidence currently available clearly points to mycorradicin as a byproduct of C_{13} α-ionol generation, without a specific function in the AM symbiosis. The primary C_{14} dialdehyde cleavage product is instable and may degrade rapidly, unless it is modified to a dialcohol (known as rosafluene), or the dioic acid mycorradicin, which are more stable.

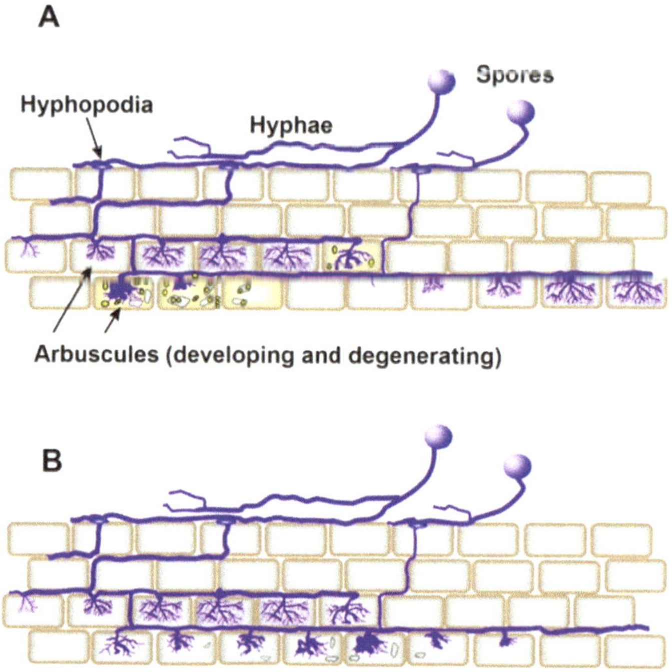

FIGURE 10.3 Proposed model for mode of action of C_{13} α-ionol and C_{14} mycorradicin derivatives in root cells harbouring fungal arbuscules. Fungal hyphae enter the roots through hyphopodia, colonize it and form arbuscules at the inner cortex (bottom). As depicted in two rows of root cells from left to right, arbuscules develop from hyphae through young, developing, mature, degenerating, and dead (clumpy) stages, after which a new life cycle is initiated. Scenario A: Normal C_{13} α-ionol and C_{14} mycorradicin deposition at late mature and early degenerating arbuscules accelerates degradation and allows for rapid clearance, recolonization and repeated hosting of arbuscules. Droplets of the 'yellow pigment complex' (yellow vesicles) appear in the cytosol and also localize to vacuoles (grey). They may eventually also spread to vicinal apoplast and cells bringing about a yellow colouration of roots (Fester et al., 2002). Scenario B: Depletion or loss of C_{13} α-ionol and C_{14} mycorradicin accumulation by suppressing biosynthesis leads to an extended lifespan of arbuscules, but also to an accumulation of older and dead stages (appearing as clumpy individuals) disturbing clearance and the emergence of new arbuscules.

As for the C_{13} compounds, several 3-hydroxy-β-ionone and 3-hydroxy-α-ionone derivatives, similar in structure to AM-induced C_{13} compounds and potential intermediates for glycosylated C_{13} derivatives, have been shown to exhibit moderate but not strong phytotoxic, antimicrobial, and allelopathic activities (Park et al., 2004; Kato-Noguchi et al., 2010, 2012, 2015), compatible with a function as allelopathic agents against specific structures of a symbiont but not a pathogen. Plant cell death is to be avoided in the process, as during turnover the arbuscule is degraded and dies, but the plant cell must stay alive and does not experience any obvious damage (in contrast to pathogenic interactions).

Based on the observations resulting from the *DXS2* silencing, a model has been proposed, which is explained here in more detail (Figure 10.3). It assigns the moderately antimicrobial C_{13} α-ionol aglycons a role in a plant-controlled abrogation and degradation of arbuscules, together with other components such as hydrolytic enzymes. This accelerates arbuscule turnover, and re-emergence of new arbuscules in cells freed from old arbuscules. It thus ultimately leads to an arbuscule population that contains more mature and fully functional individuals (Figure 10.3; Walter et al., 2010; Walter and Strack, 2011; Walter, 2013). The model is compatible with the observed nonlinear relationship between C_{13} apocarotenoid levels and the arbuscule functionality marker, the phosphate transporter gene *MtPT4* showing a significant shift in arbuscule stages only upon strong C_{13} apocarotenoid depletion, while moderate reduction rather invoked an upward trend on *MtPT4* expression (Floss et al., 2008a).

According to the model, an inhibition of C_{13} apocarotenoid-mediated arbuscule degradation can initially extend arbuscule lifetime and mycorrhizal functionality. Only strong and extended inhibition of degradation will ultimately lead to an over-ageing, that is, an accumulation of many old, non-functional arbuscules, and prevent or slow down their degradation and removal (Figure 10.3). This in turn will block the re-emergence of new arbuscules in those cells (the turnover), inhibit a new generation of arbuscules to appear, and impair mycorrhizal functionality in subsequent arbuscule generations. As introduced above, plant-controlled arbuscule degradation as a sort of punishment for poor or lack of phosphate delivery is now an accepted mechanism based on a number of independent findings (Javot et al., 2007; Floss et al., 2017; Gutjahr and Parniske, 2017). The current model of C_{13} apocarotenoid action fits well into these newly developed scenarios. Penalizing a poorly-performing or non-performing symbiotic fungal partner by degrading its symbiotic organs is part of the concept, while on the other hand, rewards, for example, energy supply to fungi, can be given in the form of plant sugars and lipids.

The regular arbuscule degradation and turnover process itself is expected to generate a lot of old arbuscule debris (e.g., old membrane components) that may accumulate in vacuoles and the apoplast and may be part of the 'yellow pigment complex'. Whether and how this material can be recycled in the root cortex cells is unknown. The hitherto detectable parts of this process are likely the C_{13}/C_{14} apocarotenoid metabolic markers, which may hence constitute a sort of metabolic record of past arbuscule degradation events in roots.

USE OF C_{13} α-IONOL DERIVATIVES AS SHOOT MARKERS INDICATIVE OF THE MYCORRHIZATION STATUS OF ROOTS

Recent work in the lab of Ian Baldwin has given new application perspectives to the massive accumulation C_{13} apocarotenoids in mycorrhizal roots (Wang et al., 2018). This group has applied advanced and sophisticated analytical methods and detection systems such as state-of-the-art triple quadrupole technology and enhanced sample preparations (e.g., by solid-phase-extraction based purification and concentration) to the analysis of α-ionol glycosides. In a combination of targeted and untargeted metabolomics they first identified by using liquid chromatography (LC) coupled to time-of-flight mass spectrometry (qTOF-MS) five glycosylated C_{13} α-ionol apocarotenoid derivatives (called blumenols according to the old literature) in roots of mycorrhizal *Nicotiana attenuata.* The five compounds were:

1. 11-Hydroxyblumenol C-9-O-Glc
2. 11-Carboxyblumenol C-9-O-Glc
3. 11-Hydroxyblumenol C-9-O-Glc-Glc
4. Blumenol C-9-O-Glc-Glc
5. Blumenol C-9-O-Glc

The structures of compounds 1 and 2 as shown by Wang et al. (2018) correspond to 13-hydroxyblumenol C-9-O-Glc and 13-carboxyblumenol C-9-O-Glc in our nomenclature (Figure 10.1D). Carboxy groups at the C-13 position of the ring (compound 2) have been found previously only in few cases (Strack and Fester, 2006) but the other structures are well in line with what is already known. These five compounds were highly clustered, and had occupied a unique metabolic space in their metabolomic analysis of roots. In particular, these compounds appeared only in empty vector transgenic plants but not in mutant transgenic lines silenced in a calcium- and calmodulin-dependent protein kinase gene (irCCaMK), which rendered them unable to establish a functional AM symbiosis. This result established the α-ionol (blumenol) derivatives again as metabolic markers of mycorrhization and arbuscule abundance in roots.

The novelty of this analysis then popped up only when shoots were analysed: Compounds 1 and 2 (hydroxy- and carboxy-blumenol C glucosides) showed a similar AM-specific accumulation in the leaves as in the roots. Their identity in leaves was verified by high resolution qTOF-MS. The other blumenols or their glycosides, respectively, detected previously in roots were not detected in leaves (compounds 3 and 4) or showed a less consistent AM-specific accumulation (compound 5) due to constitutive background levels. Constitutive ionol and ionone derivatives in tobacco leaves are well known and can serve as aroma compounds (Mookherjee and Wilson, 1990). In the aroma literature these compound class is often referred to as norisoprenoids including 3 oxo-α-ionol (Slaghenaufi and Ugliano, 2018). Structurally similar megastigmanes can also have aroma potential (Takeda et al., 1997). The crucial point in the analysis of Wang et al. (2018) was that such constitutive ionol or ionone derivatives in leaves could be separated from the two AM-related ones (compounds 1 and 2). The accumulation of these potentially AM-indicative foliar compounds was then tested for correlation with root colonization rates. From three weeks after inoculation onwards, the abundance of the two compounds (1 and 2) was sufficient to represent colonization levels in both greenhouse and field experiments. In fact, the foliar content of these compounds strictly followed the percentage of arbuscules in roots. Classical AM-marker genes, which are known to be strongly upregulated in roots, did not respond in leaves. The work of Wang et al. (2018) therefore provides for the first time a method to determine the mycorrhizal status of a plant without digging out the roots.

These authors (Wang et al., 2018) also used a couple of experimental approaches and results (chemical inhibition of carotenoid formation in shoots but not roots, dipping roots into solutions of compounds 1 and 2, and analysing shoots) to argue that there is most likely a transport of these compounds from the root to the shoot. A number of other researchers have tried in the past to find metabolites indicative of mycorrhizal colonization in leaves but were unsuccessful (Schweiger and Müller, 2015; Adolfsson et al., 2017). The work of Wang et al. (2018) now paves the way for high-throughput analysis of the leaf material from both greenhouse- and field-grown plants to detect and quantify a functional AM symbiosis via the α-ionol glycoside (blumenol glycoside) metabolic marker indicative of the presence of arbuscules and their turnover. The methodology and the equipment needed is still sophisticated, requiring advanced sensitivity and selectivity, but can be used in a wide range of applications. The method was shown to be robust and to be applicable to other plants including crop plants (*Solanum tuberosum*, *Brachypodium distachyon*, and *M. truncatula*). The specific types of α-ionol glycoside usable appears to vary between plant species, but at least in the plants mentioned the method could be verified. It is also robust with respect to different AM fungal species (*Rhizophagus irregularis*, *Funneliformis mosseae*, and *Glomus versiforme*), that is, the observed effects were not restricted to specific AMF taxa (Wang et al., 2018). These authors argue also that the new method may enable breeding programmes to select crop varieties and specific lines superior in engaging in and maintaining a functional AM symbiosis resulting in higher yields with less fertilizer input. Such methods can therefore contribute to a transition towards a more sustainable agriculture.

ACKNOWLEDGEMENTS

I would like to thank Juan Antonio López-Ráez, CSIC, Granada, Spain for critically reading the manuscript.

REFERENCES

Abuauf H, Haider I, Jia, KP, Ablazov A, Mi J, Blilou I, Al-Babili S. 2018. The Arabidopsis DWARF27 gene encodes an all-*trans*-/9-*cis*-β-carotene isomerase and is induced by auxin, abscisic acid and phosphate deficiency. *Plant Science*. 277:33–42.

Adolfsson L, Nziengui H, Abreu IN, Šimura J, Beebo A, Herdean A, Aboalizadeh J, Široká J, Moritz T, Novák O, et al. 2017. Enhanced secondary- and hormone metabolism in leaves of arbuscular mycorrhizal *Medicago truncatula. Plant Physiol*. 175(1):392–411.

Ahrazem O, Diretto G, Argandoña J, Fiore A, Rubio-Moraga A, Rial C, Varela RM, Macias FM, Castillo R, Romano E, Gómez-Gómez L. 2019. The specialised roles in carotenogenesis and apocarotenogenesis of the phytoene synthase gene family in saffron. *Front Plant Sci*. 10:249.

Al-Babili S, Bouwmeester HJ. 2015. Strigolactones, a novel carotenoid-derived plant hormone. *Annu Rev Plant Biol.* 66:161–186.

Alder A, Jamil M, Marzorati M, Bruno M, Vermathen M, Bigler P, Ghisla S, Bouwmeester H, Beyer P, Al-Babili S. 2012. The path from β-carotene to carlactone, a strigolactone-like plant hormone. *Science.* 335(6074):1348–1351.

Bothe H, Klingner A, Kaldorf M, Schmitz O, Esch B, Hundeshagen B, Kernebeck H. 1994. Biochemical approaches to the study of plant-fungal interactions in arbuscular mycorrhiza. *Experientia.* 50:919–925.

Brundrett MC. 2002. Coevolution of roots and mycorrhizas of land plants. *New Phytol.* 14(2):275–304.

Bruno M, Hofmann M, Vermathen M, Alder A, Beyer P, Al-Babili S. 2014. On the substrate-and stereospecificity of the plant carotenoid cleavage dioxygenase 7. *FEBS Lett.* 588(9):1802–1807.

Bucher M. 2007. Functional biology of plant phosphate uptake at root and mycorrhiza interfaces. *New Phytol.* 173(1):11–26.

Carbonnel S, Gutjahr C. 2014. Control of arbuscular mycorrhiza development by nutrient signals. *Front Plant Sci.* 5:462.

Fester T, Maier W, Strack D. 1999. Accumulation of secondary compounds in barley and wheat roots in response to inoculation with an arbuscular mycorrhizal fungus and co inoculation with rhizosphere bacteria. *Mycorrhiza.* 8(5):241–246.

Felemban A, Braguy J, Zurbriggen MD, Al-Babili S. 2019. Apocarotenoids involved in plant development and stress response. *Front Plant Sci.* 10:1168.

Fester T, Hause B, Schmidt D, Halfmann K, Schmidt J, Wray V, Hause G, Strack D. 2002. Occurrence and localization of apocarotenoids in arbuscular mycorrhizal plant roots. *Plant Cell Physiol.* 43:256–265.

Fiorilli V, Wan JY, Bonfante P, Lanfranco L, Al-Babili S. 2019. Apocarotenoids: Old and new mediators of the arbuscular mycorrhizal symbiosis. *Front Plant Sci.* 10:1186.

Floss DS, Hause B, Lange PR, Kuster H, Strack D, Walter MH. 2008a. Knock-down of the MEP pathway isogene 1-deoxy-D-xylulose 5-phosphate synthase 2 inhibits formation of arbuscular mycorrhiza-induced apocarotenoids, and abolishes normal expression of mycorrhiza-specific plant marker genes. *Plant J.* 56:86–100.

Floss DS, Schliemann W, Schmidt J, Strack D, Walter MH. 2008b. RNA interference-mediated repression of MtCCD1 in mycorrhizal roots of *Medicago truncatula* causes accumulation of C-27 apocarotenoids, shedding light on the functional role of CCD1. *Plant Physiol.* 148:1267–1282.

Floss DS, Gomez SK, Park HJ, MacLean AM, Müller LM, Bhattarai KK, Lévesque-Tremblay V, Maldonado-Mendoza E, Harrison MJ. 2017. A transcriptional program for arbuscule degeneration during AM symbiosis is regulated by MYB1. *Curr Biol.* 27:1206–1212.

Giger A. 2002. Chemical synthesis project. A new yellow carotenoid. *Pure Appl Chem.* 74(8):1383–1390.

Gutjahr C, Parniske M. 2013. Cell and developmental biology of arbuscular mycorrhiza symbiosis. *Annu Rev Cell Dev Biol.* 29:593–617.

Gutjahr C, Parniske M. 2017. Cell biology: Control of partner lifetime in a plant–fungus relationship. *Curr Biol.* 27(11):R420–R423.

Hans J, Hause B, Strack D, Walter MH. 2004. Cloning, characterization, and immunolocalization of a mycorrhiza-inducible 1-deoxy-D-xylulose 5-phosphate reductoisomerase in arbuscule-containing cells of maize. *Plant Physiol.* 134:614–624.

Harrison MJ, Dewbre GR, Liu J. 2002. A phosphate transporter from *Medicago truncatula* involved in the acquisition of phosphate released by arbuscular mycorrhizal fungi. *Plant Cell.* 14(10):2413–2429.

Harrison MJ. 2012. Cellular programs for arbuscular mycorrhizal symbiosis. *Curr Opin Plant Biol.* 15:691–698.

Hou X, Rivers J, Leon P, McQuinn RP, Pogson BJ. 2016. Synthesis and function of apocarotenoid signals in plants. *Trends Plant Sci.* 21:792–803.

Javot H, Penmetsa RV, Terzaghi N, Cook DR, Harrison MJ. 2007. A *Medicago truncatula* phosphate transporter indispensable for the arbuscular mycorrhizal symbiosis. *Proc Natl Acad Sci USA.* 104:1720–1725.

Jones FR. 1924. A mycorrhizal fungus in the root of legumes and some other plants. *J Agric Res.* 29:459–470.

Kato-Noguchi H, Seki T, Shigemori H. 2010. Allelopathy and allelopathic substance in the moss *Rhynchostegium pallidifolium. J Plant Physiol.* 167:468–471.

Kato-Noguchi H, Hamada N, Clements DR. 2015. Phytotoxicities of the invasive species *Plantago major* and non-invasive species *Plantago asiatica. Acta Physiol Plant.* 37(3):1794.

Kato-Noguchi H, Tamura K, Sasaki H, Suenaga K. 2012. Identification of two phyto-toxins, blumenol A and grasshopper ketone, in the allelopathic Japanese rice variety Awaakamai. *J Plant Physiol.* 169(7):682–685.

Keymer A, Pimprikar P, Wewer V, Huber C, Brands M, Bucerius SL, Delaux P-M, Klingl V, von Röpenack-Lahaye E, Wang TL, et al. 2017. Lipid transfer from plants to arbuscular mycorrhiza fungi. *eLife.* 6:e29107.

Kiers ET, Duhamel M, Beesetty Y, Mensah JA, Franken O, Verbruggen E, Fellbaum CR, Kowalchuk GA, Hart MM, Bago A, et al. 2011. Reciprocal rewards stabilize cooperation in the mycorrhizal symbiosis. *Science*. 333:880–882.
Kobae Y, Hata S. 2010. Dynamics of periarbuscular membranes visualized with a fluorescent phosphate transporter in arbuscular mycorrhizal roots of rice. *Plant Cell Physiol*. 51:341–353.
Klingner A, Bothe H, Wray V, Marner FJ. 1995a. Identification of a yellow pigment formed in maize roots upon mycorrhizal colonization. *Phytochemistry* 38:53–55.
Klingner A, Hundeshagen B, Kernebeck H, Bothe H. 1995b. Localization of the yellow pigment formed in roots of gramineous plants colonized by arbuscular fungi. *Protoplasma*. 185(1–2):50–57.
Lanfranco L, Fiorilli V, Gutjahr C. 2018. Partner communication and role of nutrients in the arbuscular mycorrhizal symbiosis. *New Phytol*. 220(4):1031–1046.
Laparre J, Malbreil, M, Letisse F, Portais JC, Roux C, Bécard G. Puech-Pagès V. 2014. Combining metabolomics and gene expression analysis reveals that propionyl- and butyryl-carnitines are involved in late stages of arbuscular mycorrhizal symbiosis. *Mol Plant*. 7(3):554–566.
López-Ráez JA, Shirasu K, Foo E. 2017. Strigolactones in plant interactions with beneficial and detrimental organisms: The Yin and Yang. *Trends Plant Sci*. 22:527–537.
Luginbuehl LH, Oldroyd GED. 2017. Understanding the arbuscule at the heart of endomycorrhizal symbioses in plants. *Curr Biol*. 27:R952–R963.
MacLean AM, Bravo A, Harrison MJ. 2017. Plant signaling and metabolic pathways enabling arbuscular mycorrhizal symbiosis. *Plant Cell*. 29:2319–2335.
Maier W, Peipp H, Schmid, J, Wray V, Strack D. 1995. Levels of a terpenoid glycoside (Blumenin) and cell wall-bound phenolics in some cereal mycorrhizas. *Plant Physiol*. 109:465–470.
Maier W, Hammer K, Dammmann U, Schulz B, Strack D. 1997. Accumulation of sesquiterpenoid cyclohexenone derivatives induced by an arbuscular mycorrhizal fungus in members of the Poaceae. *Planta*. 202:36–42.
Maier W, Schneider B, Strack D. 1998. Biosynthesis of sesquiterpenoid cyclohexenone derivatives in mycorrhizal barley roots proceeds via the glyceraldehyde 3-phosphate/pyruvate pathway. *Tetrahedron Lett*. 39:521–524.
Maier W, Schmidt J, Nimtz M, Wray V, Strack D. 2000. Secondary products in mycorrhizal roots of tobacco and tomato. *Phytochemistry*. 54(5):473–479.
Massalha H, Korenblum E, Tholl D, Aharoni A. 2017. Small molecules below-ground: The role of specialized metabolites in the rhizosphere. *Plant J*. 90(4):788–807.
Martin FM, Uroz S, Barker DG. 2017. Ancestral alliances: Plant mutualistic symbioses with fungi and bacteria. *Science* 356(6340):eaad4501.
Matsunami K, Otsuka H, Takeda Y. 2010. Structural revisions of blumenol C glucoside and byzantionoside B. *Chem Pharm Bull*. 58(3):438–441.
Mookherjee BD, Wilson RA. 1990. Tobacco constituents – Their importance in flavor and fragrance chemistry. *Perfumer Flavorist*. 15(1):27–49.
Nisar N, Li L, Lu S, Khin NC, Pogson BJ. 2015. Carotenoid metabolism in plants. *Mol Plant*. 8:68–82.
Parniske M. 2008. Arbuscular mycorrhiza: The mother of plant root endosymbioses. *Nat Rev Microbiol*. 6:763–775.
Park S, Takano Y, Matsuura H, Yoshihara T. 2004. Antifungal compounds from the root and root exudate of *Zea mays*. *Biosci Biotechnol Biochem*. 68(6):1366–1368.
Paszkowski U, Jakovleva L, Boller T. 2006. Maize mutants affected at distinct stages of the arbuscular mycorrhizal symbiosis. *Plant J*. 47:165–173.
Plett JM, Martin FM. 2018. Know your enemy, embrace your friend: Using omics to understand how plants respond differently to pathogenic and mutualistic microorganisms. *Plant J*. 93(4):729–746.
Recorbet G, Abdallah C, Renaut J, Wipf D, Dumas-Gaudot E. 2013. Protein actors sustaining arbuscular mycorrhizal symbiosis: Underground artists break the silence. *New Phytol*. 199:26–40.
Remy W, Taylor TN, Hass H, Kerp H. 1994. Four hundred-million-year-old vesicular arbuscular mycorrhizae. *Proc Nat Acad Sci*. 91(25):11841–11843.
Rohmer M, Knani M, Simonin P, Sutter B, Sahm H. 1993. Isoprenoid biosynthesis in bacteria: A novel pathway for the early steps leading to isopentenyl diphosphate. *Biochem J*. 295(2):517–524.
Scannerini S, Bonfante-Fasolo P. 1977. Unusual plastids in an endomycorrhizal root. *Can J Bot*. 55:2471–2474.
Schliemann W, Ammer C, Strack D. 2008a. Metabolite profiling of mycorrhizal roots of *Medicago truncatula*. *Phytochemistry* 69:112–146.
Schliemann W, Kolbe B, Schmidt J, Nimtz M, Wray V. 2008b. Accumulation of apocarotenoids in mycorrhizal roots of leek (*Allium porrum*). *Phytochemistry*. 69:1680–1688.

Schliemann W, Schmidt J, Nimtz M, Wray V, Fester T, Strack D. 2006. Accumulation of apocarotenoids in mycorrhizal roots of *Ornithogalum umbellatum*. *Phytochemistry*. 67:1196–1205.

Schröder S, Hildebrandt U, Bothe H, Niehaus K. 2001. Suppression of an elicitor-induced oxidative burst reaction in *Nicotiana tabacum* and *Medicago sativa* cell culture by corticrocin but not by mycorradicin. *Mycorrhiza*. 11:101–106.

Schweiger R, Müller C. 2015. Leaf metabolome in arbuscular mycorrhizal symbiosis. *Curr Opin Plant Biol*. 26:120–126.

Serra S. 2015. Recent advances in the synthesis of carotenoid-derived flavours and fragrances. *Molecules*. 20(7):12817–12840.

Slaghenaufi D, Ugliano M. 2018. Norisoprenoids, sesquiterpenes and terpenoids content of valpolicella wines during aging: Investigating aroma potential in relationship to evolution of tobacco and balsamic aroma in aged wine. *Front Chem*. 6:66.

Stauder R, Welsch R, Camagna M, Kohlen W, Balcke GU, Tissier A, Walter MH. 2018. Strigolactone levels in dicot roots are determined by an ancestral symbiosis-regulated clade of the *PHYTOENE SYNTHASE* gene family. *Front Plant Sci*. 9:255.

Strack D, Fester T. 2006. Isoprenoid metabolism and plastid reorganization in arbuscular mycorrhizal roots. *New Phytol*. 172:22–34.

Takeda Y, Zhang H, Masuda T, Honda G, Otsuka H, Sezik E, Yesilada E, Handong S. 1997. Megastigmane glucosides from *Stachys byzantina*. *Phytochemistry* 44(7):1335–1337.

Vogel JT, Walter MH, Giavalisco P, Lytovchenko A, Kohlen W, Charnikhova T, Simkin AJ, Goulet C, Strack D. Bouwmeester HJ, Fernie AR, Klee HJ. 2010. SlCCD7 controls strigolactone biosynthesis, shoot branching and mycorrhiza-induced apocarotenoid formation in tomato. *Plant J*. 61:300–311.

Walter MH. 2013. Role of carotenoid metabolism in the arbuscular mycorrhizal symbiosis. In: *Molecular Microbial Ecology of the Rhizosphere*, F. de Bruijn, ed. (Chichester: Wiley-Blackwell), pp. 513–524.

Walter MH, Fester T, Strack D. 2000. Arbuscular mycorrhizal fungi induce the non-mevalonate methylerythritol phosphate pathway of isoprenoid biosynthesis correlated with accumulation of the 'yellow pigment' and other apocarotenoids. *Plant J*. 21:571–578.

Walter MH, Hans J, Strack D. 2002. Two distantly related genes encoding 1-deoxy-D-xylulose 5-phosphate synthases: Differential regulation in shoots and apocarotenoid-accumulating mycorrhizal roots. *Plant J*. 31:243–254.

Walter MH, Floss DS, Hans J, Fester T, Strack D. 2007. Apocarotenold biosynthesis in arbuscular mycorrhizal roots: Contributions from methylerythritol phosphate pathway isogenes and tools for its manipulation. *Phytochemistry*. 68:130–138.

Walter MH, Floss DS, Strack D. 2010. Apocarotenoids: Hormones, mycorrhizal metabolites and aroma volatiles. *Planta*. 232:1–17.

Walter MH, Strack D. 2011. Carotenoids and their cleavage products: Biosynthesis and functions. *Nat Prod Rep*. 28:663–692.

Walter MH, Stauder R, Tissier A. 2015. Evolution of root-specific carotenoid precursor pathways for apocarotenoid signal biogenesis. *Plant Science*. 233:1–10.

Wang M, Schäfer M, Li D, Halitschke R, Dong C, McGale E, Paetz C, Song Y, Li S, Dong J, Heiling S, Groten K, Franken P, Bitterlich M, Harrison MJ, Paszkowski U, Baldwin IT. 2018. Blumenols as shoot markers of root symbiosis with arbuscular mycorrhizal fungi. *eLife*. 28 (7):e37093.

11 Strigolactones: Nutrient Stress Hormones Affecting Plant Development and Biotic Interactions

Michael H. Walter

CONTENTS

INTRODUCTION

The first description and subsequent structural elucidation of strigolactones (SLs) dates back to the 1960s and 1970s (Cook et al., 1966, 1972). Their discovery was completely independent of any plant mineral nutrition studies, since at this time they were characterized as seed germination stimulants for parasitic weeds. Scientists were in search of compounds exuded from cotton roots into the rhizosphere to act as signals for the parasite. Consequently, the trivial name strigol for the compound identified in this search was derived from the witchweed root parasite *Striga lutea* Lour, on which it acted in minute concentrations. To isolate the compound cotton plants (actually a false host) were grown hydroponically, and the nutrient fluid was circulated periodically through charcoal columns, which were then eluted with acetone (Cook et al., 1966). The complete structural elucidation was accomplished based on an X-ray crystallographic analysis of a single crystal, further supported by NMR and mass spectra (Cook et al., 1972; Figure 11.1). The structure was similar to other unsaturated lactones, thus determining the name for the compound class of strigolactones. With these root parasites becoming an increasingly serious threat to agriculture, various SLs have been characterized over the years as universal seed germination stimulants of parasitic weeds, exuded from the roots of both host and non-host plants into the rhizosphere, and acting on a variety of *Striga* and *Orobanche* parasite species (Bouwmeester et al., 2003; Cardoso et al., 2011; Clarke et al., 2019).

The next major step in the history of SL discoveries occurred in a search of a branching factor for hyphae of arbuscular mycorrhizal (AM) fungi. These fungi live in an obligate endosymbiosis with plant roots, where they form tiny intracellular organs called arbuscules (Parniske, 2008; Martin et al., 2017). Arbuscules are encased by a plant periarbuscular membrane, which constitutes an important

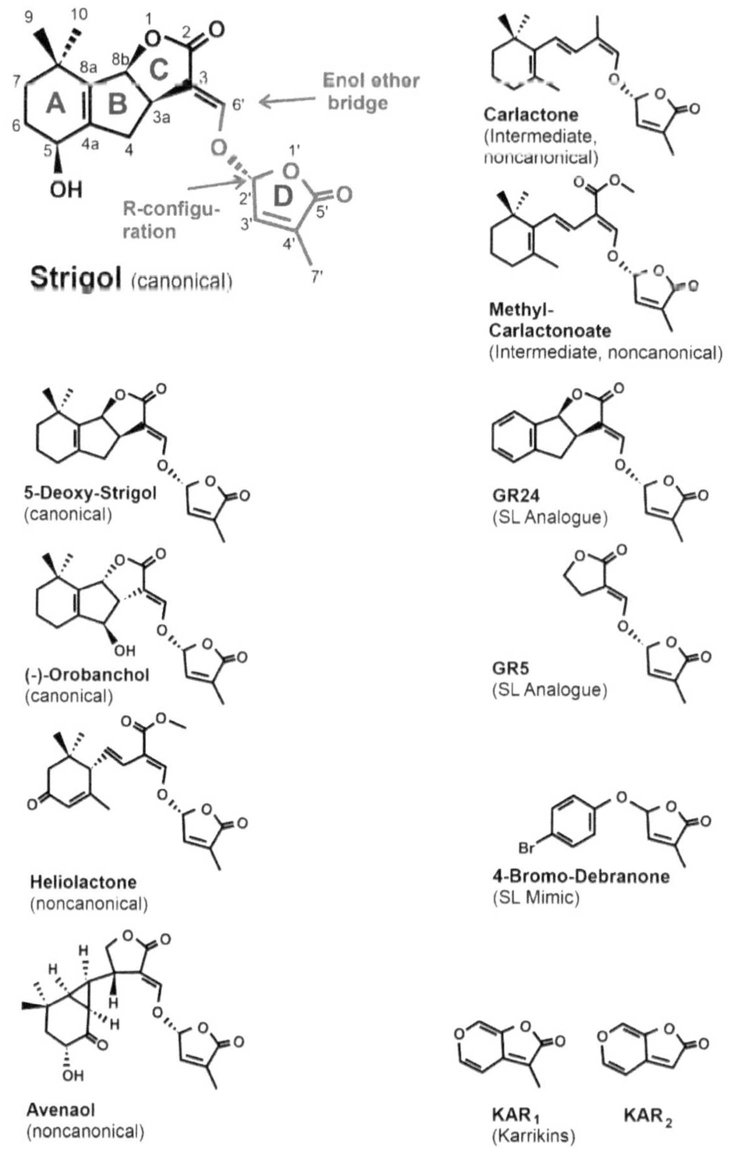

FIGURE 11.1 Structural diversity of strigolactones (SLs). Strigol (top left): prototype SL (canonical), first one isolated from root exudates, with SL-typical features such as a system of 4 rings (A, B, C, D), of which the methylbutenolide D-ring connected via an enol–ether bridge to the C-ring (or other moieties) is essential for activity. 5-Deoxy-Strigol (beneath): variant characterized first as inducer of hyphal branching of AM fungi (canonical). (−)-Orobanchol (beneath): Representative of orobanchol-type canonical SL, which differ in stereochemistry at the B–C-ring junction from strigol-type SLs. Heliolactone and avenaol (beneath): two representatives of natural noncanonical SLs missing an intact A–C-ring configuration. Carlactone and methylcarlactone (top right): two pathway intermediates identified during elucidation of the SL biosynthetic pathway, both noncanonical but exhibiting SL activity. GR24 and GR5: two synthetic SL analogues of which the first one is the classical SL analogue used frequently. 4-Bromo-Debranone (beneath): synthetic SL mimic with little more than the essential structure. KAR_1, KAR_2 (bottom right): Two karrikins with a butenolide ring structure and seed germination promoting activity but not carotenoid-derived, discovered in the smoke of wildfires.

interface for compound exchanges between the two partners of this mutualistic symbiosis. The trade is mainly inorganic phosphate and other mineral nutrients for the plant from the fungus, against sugars and lipids for the fungus from the plant (Keymer et al., 2017). In order to reach the inner root cortex of the plant, where arbuscules are formed and phosphate is delivered, the fungus must colonize the root first extraradically, and then intraradically. Root colonization involves the ramification of fungal hyphae, and this process is enabled – or at least supported – by plant signals. In a search of the still elusive hyphal branching factor, a strigolactone was isolated from root exudates of *Lotus japonicus*. Its identification as 5-deoxy-strigol (Figure 11.1) was based on spectroscopic analyses and chemical synthesis. Both this natural SL and a synthetic analogue (GR24, Figure 11.1) induced extensive hyphal branching in germinating spores of the AM fungus *Gigaspora margarita* at very low concentrations (Akiyama et al., 2005).

Initially, the SL hyphal branching signals for AM fungi were described as sesquiterpenes (Akiyama et al., 2005). At about the same time, the first arguments came up that SLs are not classical sesquiterpene lactones synthesized directly, but that their biogenesis rather involves an indirect pathway through carotenoid cleavage in analogy to ABA. The key arguments for the apocarotenoid nature of SLs came from the use of carotenoid mutants and inhibitors of isoprenoid pathways in maize, sorghum and cowpea (Matusova et al., 2005). These results were subsequently further supported by work on tomato (López-Ráez et al., 2008).

A breakthrough in SL research opening up new perspectives came in 2008 from a completely different direction. Other researchers had been looking for the structure of a shoot branching inhibitor hormone for a long time. Some parts of its biogenesis in roots, and propagation into shoots to regulate shoot branching architecture, was well established from highly branched/tillered mutants and from grafting studies (reviewed by Mouchel and Leyser, 2007). The apocarotenoid nature of the inhibitor was suggested from studies with mutants in two genes encoding carotenoid cleavage dioxygenase enzymes (CCDs): CCD7 and CCD8. Intriguingly, both mutants exhibit an increased shoot branching phenotype in several plant species (Sorefan et al., 2003; Booker et al., 2004; Schwartz et al., 2004). Eventually, the research directions united, when known or suspected carotenoid-derived compounds were tested, including an application of the synthetic strigolactone analogue GR24 to shoot branching CCD7 and CCD8 mutants of pea, rice and Arabidopsis. The resulting phenotypical changes proved the ability of SLs – or a compound derived from them – to at least partially restore wild-type branching phenotypes and bud outgrowth in the mutants (Gomez-Roldan et al., 2008; Umehara et al., 2008). The story of SL identification as the long-sought shoot branching inhibitor hormone has been covered in several reviews from various perspectives (Leyser, 2008; Dun et al., 2009; Tsuchiya and McCourt, 2009).

SL research has become a very popular topic since then. Many studies have been initiated on biosynthesis, regulation of biosynthesis, hormone transport, hormone perception mechanisms in the plant as well as interactions with other hormones (reviewed recently in Jia et al., 2018; Zwanenburg and Blanco-Ania, 2018; Felemban et al., 2019; Machin et al., 2020). The role of SLs in the AM symbiosis is also being investigated (reviewed in Lanfranco et al., 2018; Fiorilli et al., 2019). Additional roles in other symbioses and generally in biotic interactions and also abiotic stresses is emerging. Secondary growth (Agusti et al., 2011), leaf senescence (Felemban et al., 2019), stomata closure (Lv et al., 2018) and drought response regulation together with ABA (Haider et al., 2018) are processes where SLs appear to play a role. Furthermore, the exploration of SL action as a germination stimulant of parasitic weeds and its perception by the parasite has been carried on extensively including various approaches to combat the threat to agriculture by these parasites using novel strategies (Clarke et al., 2019).

INDUCIBILITY BY MINERAL NUTRIENT STARVATION

SLs have been associated with nutrient stress responses because SL levels in roots and their biosynthesis is induced by mineral nutrient starvation (reviewed in Umehara, 2011; Marzec et al., 2013). Similarly, they have been characterized as a "cry for help in the rhizosphere" calling symbiotic microbes such as AM fungi to the rescue, which can assist in uptake during situations of poor mineral nutrient availability (López-Ráez et al., 2011b; Andreo-Jimenez et al., 2015). Inhibiting branching/tillering and focusing growth on single shoots leading to more rapid production of progeny in cases of nutrient shortage can also be seen as a SL-driven measure to adapt plant shoot architecture to reduced nutrient availability.

Phosphorus (P) and Nitrogen (N) are key macronutrients that sustain plant growth and crop yield. Inorganic phosphate (Pi), in particular, is often very limited in soils in a form accessible by plant roots. Plants have developed sophisticated strategies to enhance Pi acquisition on their own (Lambers et al., 2015; Abel, 2017), but in terrestrial ecosystems they often rely on soil microbes to assist them. SLs seem to be involved both in direct measures, remodelling root system architecture, and in calling/supporting symbiotic microbes. To achieve these modifications to enable coping with nutrient stress, SL biosynthesis has to be enhanced in situations of nutrient deprivation. There is ample documentation on elevated SL levels in roots under phosphate starvation. Early work was done in red clover and sorghum, which showed that phosphate deficiency alone or deficiency in either phosphate or nitrogen can lead to elevated levels of SLs (Yoneyama et al., 2007a, b). A wider range of species was investigated later, indicating that there are differences between species, but that a basic response can be generally observed (Yoneyama et al., 2012). This was followed up in studies on phosphate and nitrogen starvation responses in pea (Foo et al., 2013). Tomato, Arabidopsis and rice were other species, in which Pi-deprivation led to enhanced SL levels in roots, resulting in changes in shoot architecture (López-Ráez et al., 2008; Umehara et al., 2010; Kohlen et al., 2011). SL-insensitive or SL-deficient mutants did not display this response (Umehara et al., 2010; Kohlen et al., 2011). Regulation of additional genes (TIR1 auxin receptor, phosphate transporters) was shown to be affected by phosphate starvation in Arabidopsis, and this response was compromised in the *max2-1* SL receptor component mutant, indicating that it is dependent on SLs (Mayzlish-Gati et al., 2012). Among many other genes, SL biosynthetic genes (*CCD7* and *CCD8*) are upregulated by phosphate deficiency in *M. truncatula* (Bonneau et al., 2013). Recent results from both *M. truncatula* and tomato have confirmed this finding, and extended it by demonstrating that the presence of AM fungi further upregulates these and other SL biosynthetic genes (Stauder et al., 2018). Similarly, an ATP-binding cassette (ABC) transporter gene specific for SLs (*PDR1*) from *Petunia hybrida* was transcriptionally upregulated by phosphate starvation and by the presence of the AM fungus *Rhizophagus irregularis* (Kretzschmar et al., 2012). Taken together, there is ample evidence for SL upregulation at various levels by mineral nutrient deficiency. Surprisingly, this upregulation seems to be further enhanced by the presence of the symbionts, although phosphate deprivation should be alleviated by their activities. It further seems to be continued into the later stages of the interaction arguing that SLs are not only needed to lure the microbial symbiont from the rhizosphere to the root upon first encounter but also plays a role in supporting the symbiosis later on. Further details on the action of SLs on AM fungi will be discussed below.

STRUCTURAL DIVERSITY AND BIOSYNTHESIS

Naturally occurring SLs, now amounting to about 30 variants, are structurally characterized by four rings (A–D), of which the methylbutenolide D-ring seems to be the most important and invariably essential part for biological activity (Figure 11.1; Al-Babili and Bouwmeester, 2015; Jia et al., 2018; Wang and Bouwmeester, 2018; Yoneyama et al., 2018b; Machin et al., 2020). The D-ring is linked by an enol–ether bridge in R-configuration to the ABC rings, which are more structurally variable. In fact, the enol–ether-D-ring moiety is essential for biological activity. SLs can be separated into strigol or orobanchol types. While the D-ring is always in R-configuration, the two types differ in the stereochemistry of the BC-ring junction (3a*R*, 8b*S* for strigol, 3a*S* 8b*S* for orobanchol, Figure 11.1).Together, they are known as the canonical SLs, having the typical tricyclic lactone signature (ABC rings), while the noncanonical SLs, sometimes also called SL-like, display other or partly missing ABC ring structures (Figure 11.1). Revised definitions for SLs have been proposed, whereby SLs are defined as "a carotenoid-derived molecule with a butenolide D-ring", or "products of CCD7 and CCD8 activity in plants that have a 2′R-configured D-ring", which would include both canonical and noncanonical SLs (Wang and Bouwmeester, 2018; Machin et al., 2020). In canonical SLs, the A and B rings can be modified/decorated by hydroxylation, demethylation, epoxidation, acetoylation, or ketolation (Wang and Bouwmeester, 2018). Plants typically produce a complex blend of SLs, both canonical and noncanonical, and both types seem to have important functions (Wang and Bouwmeester, 2018; Yoneyama et al., 2018b). As an example, the tomato root exudate contains orobanchol, solanacol, hydroxyorobanchol, oxoorobanchol, orobanchyl acetate, and didehydro-orobanchol isomers (Kohlen et al., 2013; Wang and Bouwmeester, 2018). The chemistry and the distribution of canonical and noncanonical SLs in the plant kingdom has recently been summarized in a review. It assigns various SL-typical functions in the rhizosphere or in plant-internal signalling to both types of strigolactones (Yoneyama et al., 2018b). Examples for noncanonical SLs are carlactone (a biosynthetic intermediate of SLs), methyl carlactonoate, heliolactone, zealactone, zeapyranolactone, and avenaol (Jia et al., 2018; Felemban et al., 2019; Figure 11.1). Akiyama and coworkers searched for structural requirements determining hyphal branching of AM fungi. Truncation of the A- and AB-rings in tricyclic canonical SLs resulted in a drastic reduction of such activities. Although the connection of the C-ring in the tricyclic lactone to the methylbutenolide D-ring was shown to be essential for hyphal branching, the bridge structure in the C–D part was found not necessarily to be enol–ether, being replaceable with either alkoxy or imino ethers. These structural requirements for acting on AM fungi were very similar, but not identical, to those observed for their action on root-parasitic weeds (Akiyama et al., 2010).

In addition, there are even smaller molecules called karrikins, which contain a butenolide ring structure but not the ABC rings (Figure 11.1). These are smoke-derived compounds occurring after wild fires by thermal degradation. They can have seed germination promoting activity. Karrikins are not considered SLs as they do not conform to the definition of carotenoid-derived. Nevertheless their signalling appears to go through a signalling cascade, which uses similar components (D14/KAI2) as do SLs (Flematti et al., 2015; De Cuyper et al., 2017, see below).

Synthetic analogues of SLs can be perceived by the SL perception system and can mimic many biological effects of SLs. The most important and most frequently used analogue is GR24, having an aromatic A ring (Figure 11.1). Known for a long time (Johnson et al., 1981), it was initially developed for its high parasitic weed seed germination activity and its increased stability compared to natural SLs. The question has been raised, whether all the effects determined over the years are specific to SLs when using racemic GR24 as most researchers commonly did. The mixture contains four isomers, two of which mimic natural strigolactones. Zwanenburg and Blanco-Ania (2018) compared the bioactivities on germination of parasitic weeds (*Striga hermontica*) of enantiopure GR24 forms with that of racemic GR24 and found little if any differences. However, when investigating hormonal effects in the plant, the use of specific isomers only is recommended, since the perception complex is strictly stereospecific. The chemical biology of further SL analogues will be discussed in the last subchapter.

The biogenesis of SLs was elusive for a long time. Only when the apocarotenoid nature of these compounds was becoming likely, was a hypothetical scheme for its biosynthesis proposed, with some analogy to ABA formation (Figure 11.2), but with a late coupling of the D-ring to an existing tricyclic ABC molecule (Matusova et al., 2005). After SLs were identified as shoot branching inhibitor hormones in 2008, research on SL biogenesis picked up rapidly. Apart from the two carotenoid cleavage steps encoding CCD7 and CCD8, which now clearly established the apocarotenoid nature of SLs, little was known about the strigolactone biosynthetic pathway at the time. A series of high branching, tillering, and dwarf mutants from *A. thaliana*, pea, rice and petunia defective in strigolactone biosynthesis or perception were available (Stirnberg et al., 2002; Sorefan et al., 2003; Ishikawa et al., 2005; Johnson et al., 2006; Simons et al., 2007; Drummond et al., 2009). Four *A. thaliana max* mutants and eight *d/htd* rice mutants had been characterized at the molecular level, similar to several pea *rms* and petunia *dad* mutants. The *MAX3-*, *RMS5-*, *D17/HTD1-*, and *DAD3*-defective mutants have been

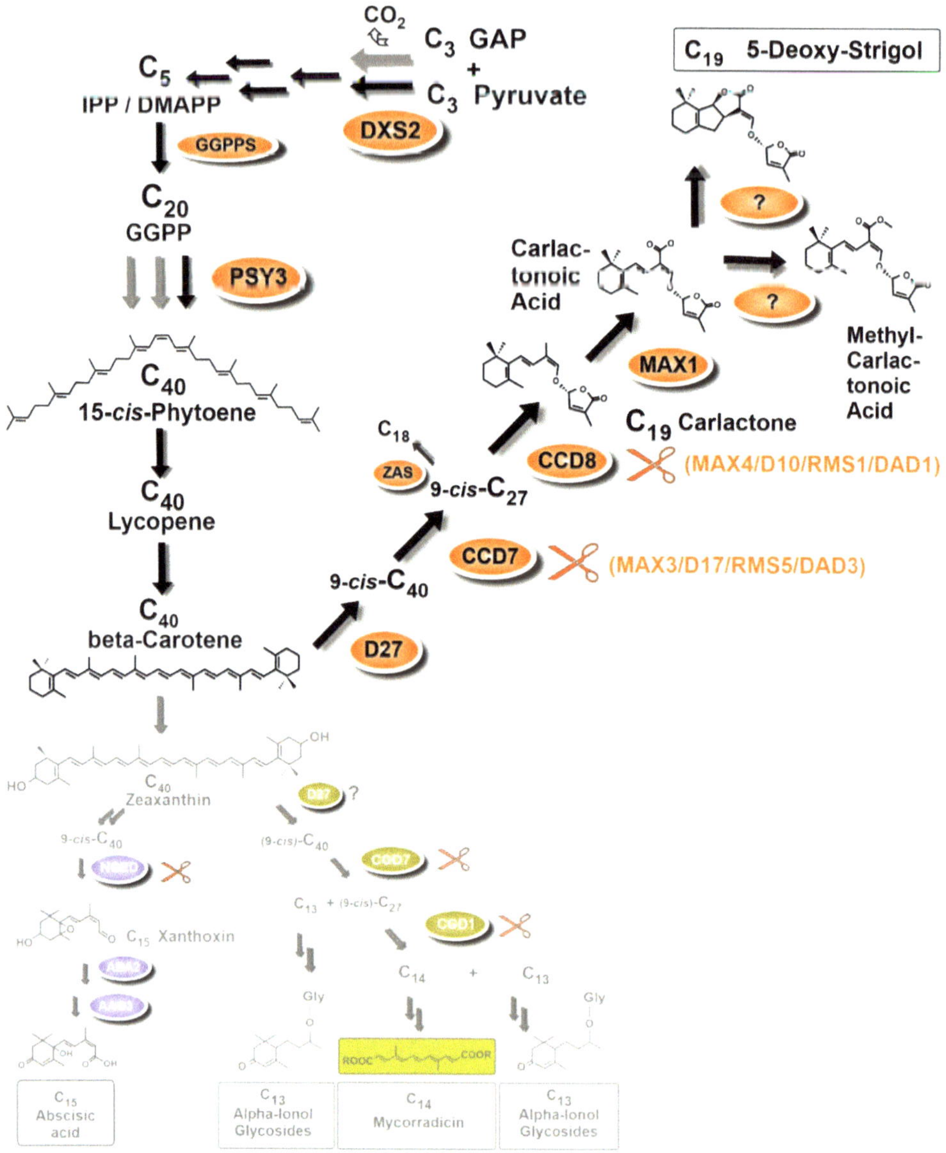

FIGURE 11.2 Apocarotenoid biosynthetic pathways in roots and their C_{40} carotenoid and C_5 IPP precursor supply. The pathway to C_{19} strigolactones is shown in full print while potentially competing side pathways to C_{13} α-ionol glycosides/C_{14} mycorradicin derivatives and ABA are faded out. The simplified scheme indicates key enzymatic step of the MEP pathway (horizontal to the left, top), of carotenoid biosynthesis (vertical) and of carotenoid cleavage steps (inclined and going up for strigolactone end products). Precursor supply in roots is regulated through specialized isoforms of 1-deoxy-D-xylulose 5-phosphate synthase (DXS2) and phytoene synthase (PSY3) in most plants except Arabidopsis, which has single copy genes in these steps. Cleavage by CCDs or 9-*cis*-epoxy-carotenoid cleavage dioxygenases (NCEDs) typically starts from 9-*cis* isomerized substrates (brought about by D27 *cis–trans* isomerase) and is carried out by a single (ABA) or by sequential double cleavage reactions followed by modification steps leading to various apocarotenoid products (boxed). Various other designations of CCD7 and CCD8 resulting from naming mutants are indicated in orange print. Recently, an enzyme called zaxinone synthase (ZAS) has been described, which diverts C_{27} intermediates to a C_{18} compound. Steps downstream of carlactone (top right) are not fully elucidated and are indicated by question marks.

shown to be orthologous and encode non-functional *CCD7* genes. Similarly, *MAX4*, *RMS1*, *D10*, and *DAD1* are compromised in the *CCD8* gene function (Figure 11.2). Another *A. thaliana* branching mutant (*MAX1*) has led to the identification of an oxidative modification step downstream of CCD7/CCD8 catalysed by a cytochrome P450 enzyme (CYP711A1-related). The *MAX2* mutant is not a biosynthesis mutant, but rather defective in strigolactone perception. *MAX2* encodes an F-box leucine-rich repeat protein, which is involved in the SL perception protein complex. An iron-containing rice protein (DWARF27, D27) was shown to be required for strigolactone biosynthesis and its mutation caused a phenotype similar to the *D10* (*CCD8*) mutant (Lin et al., 2009).

Tremendous progress in elucidating strigolactone biosynthesis was obtained in 2012. The group of Salim Al-Babili established the path from β-carotene to a new SL biosynthetic intermediate by characterizing two important enzymic steps (Alder et al., 2012). The first one was the identification of D27 as a *cis/trans* isomerase catalysing the conversion of all-*trans*-β-carotene into 9-*cis*-β-carotene. Only the 9-*cis*-β-carotene is a proper substrate for the subsequent first carotenoid cleavage step catalysed by CCD7 producing a 9-*cis*-configured aldehyde (9-*cis*-β-apo-10′-carotenal, C_{27}) and β-ionone (C_{13}) (Figure 11.2). The second step characterized biochemically in this paper was the further conversion of this C_{27} apocarotenal by the second cleavage enzyme CCD8. CCD8 was shown to be an unusual CCD, catalysing the incorporation of three oxygens and a molecular rearrangement in a combination of different reactions, leading to a newly identified C_{19} SL intermediate that was called carlactone (Alder et al., 2012; Figure 11.2). The catalytic mechanism of CCD8 has later been discussed in quite some detail (Bruno et al., 2017). The presence of carlactone *in planta* could not be proven at the time of carlactone identification *in vitro* (Alder et al., 2012) but was later confirmed to be a native SL intermediate detectable in plants. This identification was facilitated by an approximately 700-fold higher content of carlactone in the Arabidopsis *MAX1* mutant suggesting that the enzyme encoded may directly convert carlactone (Seto et al., 2014). In addition, it was shown that recombinant CCD7 and CCD8 thioredoxin fusion proteins from Arabidopsis, rice or pea can also utilize hydroxylated substrates (xanthophylls) producing 4-hydroxy-carlactone, which might give rise to further yet unidentified SLs (Baz et al., 2018). Most recently, a new player acting on SL biosynthesis emerged. It is zaxinone synthase (ZAS), a member of a hitherto overlooked class of CCDs acting on C_{27} apocarotenoid intermediates similar to CCD8. It appears to be a negative regulator of SL biosynthesis, since it converts C_{27} apocarotenoid intermediates to a C_{18} compound called zaxinone, and thus removes C_{27} intermediates from the pathway. Consequently, ZAS rice mutants have more C_{27} and thus produce more SLs (Wang et al., 2019; Figure 11.2).

There are many reports on the regulation of SL biosynthesis in roots during root colonization by AM fungi. These data have been reviewed recently (Lanfranco et al., 2018) and therefore only the most pertinent and the latest original work will be cited here. *CCD7* transcript level have been found to be upregulated in many species, and in early as well as late stages of the symbiosis; for example, in *Lotus japonicus* (Guether et al., 2009; Fiorilli et al., 2015). Recent confirmations come from tomato and *M. truncatula* (López-Ráez et al., 2015; Stauder et al., 2018). *CCD8* transcripts do not exhibit such a clearcut upregulation, showing induction in only a few species. This has been attributed to rather high constitutive levels and to the role of CCD7, but not CCD8, in the production of other apocarotenoids, which can accumulate in mycorrhizal roots in rather massive amounts (Vogel et al., 2010; Walter et al., 2010; Walter and Strack, 2011; Walter, 2013). These other classes of C_{13}/C_{14} apocarotenoids, which share some steps with SL biosynthesis (Figure 11.2), are covered in detail in Chapter 10.

Several of the reactions downstream of carlactone have been elucidated by now (Figure 11.2). While CCD8 is located in plastids, subsequent steps are located in the cytosol, suggesting an unknown mechanism of carlactone export from plastids. One or more cytochrome P450 enzymes (CYPs) in the cytoplasm convert carlactone into further oxidized compounds. Since from Arabidopsis there is only a single mutant known (*MAX1*), the complexity of these reactions is probably lower than initially expected, since CCD8 already introduces several of the structural modifications needed. The MAX1 protein belongs to the CYP711 clade and analysis of the Arabidopsis enzyme AtMAX1

(CYP711A1) indicated that it oxidizes the C-19 methyl group of carlactone to produce carlactonoic acid (Abe et al., 2014). In rice, it was reported that only one out of five MAX1 homologs (Os900/CYP711A2) converts carlactone to 4-deoxy-orobanchol, which already contains the typical four-ring SL structure (Zhang et al., 2014). Another member of this clade Os1400/CYP711A3 converts 4-deoxy-orobanchol to orobanchol (Zhang et al., 2014). Recent studies into the distribution of carlactone converting enzymes and genes across plant species (Arabidopsis, rice, maize, tomato, poplar, and *Selaginella moellendorffii*) have shown that this step is a common reaction, catalysed by MAX1 homologs and that these homologs can be classified into three types: (i) A1, converting carlactone to carlactonoic acid; (ii) A2, converting carlactone to 4-deoxy-orobanchol through a carlactonoic acid intermediate, and (iii) A3, converting carlactone to carlactonoic acid and 4-deoxy-orobanchol to orobanchol (Yoneyama et al., 2018a). Carlactonoic acid is also a substrate for an unidentified methyltransferase that produces methyl carlactonoate (Abe et al., 2014; Jia et al., 2018; Yoneyama et al., 2018a). Another step downstream of MAX1 was described as a lateral branching oxidoreductase (LBO), a member of the 2-oxoglutarate and Fe-II-dependent dioxygenase family. It converts methyl-carlactonoic acid to an unknown product with an additional oxygen atom (Brewer et al., 2016).

Biosynthesis of SLs, the C_{13}/C_{14} apocarotenoids, and ABA depends on carotenoid precursor supply. However, despite the importance of precursor availability, reactions upstream of β-carotene, their regulation by phosphate starvation and mycorrhizal colonization and their potential role in regulating precursor levels for SL formation, have received rather little attention. My group, in collaboration with others, has investigated several steps within the methylerythritol phosphate (MEP) pathway of C_5-IPP precursor supply and in the carotenoid core pathway for transcriptional regulation during root colonization by AM fungi. First, an isogene of *1-DEOXY-D-XYLULOSE 5-PHOSPHATE SYNTHASE* (*DXS*) was identified, called *DXS2*, which was strongly upregulated, particularly in the later stages of the AM symbiosis, whereas the housekeeping *DXS1* isogene was not (Walter et al., 2002; Figure 11.2). Silencing *DXS2* expression in mycorrhizal transgenic plants led to strongly reduced C_{13}/C_{14} apocarotenoid accumulation, which correlated with a strong down-regulation of AM marker genes and a remarkable change in arbuscule population structures leading to more older and fewer younger arbuscules. SL levels were not determined at the time, but in retrospect it is quite likely that SL levels were also reduced, since we observed a moderate but not strong reduction in root colonization (Floss et al., 2008). Recently, we have targeted the core step of carotenoid biosynthesis catalysed by PHYTOENE SYNTHASE (PSY) in a search for AM-regulated isogenes. We have identified a new class of *PSY3* isogenes called dicot *PSY3* (d*PSY3*), since it is absent in cereals. It is strongly regulated in roots by AM fungal colonization and moderately by Pi-starvation in the absence of the fungi (Stauder et al., 2018; Figure 11.2). Silencing this gene in transgenic *M. truncatula* roots led to significantly reduced levels of SL and also of C_{13}/C_{14} apocarotenoids establishing an important role for carotenoid precursor supply in both SL and C_{13}/C_{14} biosynthesis (Stauder et al., 2018). This root symbiosis-regulated *PSY3* class has now also been discovered in the monocot *Crocus sativus* (Ahrazem et al., 2019). Taken together, the results on the precursor pathways establish a prominent role for carotenoid precursor supply in SL biosynthesis. Interestingly, the two AM-regulated isogenes *DXS2* and *PSY3* governing fortification of carotenoid precursor production for SL biosynthesis are not found in the genome of the AM fungi-nonhost plant Arabidopsis. Whether these gene losses are cause or effect of the nonhost property is unknown.

HORMONE TRANSPORT AND PERCEPTION

The SL internal phytohormone and the SL cue for AM fungi both need to be transported within the plant and exported into the rhizosphere or to the site of fungal entry in roots. Orobanchol was detected in the xylem sap of Arabidopsis consistent with the assumption that strigolactones are mainly produced in roots and transported to the shoot to regulate shoot branching (Kohlen et al., 2011; Figure 11.3). Such a direction of SL fluxes in the plant was already suggested by grafting

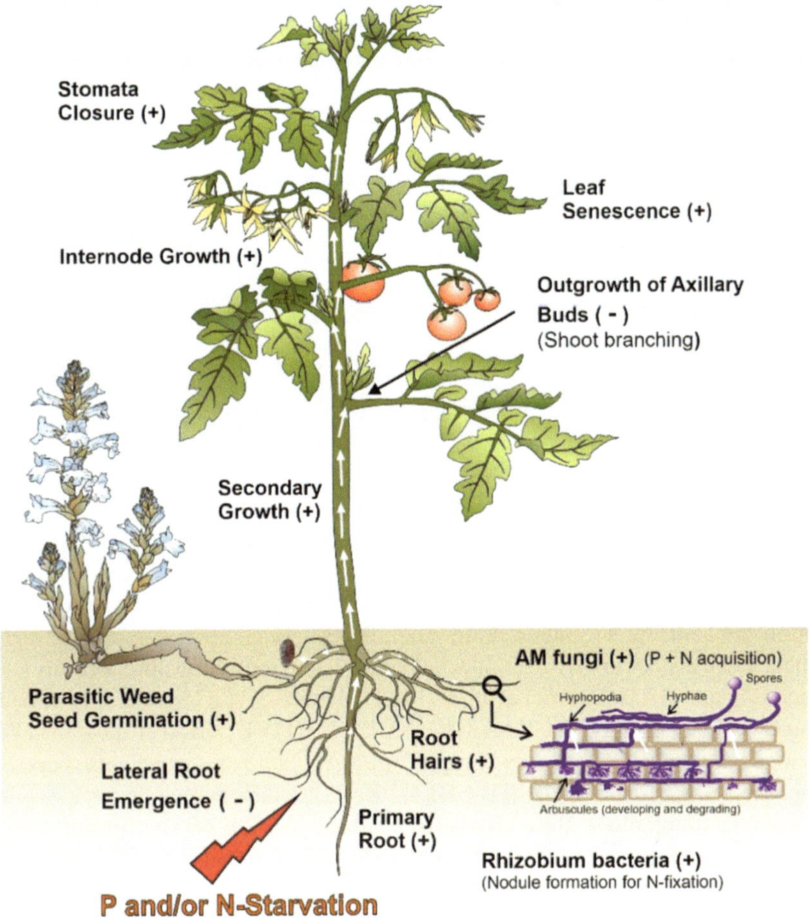

FIGURE 11.3 Multiple actions (promoting + or inhibiting –) of SLs on plant development and biotic interactions. SLs transmit nutrient stress information (P- and/or N-starvation) through elevated SL production in the roots followed by exudation as well as by transport of the hormone to the shoot (white arrows). Exudation of SLs to AM fungal nutrient-providers supports their hyphal branching as well as the formation and function of fungal hyphopodia entry structures (belowground, right). Rhizobium bacteria fixing atmospheric nitrogen are also supported (beneath). The SL signal in the rhizosphere is exploited by root-parasitic plants to detect host roots and start seed germination (belowground left) leading to harmful root parasite growth (witchweeds, broomrapes) at the expense of the host plant (aboveground, left).

experiments, where SL biosynthesis mutants could be rescued by wild-type root stocks (Leyser, 2009). The upwards translocation of SLs has been confirmed, but it has been questioned whether or not this occurs through the xylem (Xie et al., 2015).

ATP-binding cassette (ABC)-transporter proteins are known to have roles in phytohormone translocations. In 2012 an ABC transporter (PDR1) specific for SLs was identified from *P. hybrida*, which appeared to function as a cellular exporter (Kretzschmar et al., 2012). It has a key role in both regulating shoot branching and root colonization by AM fungi. *PDR1* mutants exhibit an enhanced shoot branching phenotype and reduced AM fungal colonization due to a deficiency in SL exudation (Kretzschmar et al., 2012). Later studies by the same group showed that *PDR1* overexpression causes reduced shoot branching and induced seed germination of root-parasitic weeds. There is directional transport in roots, where PDR1 is apically localized in root tips, and laterally in hypodermal passage cells, thus facilitating the entry of AM fungi into roots (Sasse et al., 2015).

While the perception of SLs by AM fungi and their action on the fungi is still largely a mystery (see below), perception in the plant has been elucidated in quite some detail. The full, highly complex picture currently known is presented in a recent review (Machin et al., 2020), which can be discussed here only to a limited extent. SL perception starts by binding to an SL receptor, which is part of a large receptor/F-box protein complex. *DWARF14* (*D14*), first defined from a rice mutant (homologous to *DAD2* from *P. hybrida*) encodes an α/β-hydrolase-fold protein with an internal cavity capable of accommodating SLs. However, D14/DAD2 is not only a receptor but also modifies its ligand, that is, it has an enzyme function (Janssen and Snowden, 2012). It is thus an unusual receptor, which has been called noncanonical (Yao et al., 2016). It contains a highly conserved catalytic triad (Ser-His-Asp) that is required for SL hydrolysis and signal transduction (Hamiaux et al., 2012; Waters et al., 2017). Integration of SL in the open solvent-exposed binding pocket triggers its hydrolysis into the ABC ring and an intermediate D-ring moiety. The next step is the release of the ABC ring, which is then covalently bound, forming an intermediate called CLIM (covalently linked intermediate molecule) (Waters et al., 2017; Machin et al., 2020). The process continues with binding of D14 to the F-box protein MAX2 (homologous to D3 in rice), and with interacting with other components of the receptor complex. These include rice D53 proteins or Arabidopsis SMXL proteins to eventually result in ubiquitination and proteasome-mediated degradation (Waters et al., 2017; Machin et al., 2020). Recent studies have described the interaction of D14/D3/D53 and the mechanisms of SL hydrolysis in more detail, ascribing D3/MAX2 a key role in the coordination of SL perception and hydrolysis (Shabek et al., 2018; Barbier et al., 2019).

One aspect that has only briefly been touched on above in the context of karrikins is the relationship between D14 and KAI2 (Scaffidi et al., 2014). D14 belongs to a larger family of α/β hydrolases in land plants, which includes the KARRIKIN INSENSITIVE2 (KAI2) receptor (Waters et al., 2017). Phylogenetic analysis has shown that the D14/KAI2 family in land plants is derived from two ancient lineages, eu-KAI2 and DDK (D14/DLK2/KAI2), which evolved from a single KAI2-like lineage, present in streptophyte algae (Bythell-Douglas et al., 2017). The ligands for the eu-KAI2 and KAI2-like receptors remain largely elusive, but it is known that in angiosperms, eu-KAI2 acts in the perception of karrikins. Adding to the confusion about the role of KAI2 it was reported that perception of AM fungi requires a karrikin receptor (KAI2) complex and that a rice *KAI2/D14*-like mutant loses responsiveness to AM fungi (Gutjahr et al., 2015). The unidentified molecule perceived by rice roots might be a KAI2 ligand, which might mimic endogenous plant signals. Recent work has shown that KAI2 is duplicated and has been subfunctionalized in the legume *Lotus japonicus* having different binding preferences to various synthetic ligands. KAI2a and KAI2b respond organ-specifically to different karrikins (KAR1, KAR2) (Carbonnel et al., 2019). Taken together, the KAI2 story and the relationship between karrikin and strigolactone signalling still remains rather enigmatic.

ROLES IN SHAPING SHOOT ARCHITECTURE

Plants have an enormous phenotypic plasticity. They can adjust their shoot architecture from apical growth to extended branching in order to adapt to certain environmental conditions including the supply of mineral nutrients. Axillary buds and their meristems can remain dormant or become activated to produce a branch. Information on how such a decision is made is collected in the axillary meristem through a network of hormone and other signals that can integrate various inputs. SLs have now become a major player – but not the only one – in determining such a decision (Barbier et al., 2019). As introduced above, the discovery of SLs as a phytohormone was based on an extended investigation of a large number of mutants in different species exhibiting more shoot branching, a higher tillering or a dwarf phenotype. These mutants were named differently, but all the designations refer to the observed shoot phenotypes: more axillary branching (MAX) in Arabidopsis; *ramosus* (latin for branched, *rms*) in pea; decreased apical dominance (DAD) in *P. hybrida* and dwarf (D) in rice. Many of these mutants from different species turned out to be orthologous. Most instrumental for the final discovery of SLs as the signal causing these phenotypes (or rather, the lack of it, or the absence of perception) was the biochemical identification of the gene functions of *MAX3/RMS5/D17/DAD3* all encoding CCD7, as well as *MAX4/RMS1/D10/DAD1* all encoding CCD8, suggesting an apocarotenoid hormone, which was still elusive before 2008.

The SL signal produced in the roots travels up into the shoot to axils and axillary meristems, where it participates in a cellular programme to regulate branching. Indeed, the local expression in shoots of the SL transporter *PDR1* as shown by promoter-GUS fusions was confined to the vasculature and nodal tissues adjacent to the leaf axils, but was absent from dormant buds (Kretzschmar et al., 2012). In roots, GUS staining was enhanced in roots grown under phosphate-deficient conditions. This finding argues that there is a phosphate/nitrogen starvation-driven programme of SL biosynthesis in roots and translocation into shoots finalized by SL deposition at axils to suppress bud outgrowth and branching (Figure 11.3). On the other hand, while the grafting experiments (wild-type rootstocks rescue entirely the shoot branching phenotype of mutant scions) underline the importance of roots in producing SLs, or at least their precursors, SLs appear not to be exclusively produced in roots. The reverse grafting of SL biosynthesis deficient rootstocks to wild-type scions did not lead to an increased branching phenotype in the scions, which is what one would expect when SL supply from roots is cut off (Turnbull et al., 2002; Sorefan et al., 2003).

Clearly, at the axils in the shoot, SLs are not the only regulators of shoot branching and bud outgrowth. At least three other hormonal signals have been identified as important regulators, namely auxin, cytokinin, and gibberellins. Other players may also contribute (Rameau et al., 2015). Auxins have long been known to be involved in the process. Many studies have shown that SLs interact with auxins, and it is known that SLs regulate auxin homeostasis by modulating polar auxin transport (PAT) (Barbier et al., 2019; Tan et al., 2019). The apically-derived auxin transported in stems induces expression of SL biosynthetic genes and represses cytokinin levels, thus balancing bud outgrowth (Domagalska and Leyser, 2011; Rameau et al., 2015). Sucrose is another player, mediating bud outgrowth and interacting with the hormones. Sucrose upregulates the expression of auxin biosynthetic genes and promotes the accumulation of cytokinin accumulation, which in turn triggers bud outgrowth (Barbier et al., 2019). Bud outgrowth correlates well with the expression of cytokinin- and SL-signalling genes in axillary buds, which indicates that buds may be a site of action of these hormones (Barbier et al., 2019). SLs have an inhibitory effect on the polar subcellular localization of PIN1 in axillary buds that differs from the largely SL-independent regulation of auxin transport and cytokinin biosynthesis in stems (Tan et al., 2019). Local auxin transport between PAT and surrounding tissues called connective auxin transport, rather than PAT itself, was shown to be essential for the initial bud outgrowth, while PAT is necessary for sustainable bud growth both in Arabidopsis and pea (Leyser, 2009; Barbier et al., 2019). A recent genetic analysis of connective auxin transport involving additional PINs in Arabidopsis indicated that it plays an important role in SL-mediated control of shoot branching with a triple *pin3pin4pin7* mutant being able to suppress at least partially the highly branched phenotype of Arabidopsis SL-deficient mutants (van Rongen et al., 2019). We are only beginning to understand the highly complex networks of signals, signal transduction and target genes determining the ensuing cellular actions. SLs are a rather new player in this research field of molecular components of hormone and sugar signalling bringing in new inputs from the mineral nutrient supply situation in the soil.

ROOT DEVELOPMENT AND SUPPORT OF BENEFICIAL FUNGAL SYMBIONTS ENABLING IMPROVED NUTRIENT ACQUISITION

SLs play important roles in supporting AM fungi (see below), but apart from this well-known action they can also determine root architecture and root development in the absence of symbionts, or in symbiont-nonhost plants such as Arabidopsis (Koltai, 2011; Figure 11.3). Much of the work on this topic has been done on the latter species, starting with investigating the effects of GR24 on root architecture. GR24 application reduced the number of lateral roots in wild-type plants, and in the biosynthetic mutants *MAX3* and *MAX4*, but not in the *MAX2* signalling mutant. Moreover GR24 led to increased length of root hairs but again *MAX2* did not display this response, suggesting that both effects of SL on root architecture is mediated via the MAX2/D14 receptor complex (Kapulnik et al., 2011). Ruyter-Spira et al. (2011) did similar experiments, and also found suppression by GR24 application of lateral root primordia development, leading to reduced lateral root density under phosphate-sufficient conditions. Moreover, primary root lengths of SL-deficient and SL-insensitive Arabidopsis plants were shorter than those of wild-type plants (Ruyter-Spira et al., 2011). SLs also suppress the formation of adventitious roots, in both Arabidopsis and pea (Rasmussen et al., 2012).

For rice, comparable effects have been reported, although the seminal root of SL mutants was shorter than that of the wild-type only under low phosphate and low nitrate conditions, demonstrating that the effect of endogenous SLs is affected by the growth conditions (Sun et al., 2014). Under both normal and nutrient-poor conditions, application of GR24 resulted in an elongation of the primary root of wild-type and SL mutants, except for the signalling mutant *D3* (Sun et al., 2014). For tomato, only a combined treatment with auxin uncovered an impact on the primary root length: the inhibitory effect of auxin on root growth diminished with the addition of increased concentrations of GR24 (Koltai et al., 2010). It was concluded that general statements are difficult to make on SL impact on primary root length, probably because of subtle phenotypes, of influence of the growth conditions, or, possibly, due to varying endogenous hormonal backgrounds between different species (reviewed by Matthys et al., 2016). Taken together, of the four root-specific GR24-triggered phenotypes that have been observed, two seem directly related to SLs, specifically increased primary root length and adventitious rooting. For the effect on root hairs and lateral root development, more research needs to be done. For now, it cannot be excluded that an as yet unknown signal, mimicked by GR24 and signalling through MAX2/D14, might play a role (Matthys et al., 2016).

Further questions on to the where and how of SL action in root development have been addressed in the review by Matthys et al. (2016). As mentioned above, the results of reciprocal grafting experiments argue for a predominant but not exclusive production of SL in roots (Turnbull et al., 2002; Sorefan et al., 2003). Furthermore, many SL biosynthetic genes are preferentially expressed in roots, but not all of them, for example *MAX1*, whose expression is spread all over the plant. Also, expression levels of *CCD7* from tomato were high in roots but also high in immature green fruit (Vogel et al., 2010). It is obvious that again there is a crosstalk of SL with other phytohormones as in the shoot; namely auxin, cytokinin and also ethylene. This crosstalk might be differentially regulated between root and shoot. Alterations in lateral root development by GR24 application are dependent on the auxin status of the root (Ruyter-Spira et al., 2011). Also, root hair elongation seems to be a process dependent on both auxin and SLs. GR24 induced PIN2 abundance at the epidermal plasma membrane during root hair elongation, indicating that GR24 might affect PIN2 endocytosis and endosomal trafficking via actin dynamics in a MAX2-dependent manner (Pandya-Kumar et al., 2014).

SLs initially evolved as regulators of developmental processes in ancient plant lineages, and then adopted new roles to support plant interactions with microbes (Delaux et al., 2012). One major involvement of SLs today is in the AM symbiosis of plant roots (reviewed recently by Lanfranco et al., 2018; Fiorilli et al., 2019). This topic has been investigated extensively, but the roles of SLs in this interaction are still far from being fully understood. The work of Khoki Akiyama first established a role of strigolactones as an inducer of hyphal branching of AM fungi (Akiyama et al., 2005). Early on, and still now in most opinions, SLs were thought to function as a pre-symbiotic host plant signal for recognition by the fungi. The context described was that of a molecular dialogue of signals, in which SLs act at picomolar concentrations to trigger hyphal branching at the onset of root colonization. However, already in 2006 it was described that strigolactones affect the shape, movement, and density of fungal mitochondria and fungal respiration (Besserer et al., 2006). GR24 application stimulated mitosis, growth, and ATP production of AM fungi by boosting their energy metabolism (Besserer et al., 2008). RNA sequencing of germinated fungal spores after GR24 treatment later confirmed these findings on the transcriptional level by showing fungal mitochondrial genes being upregulated. The authors also identified more upregulated genes, which suggested that also other cellular compartments of the fungi are responsive to SLs (Salvioli et al., 2016). SLs were further shown to induce the release of fungal chitin oligomer signals (Genre et al., 2013) and to enhance fungal spore germination, hyphal elongation and hyphopodia formation (reviewed by Lanfranco et al., 2018). These results imply that SLs profoundly modify fungal metabolism.

To this day it is still not clear how SLs are perceived by the fungus, and how exactly they affect the fungi. Already in the first SL hormone identification paper it was shown that SL-deficient *CCD8* mutants of pea exhibit a root colonization rate reduced to about 30% of wild-type (Gomez-Roldan et al., 2008). The colonization was not fully abolished though, pointing to a supportive but not essential function in the AM symbiosis. Colonization was also reduced moderately in the *PDR1* SL transporter mutant, but arbuscules did not exhibit any aberrant morphology (Kretzschmar et al., 2012). Recently, Kobae et al. (2018) investigated in detail both the *D10* (CCD7) and the *D17* (CCD8) mutants of rice with respect to their root colonization behaviour relative to wild-type rice plants. Most of the colonization process was normal, that is, AM fungal hyphae approached the roots and clung around them. Epidermal penetration, arbuscule size, arbuscule number per hyphopodium, and metabolic activity of the intraradical mycelium, were not different from wild-type. However, the formation of hyphopodia was severely affected, suggesting that SLs are required for the proper generation of these entry points, and for the punctual internalization of hyphae (Kobae et al., 2018; Figure 11.3). These conclusions are supported by earlier observations when localizing expression sites of the *PDR1* SL transporter in mycorrhizal roots. *PDR1* promoter activity was present in hypodermal passage cells above the root tip that form gates for the entry of AM fungi. It was also detected in the outer-lateral membrane compatible with a postulated function as an SL exporter to the site of fungal entry (Sasse et al., 2015). In contrast, the C_{13}/C_{14} apocarotenoids accumulating in late stages of the AM symbiosis appear to be involved in arbuscule functions, particularly in the regulation of the arbuscule degeneration process, and thus determining arbuscule lifetimes (Figure 11.3; Walter, 2013; Walter et al., 2015; see also Chapter 10).

As mentioned above, SL-deficient plants mutated in biosynthetic or transporter genes have lower colonization levels than wild-type plants, but are still colonized at low to intermediate levels. This has now been shown for many different plant species (reviewed in Lanfranco et al., 2018). The effect can be overcome to some extent by feeding GR24 (Gomez-Roldan et al., 2008; Breuillin et al., 2010). However, some mutants in SL/karrikin signalling and perception components behave in a more severe and unexpected manner upon mycorrhization attempts. The *D14* rice mutant is still only weakly affected and shows slightly higher colonization, probably due to higher SL exudation (Yoshida et al., 2012). Remarkably, mutants defective in *D3* or *RMS4* show several aborted infection attempts, and a significant reduction in arbuscule and vesicle formation (Yoshida et al., 2012; Foo et al., 2013). D3/RMS4 is present in both the SL and the karrikin signalling pathway, and thus it has been hypothesized that the severe phenotype might be related to defects in karrikin signalling (Waters et al., 2017). In agreement with such ideas are the results of Gutjahr et al., (2015) already mentioned above, who showed that a mutation in the *D14*-like/*KAI2* gene of rice has a very severe symbiotic phenotype and completely abolishes the formation of hyphopodia thus resulting in a complete inability of the *hebiba* mutant to establish any internal symbiotic structures. The signal to which this mutant is insensitive is still elusive, but it may have some structural similarities to either SLs or karrikins. It is tempting to speculate that such a signal might be generated from SLs by an SL-converting activity in the fungus similar to the hydrolytic activity of D14 in the plant. Indeed, the sequence of a putative fungal α/β hydrolase (RiSP811) potentially hydrolysing SLs has recently been described. Its transcripts were induced upon GR24 treatment and during root colonization (Kamel et al., 2017).

The idea of a fungal hydrolase or any other fungal activity processing SLs has already been put forth earlier to explain the observation that SL levels in mycorrhizal roots are lower than in nonmycorrhizal roots (Koltai, 2014; Walter et al., 2015). This remarkable observation has been made previously by a considerable number of scientists (Lendzemo et al., 2009; Fernàndez-Aparicio et al., 2010; López-Ráez et al., 2011a, 2015; Aroca et al., 2013) and has been confirmed in our work with mature mycorrhizal vs nonmycorrhizal roots of *M. truncatula*. The reduced levels of SLs are in contrast with a still highly active SL production machinery as indicated by high transcript levels of SL biosynthetic genes in later stages of the symbiosis including *CCD8* in this case (Stauder et al., 2018). This discrepancy thus argues for an AM fungus-induced production of SLs by the plant, but accompanied by SL metabolization/catabolism presumably brought about by the fungus (Walter et al., 2015; Stauder et al., 2018). The drain on SL from such fungal activities or simply from continuously exuding SLs into the rhizosphere may necessitate maintaining a permanently high level of SL production in roots in both pre-symbiotic and symbiotic stages dependent on the timely production of its carotenoid precursors.

SLs also influence other plant–microbe interactions. The legume roots–Rhizobium bacteria symbiosis resulting in root nodules that can fix atmospheric nitrogen and provide nitrogen sources to roots also seems to be positively affected by SLs (Figure 11.3; reviewed in López-Ráez et al., 2017). How this happens is even less clear than for the AM symbiosis. The first reports came from studies on *Medicago sativa* (Soto et al., 2010) and pea (Foo and Davies, 2011). GR24 treatment elevated the nodule number in wild-type pea plants. SL-deficient mutants (*rms1*) had about 40% fewer nodules (Foo and Davies, 2011). Another parameter potentially affected by SLs was found to be rhizobial motility by affecting swarming behaviour (Peláez-Vico et al., 2016). Several SL biosynthetic genes are upregulated during nodulation, albeit not at the same magnitude as elicited by AM fungi (Stauder et al., 2018). Both *MtD27* and *MtCCD8* were significantly upregulated in the root hair infectome after rhizobium inoculation (Breakspear et al., 2014). Moreover, the expression of a *MtD27* gene was co-opted in the rhizobium symbiosis, being highly expressed locally in nodule primordial and infection zones (van Zeijl et al., 2015). Taken together, these reports argue for a role of SLs in rhizobial infection structure and nodule development.

There are also reports on SL effects on detrimental pathogenic interactions. One is a positive effect of MAX2 on plant disease resistance in leaves, since in the absence of its function by mutation Arabidopsis plants were more susceptible to the bacterial hemi-biotrophic pathogens *Pseudomonas syringae* and *Pseudomonas carotovorum*. The authors propose elevated stomatal conductance in the mutant as a potential cause for this phenotype (Piisilä et al., 2015). Other reports in the context of SL effects on microbial pathogens have been reviewed (De Cuyper and Goormachtig, 2017; López-Ráez et al., 2017). The most important detrimental interaction affected by SLs are the root-parasitic weeds. There is extensive information available on this plant-host/root-parasite relationship, and it will be dealt with more extensively next.

FRAUDULENT EXPLOITATION OF SL SIGNALS BY HARMFUL ROOT PARASITES

The obligate root-parasitic weeds of the families *Orobanchaceae* (*Aeginetia*, *Orobanche*, broomrape) and *Scrophulariaceae* (*Alectra*, *Striga*, witchweed) are considered to be among the most serious and devastating agricultural pests of economic importance in many parts of the world, particularly in Africa. These parasites form specialized organs called haustoria that attach to host roots enabling the parasite to benefit from water, sugar and other nutrients drawn from the host, which can severely damage host growth and reduce crop yield. The parasites produce a large number of minute seeds that can remain viable in the soil for more than a decade. Only upon receiving chemical signals from the roots of a potential host plant, parasite seed germination is initiated (Figure 11.3; Cardoso et al., 2011; Clarke et al., 2019).

As introduced above, SLs were identified as highly potent seed germination stimulants of these parasites more than 50 years ago, and this action of strigol was the first bioactivity known for any SL at all (Cook et al., 1966). With today's knowledge, it is obvious that with using SLs the parasites exploit ancient plant hormone and signalling molecules for symbiont support to detect their hosts. For the parasites this is an evolutionary stable, widespread, and reliable signal, that is unlikely to be lost during selection, since plants need it to establish and support rewarding relationships with beneficial microbes in the soil. The parasites thus hijack an existing and essential plant signal, und utilize it for their own benefit in a manner that could be called fraudulent. Elucidating the perception and signalling mechanism of SLs in parasitic weeds may therefore be a key to controlling these root parasites with new rational strategies. Indeed, there has been tremendous progress in this direction, designing many new SL mimics with new bioactivities. This progress will be more detailed in the 'Chemical biology and applications' section. One prerequisite – or, at least a helpful basis for such approaches – is the characterization of parasite SL perception components. One idea was to use a fluorescence turn-on probe, Yoshimura Green (YLG), which activates SL signalling in the parasites. Using this method it could be elucidated how SLs bind and act via *Striga hermontica* HTLs, encoding a diverged family of α/β hydrolases related to D14 and KAI2. A dynamic wavelike propagation of strigolactone perception appears to wake up Striga seeds (Tsuchiya et al., 2015). Remarkably, the parasite SL receptors did not evolve from D14, but from ancestral KAI2/HTL (Toh et al., 2015; Tsuchiya, 2018). The *KAI2/HTL* homologs in the parasites have many copies, for example, as many as 11 in *Striga hermonthica* (Toh et al., 2015; Tsuchiya, 2018; Tsuchiya et al., 2018). The reason for this amplification is probably to acquire the ability to perceive many different SLs, and in particular the blend of SLs produced by their preferred hosts. Based on this knowledge new synthetic agonists for parasite SL receptors can now be developed. Various SL and SL mimics have already been tested earlier for their differential action on AM fungi to find out about structure–activity relationships (Akiyama et al., 2010). Even beyond classical SLs, many efforts are now ongoing on systematic development of small molecule probes using molecular design-based approaches and/or chemical screening.

Such new developments will extend the currently ongoing classical approaches to reduce these devastating parasites (Samejima and Sugimoto, 2018). One strategy is to use suicide germination of parasite weed seeds in the absence of or away from the host roots (Toh et al., 2014). False germination in the absence of the host results in seed death within a few days, since the tiny seeds have only little reserves. Another possibility is the use of SL biosynthesis inhibitors to reduce the levels of SLs exuded and thus reduce germination (Tsuchiya, 2018). Natural germplasm with strongly reduced exudation of SLs has been identified within local pea lines, which is resistant to broomrape infection (Pavan et al., 2016). However, such approaches may negatively affect the AM fungi, and a beneficial symbiosis in the soil, which is a desired feature at least in sustainable agriculture. Such a double-edged approach requires even more knowledge on structure–activity relationships to hopefully design SL mimics, which act differently on the parasite and on the AM fungi. Many efforts in this direction are ongoing, some of which will be reported in the next section.

CHEMICAL BIOLOGY AND APPLICATIONS

Chemical biology and chemical genetics involves the development of chemical agonists and antagonists to probe biological processes. SLs are a suitable target for such approaches, since they are small molecules with even smaller core bioactivity centres (the methylbutenolide D-ring and the enol–ether bridge, Figure 11.1). Chemical biology can thus be used both to elucidate the mechanism of perception and to generate molecules, which have SL-related bioactivity and can potentially be used in agricultural or horticultural applications. The structure of most natural SLs is quite complex and unsuitable to be chemically synthesized on a larger scale. However, since the core of bioactivities resides in only a small part of the molecule (as discussed above), relatively simple molecules acting as SL analogues and mimics with the same or similar bioactivities as natural SLs can quite easily be synthesized. Older examples are the universally used GR24, as well as the less prominent GR5, GR7, or Nijmegen-1 (Figure 11.1). Nijmegen-1 has been used quite successfully in control of *Orobanche ramosa* on tobacco fields using only 8 g/ha[1] (reviewed in Zwanenburg and Blanco-Ania, 2018). However, there has always been the argument that the compounds decompose too fast in soil to be active and few further studies have been done. Five newly designed SL analogues were successfully tested on *Striga hermonthica* and *Striga asiatica* albeit only in pot experiments (Kgosi et al., 2012). Fluorescent analogues with bioactivities on the parasites and on AM fungi have also been synthesized (Prandi et al., 2011). Numerous further analogues and mimics have been developed (reviewed in Zwanenburg and Pospíšil, 2013; Zwanenburg and Blanco-Ania, 2018; Yoneyama et al., 2019). Finally, SLs have now attracted also the interest of agrochemical companies such as Syngenta, who are developing their own SL mimics for potential applications on parasitic weeds (Lumbroso et al., 2016; Bouwmeester et al., 2019).

Other potential applications might be the use of SL analogues and mimics in modifications or management of shoot architecture. It has been suggested that SLs could be used in suppressing unwanted flush of growth after pruning or in prompting branching by using inhibitors of SL biosynthesis (Beveridge et al., 2009). Some new analogues were designed rather for such hormonal applications having low activity in the rhizosphere (Boyer et al., 2014). Further mimics and analogues acting as SL biosynthesis inhibitors (Tsuchiya, 2018) or specifically designed as antagonists to interfere with receptors and signalling (Lumba et al., 2017) are coming up. Another potential application is their use as synthetic 'biostimulants' to enhance root colonization by AM fungi but again a problem of insufficient persistence in soil may exist. Other major problems still hampering their more widespread use in agriculture are potential off-target effects and regulatory issues at governmental authorities (Vurro et al., 2016).

Applications are also feasible more on the biological and ecological side. It has been mentioned above, and multiple references given, that there are reduced levels of SLs in roots and in root exudates correlated with reduced parasite infestations, when roots are colonized by AM fungi. Such effects have also been observed on maize plants even without determining SL levels (Othira et al., 2012). This phenomenon has been provisionally attributed above to a potential SL-degrading activity brought about by the fungi. Improving mycorrhization could therefore be an ecological and sustainable strategy to reduce infestation rates by root-parasitic weeds. If AM fungi are around in the soil, strengthening the plant-internal production of SLs and its exudation into the rhizosphere may thus not lead to enhanced infestation rates by more SLs as one might be afraid of, because SLs somehow disappear in the presence of the fungi.

It has turned out that carotenoid precursor supply is a major bottleneck in root SL biosynthesis, which even seems to require the maintenance in evolution of an extra Pi-starvation and symbiosis-regulated *PSY* isogene (Stauder et al., 2018). It might thus be worth considering to overexpress d*PSY3* genes or other *PSY* isogenes catalysing a rate-limiting step in carotenoid biosynthesis in a root-specific manner and/or driven by a strong AM-inducible promoter. This might overcome a likely bottleneck in SL formation and thus achieve an improvement of root colonization and mycorrhizal performance beyond wild-type plants. Another bottleneck is SL transport. Overexpression of the *PDR1* transporter in transgenic *P. hybrida* plants resulted in increased lateral root formation, extended root hair elongation, faster mycorrhization and reduced leaf senescence. These alterations translated into an improvement of plant biomass production on phosphate-poor soils (Liu et al., 2018a). Extending these studies, the authors conducted mycorrhization experiments under simulated microgravity (s0-g) and showed that elevated SL supply from either *PDR1* overexpression or via GR24 application can boost mycorrhization and plant growth even under these unnatural conditions. Microgravity is a very special environment occurring in outer space and the authors suggested the use of AM symbiotic fungi boosted by SLs for growing plants in future farming facilities on spaceships or on the moon (Liu et al., 2018b).

In summary, overcoming bottlenecks in SL production and exudation in roots can boost AM symbiosis functionality already in current agriculture at normal gravity. It would provide better access to the many benefits of the AM symbiosis (Sosa-Hernández et al., 2019). It could also help in eventually introducing sustainable practices in agriculture relying on low input plant mineral nutrition through a better support of root symbioses and at the same time reduce infestation rates by root parasitic weeds.

ACKNOWLEDGEMENTS

I would like to thank Juan Antonio López-Ráez, CSIC, Granada, Spain for critically reading the manuscript.

REFERENCES

Abe S, Sado A, Tanaka K, Kisugi T, Asami K, Ota S, Kim HI, Yoneyama K, Xie X, Ohnishi T, Seto Y, Yamaguchi S, Akiyama K, Yoneyama K, Nomura T. 2014. Carlactone is converted to carlactonoic acid by MAX1 in *Arabidopsis* and its methyl ester can directly interact with AtD14 in vitro. *Proc Natl Acad Sci USA*. 111(50):18084–18089.

Abel S. 2017. Phosphate scouting by root tips. *Curr Opin Plant Biol*. 39:168–177.

Agusti J, Herold S, Schwarz M, Sanchez P, Ljung K, Dun EA, Brewer PB, Beveridge CA, Greb T. 2011. Strigolactone signaling is required for auxin-dependent stimulation of secondary growth in plants. *Proc Natl Acad Sci*. 108(50):20242–20247.

Ahrazem O, Diretto G, Argandoña J, Fiore A, Rubio-Moraga A, Rial C, Varela RM, Macias FM, Castillo R, Romano E, Gómez-Gómez L. 2019. The specialised roles in carotenogenesis and apocarotenogenesis of the phytoene synthase gene family in saffron. *Front Plant Sci*. 10:249.

Akiyama K, Matsuzaki KI, Hayashi H. 2005. Plant sesquiterpenes induce hyphal branching in arbuscular mycorrhizal fungi. *Nature*. 435:824–827.

Akiyama K, Ogasawara S, Ito S, Hayashi H. 2010. Structural requirements of strigolactones for hyphal branching in AM fungi. *Plant Cell Physiol*. 51:1104–1117.

Al-Babili S, Bouwmeester HJ. 2015. Strigolactones, a novel carotenoid-derived plant hormone. *Annu Rev Plant Biol*. 66:161–186.

Alder A, Jamil M, Marzorati M, Bruno M, Vermathen M, Bigler P, Ghisla S, Bouwmeester H, Beyer P, Al-Babili S. 2012. The path from β-carotene to carlactone, a strigolactone-like plant hormone. *Science*. 335(6074):1348–1351.

Andreo-Jimenez B, Ruyter-Spira C, Bouwmeester HJ, López-Ráez JA. 2015. Ecological relevance of strigolactones in nutrient uptake and other abiotic stresses, and in plant-microbe interactions below-ground. *Plant Soil*. 394(1–2):1–19.

Aroca R, Ruiz-Lozano JM, Zamarreño AM, Paz JA, García-Mina JM, Pozo MJ, López-Ráez JA. 2013. Arbuscular mycorrhizal symbiosis influences strigolactone production under salinity and alleviates salt stress in lettuce plants. *J Plant Physiol*. 170:47–55.

Barbier FF, Dun EA, Kerr SC, Chabikwa TG, Beveridge CA. 2019. An update on the signals controlling shoot branching. *Trends Plant Sci*. 24(3):220–236.

Baz L, Mori N, Mi J, Jamil M, Kountche BA, Guo X, Balakrishna A, Jia KP, Vermathen M, Akiyama K, Al-Babili S. 2018. 3-Hydroxycarlactone, a novel product of the strigolactone biosynthesis core pathway. *Mol Plant*. 11:1312–1314.

Besserer A, Bécard G, Jauneau A, Roux C, Séjalon-Delmas N. 2008. GR24, a synthetic analog of strigolactones, stimulates the mitosis and growth of the arbuscular mycorrhizal fungus *Gigaspora rosea* by boosting its energy metabolism. *Plant Physiol*. 148:402–413.

Besserer A, Puech-Pages V, Kiefer P, Gomez-Roldan V, Jauneau A, Roy S, Portais JC, Roux C, Becard G, Séjalon-Delmas N. 2006. Strigolactones stimulate arbuscular mycorrhizal fungi by activating mitochondria. *PLoS Biol*. 4:1239–1247.

Beveridge CA, Dun EA, Rameau C. 2009. Pea has its tendrils in branching discoveries spanning a century from auxin to strigolactones. *Plant Physiol*. 151(3):985–990.

Bonneau L, Huguet S, Wipf D, Pauly N, Truong HN. 2013. Combined phosphate and nitrogen limitation generates a nutrient stress transcriptome favorable for arbuscular mycorrhizal symbiosis in *Medicago truncatula*. *New Phytol*. 199:188–202.

Booker J, Auldridge M, Wills S, McCarty D, Klee H, Leyser O. 2004. MAX3/CCD7 is a carotenoid cleavage dioxygenase required for the synthesis of a novel plant signaling molecule. *Curr Biol*. 14(14):1232–1238.

Bouwmeester HJ, Matusova R, Zhongkui S, Beale MH. 2003. Secondary metabolite signalling in host-parasitic plant interactions. *Curr Opin Plant Biol*. 6(4):358–364.

Bouwmeester HJ, Fonne-Pfister R, Screpanti C, De Mesmaeker A. 2019. Strigolactones: Plant hormones with promising features. *Angew Chemie Int Ed*. 58(37):12778–12786.

Boyer FD, de Saint Germain A, Pouvreau JB, Clavé G, Pillot JP, Roux A, Rasmussen A, Depuydt S, Lauressergues D, Frey N, Heugebaert TS, Stevens C, Van Geelen D, Goormachtig S, Rameau C. 2014. New strigolactone analogs as plant hormones with low activities in the rhizosphere. *Mol Plant*. 7(4):675–690.

Breakspear A, Liu C, Roy S, Stacey N, Rogers C, Trick M, Morieri G, Mysore KS, Wen J, Oldroyd GED, Downie A, Murray JD. 2014. The root hair 'infectome' of *Medicago truncatula* uncovers changes in cell cycle genes and reveals a requirement for auxin signaling in rhizobial infection. *Plant Cell*. 26:4680–4701.

Breuillin F, Schramm J, Hajirezaei M, Ahkami A, Favre P, Druege U, Hause B, Bucher M, Kretzschmar T, Bossolini E, Kuhlemeier C, Martinoia E, Franken F, Scholz U, Didier Reinhardt D. 2010. Phosphate systemically inhibits development of arbuscular mycorrhiza in *Petunia hybrida* and represses genes involved in mycorrhizal functioning. *Plant J.* 64:1002–1017.

Brewer PB, Yoneyama K, Filardo F, Meyers E, Scaffidi A, Frickey T, et al. 2016. Lateral branching oxidoreductase acts in the final stages of strigolactone biosynthesis in *Arabidopsis*. *Proc Natl Acad Sci USA*. 113(22):6301–6306.

Bruno M, Vermathen M, Alder A, Wüst F, Schaub P, van der Steen R, Beyer P, Ghisla S, Al-Babili S. 2017. Insights into the formation of carlactone from in-depth analysis of the CCD 8-catalyzed reactions. *FEBS Lett*. 591(5):792–800.

Bythell-Douglas R, Rothfels CJ, Stevenson DW, Graham SW, Wong GKS, Nelson DC, Bennett T. 2017. Evolution of strigolactone receptors by gradual neofunctionalization of KAI2 paralogues. *BMC Biol* 15(1):52.

Carbonnel S, Torabi S, Griesmann M, Bleek E, Tang Y, Buchka S, et al. 2019. Duplicated KAI2 receptors with divergent ligand-binding specificities control distinct developmental traits in *Lotus japonicus*. *bioRxiv*. 754937.

Cardoso C, Ruyter-Spira C, Bouwmeester HJ. 2011. Strigolactones and root infestation by plant-parasitic Striga, Orobanche and *Phelipanche* spp. *Plant Science*. 180(3):414–420.

Clarke CR, Timko MP, Yoder JI, Axtell MJ, Westwood JH. 2019. Molecular dialog between parasitic plants and their hosts. *Annu Rev Phytopathol*. 57:279–299.

De Cuyper C, Goormachtig S. 2017. Strigolactones in the rhizosphere: Friend or foe? *Mol Plant-Microbe Interact*. 30(9):683–690.

De Cuyper C, Struk S, Braem L, Gevaert K, De Jaeger G, Goormachtig S. 2017. Strigolactones, karrikins and beyond. *Plant Cell Environ*. 40(9):1691–1703.

Cook CE, Whichard LP, Turner B, Wall ME, Egley GH. 1966. Germination of witchweed (*Striga lutea* Lour.): Isolation and properties of a potent stimulant. *Science*. 154:1189–1190.

Cook CE, Whichard LP, Wall M, Egley GH, Coggon P, Luhan PA, McPhail AT. 1972. Germination stimulants. II. Structure of strigol, a potent seed germination stimulant for witchweed (*Striga lutea*). *J Am Chem Soc*. 94(17):6198–6199.

Delaux PM, Xie X, Timme RE, Puech-Pages V, Dunand C, Lecompte S, Delwiche CF, Yoneyama K, Bécard G, Séjalon-Delmas N. 2012. Origin of strigolactones in the green lineage. *New Phytol*. 195:857–871.

Domagalska MA, Leyser O. 2011. Signal integration in the control of shoot branching. *Nat Rev Mol Cell Biol*. 12:211–221.

Drummond RSM, Martinez-Sanchez NM, Janssen BJ, Templeton KR, Simons JL, Quinn BD, Karunairetnam S, Snowden KC. 2009. *Petunia hybrida* carotenoid cleavage dioxygenase7 is involved in the production of negative and positive branching signals in petunia. *Plant Physiol*. 151:1867–1877.

Dun EA, Brewer PB, Beveridge CA. 2009 Strigolactones: Discovery of the elusive shoot branching hormone. *Trends Plant Sci*. 14:364–372.

Felemban A, Braguy J, Zurbriggen MD, Al-Babili S. 2019. Apocarotenoids involved in plant development and stress response. *Front Plant Sci*. 10:1168

Fernández-Aparicio M, García-Garrido JM, Ocampo JA, Rubiales D. 2010. Colonization of field pea roots by arbuscular mycorrhizal fungi reduces Orobanche and Phelipanche species seed germination. *Weed Res*. 50:262–268

Fiorilli V, Vallino M, Biselli C, Faccio A, Bagnaresi P, Bonfante P. 2015. Host and non-host roots in rice: Cellular and molecular approaches reveal differential responses to arbuscular mycorrhizal fungi. *Front Plant Sci*. 6:636.

Fiorilli V, Wang JY, Bonfante P, Lanfranco L, Al-Babili S. 2019. Apocarotenoids: Old and new mediators of the arbuscular mycorrhizal symbiosis. *Front Plant Sci*. 10:1186.

Flematti GR, Dixon KW, Smith SM. 2015. What are karrikins and how were they 'discovered' by plants? *BMC Biol*. 13(1):108.

Floss DS, Hause B, Lange PR, Kuster H, Strack D, Walter MH. 2008. Knock-down of the MEP pathway isogene 1-deoxy-D-xylulose 5-phosphate synthase 2 inhibits formation of arbuscular mycorrhiza-induced apocarotenoids, and abolishes normal expression of mycorrhiza-specific plant marker genes. *Plant J*. 56:86–100.

Foo E, Yoneyama K, Hugill CJ, Quittenden LJ, Reid JB. 2013. Strigolactones and the regulation of pea symbioses in response to nitrate and phosphate deficiency. *Mol Plant*. 6:76–87.

Foo E, Davies NW. 2011. Strigolactones promote nodulation in pea. *Planta*. 243:1073–1081.

Genre A, Chabaud M, Balzergue C, Puech-Pagès V, Novero M, Rey T,... Barker DG. 2013. Short-chain chitin oligomers from arbuscular mycorrhizal fungi trigger nuclear Ca^{2+} spiking in *Medicago truncatula* roots and their production is enhanced by strigolactone. *New Phytol*. 198(1):190–202.

Gomez-Roldan V, Fermas S, Brewer PB, Puech-Pages V, Dun EA, Pillot JP, Letisse F, Matusova R, Danoun S, Portais JC, Bouwmeester H, Becard G, Beveridge CA, Rameau C, Rochange SF. 2008. Strigolactone inhibition of shoot branching. *Nature*. 455:189–194.

Guether M, Balestrini R, Hannah M, He J, Udvardi MK, Bonfante P. 2009. Genome-wide reprogramming of regulatory networks, transport, cell wall and membrane biogenesis during arbuscular mycorrhizal symbiosis in *Lotus japonicus*. *New Phytol*. 182:200–212.

Gutjahr C, Gobbato E, Choi J, Riemann M, Johnston MG, Summers W, Carbonnel S, Mansfield C, Yang SY, Nadal M, Acosta I, Takano M, Jiao WB, Schneeberger K, Kelly KA, Paszkowski U. 2015. Rice perception of symbiotic arbuscular mycorrhizal fungi requires the karrikin receptor complex. *Science*. 350(6267):1521–1524.

Haider I, Andreo-Jimenez B, Bruno M, Bimbo A, Floková K, Abuauf H,... Bouwmeester HJ. 2018. The interaction of strigolactones with abscisic acid during the drought response in rice. *J Exp Bot*. 69(9):2403–2414.

Hamiaux C, Drummond R S, Janssen BJ, Ledger SE, Cooney JM, Newcomb RD, et al. 2012. DAD2 is an α/β hydrolase likely to be involved in the perception of the plant branching hormone, strigolactone. *Curr Biol*. 22(21):2032–2036.

Ishikawa S, Maekawa M, Arite T, Onishi K, Takamure I, Kyozuka J. 2005. Suppression of tiller bud activity in tillering dwarf mutants of rice. *Plant Cell Physiol*. 46:79–86.

Janssen BJ, Snowden KC. 2012. Strigolactone and karrikin signal perception: Receptors, enzymes, or both?. *Front Plant Sci*. 3:296.

Jia KP, Baz L, Al-Babili S. 2018. From carotenoids to strigolactones. *J Exp Bot*. 69:2189–2204.

Johnson AW, Gowda G, Hassanali A, Knox J, Monaco S, Razavi Z, Rosebery G. 1981. The preparation of synthetic analogs of strigol. *J Chem Soc, Perkin Trans*. 1:1734–1743.

Johnson X, Brcich T, Dun EA, Goussot M, Haurogne K, Beveridge CA, Rameau C. 2006. Branching genes are conserved across species. Genes controlling a novel signal in pea are coregulated by other long-distance signals. *Plant Physiol*. 142:1014–1026.

Kamel L, Tang N, Malbreil M, San Clemente H, Le Marquer M, Roux C, Frei Dit Frey N. 2017. The comparison of expressed candidate secreted proteins from two arbuscular mycorrhizal fungi unravels common and specific molecular tools to invade different host plants. *Front Plant Sci*. 8:124.

Kapulnik Y, Delaux P-M, Resnick N, Mayzlish-Gati E, Wininger S, Bhattacharya C, Séjalon-Delmas N, Combier J-P, Becard G, Belausov E, Beeckman T, Dor E, Hershenhorn J, Koltai H. 2011. Strigolactones affect lateral root formation and root-hair elongation in Arabidopsis. *Planta*. 233:209–216.

Keymer A, Pimprikar P, Wewer V, Huber C, Brands M, Bucerius SL, Delaux PM, Klingl V, von Röpenack-Lahaye E, Wang TL, Eisenreich W, Dörmann P, Parniske M, Gutjahr C. 2017. Lipid transfer from plants to arbuscular mycorrhiza fungi. *eLife*. 6, e29107.

Kgosi RL, Zwanenburg B, Mwakaboko AS, Murdoch AJ. 2012. Strigolactone analogues induce suicidal seed germination of *Striga* spp. in soil. *Weed Res*. 52(3):197–203.

Kobae Y, Kameoka H, Sugimura Y, Saito K, Ohtomo R, Fujiwara T, Kyozuka J. 2018. Strigolactone biosynthesis genes of rice are required for the punctual entry of arbuscular mycorrhizal fungi into the roots. *Plant Cell Physiol*. 59(3):544–553.

Kohlen W, Charnikhova T, Liu Q, Bours R, Domagalska MA, Beguerie S, Verstappen F, Leyser O, Bouwmeester H, Ruyter-Spira C. 2011. Strigolactones are transported through the xylem and play a key role in shoot architectural response to phosphate deficiency in nonarbuscular mycorrhizal host Arabidopsis. *Plant Physiol*. 155:974–987.

Kohlen W, Charnikhova T, Bours R, López-Ráez JA, Bouwmeester H. 2013. Tomato strigolactones: A more detailed look. *Plant Signal Behav*. 8(1):e22785.

Koltai H, Dor E, Hershenhorn J, Joel DM, Weininger S, Lekalla S, et al. 2010. Strigolactones' effect on root growth and root-hair elongation may be mediated by auxin-efflux carriers. *J Plant Growth Regul*. 29(2):129–136.

Koltai H. 2011. Strigolactones are regulators of root development. *New Phytol*. 190(3):545–549.

Koltai H. 2014. Implications of non-specific strigolactone signaling in the rhizosphere. *Plant Sci*. 225:9–14.

Kretzschmar T, Kohlen W, Sasse J, Borghi L, Schlegel M, Bachelier JB, et al. 2012. A petunia ABC protein controls strigolactone-dependent symbiotic signalling and branching. *Nature*. 483(7389):341.

Lambers H, Martinoia E, Renton M. 2015. Plant adaptations to severely phosphorus-impoverished soils. *Curr Opin Plant Biol*. 25:23–31.

Lanfranco L, Fiorilli V, Venice F, Bonfante P. 2018. Strigolactones cross the kingdoms: Plants, fungi, and bacteria in the arbuscular mycorrhizal symbiosis. *J Exp Bot*. 69:2175–2188.

Lendzemo V, Kuyper TW, Vierheilig H. 2009. Striga seed-germination activity of root exudates and compounds present in stems of Striga host and nonhost (trap crop) plants is reduced due to root colonization by arbuscular mycorrhizal fungi. *Mycorrhiza*. 19:287–294.

Leyser O. 2008. Strigolactones and shoot branching: A new trick for a young dog. *Dev Cell.* 15:337–338.

Leyser O. 2009. The control of shoot branching: An example of plant information processing. *Plant Cell Environ.* 32(6):694–703.

Lin H, Wang RX, Qian Q, Yan MX, Meng XB, Fu ZM, Yan CY, Jiang B, Su Z, Li JY, Wang YH. 2009. DWARF27, an iron-containing protein required for the biosynthesis of strigolactones, regulates rice tiller bud outgrowth. *Plant Cell.* 21:1512–1525.

Liu G, Pfeifer J, de Brito Francisco R, Emonet A, Stirnemann M, Gübeli C, Hutter O, Sasse J, Matheyers C, Stelzer E, Walter A, Martinoia E, Borghi L. 2018a. Changes in the allocation of endogenous strigolactone improve plant biomass production on phosphate-poor soils. *New Phytol.* 217(2):784-798.

Liu G, Bollier D, Gübeli C, Peter N, Arnold P, Egli M, Borghi L. 2018b. Simulated microgravity and the antagonistic influence of strigolactone on plant nutrient uptake in low nutrient conditions. *Nature Microgravity.* 4(1):20.

López-Ráez JA, Charnikhova T, Gomez-Roldan V, Matusova R, Kohlen W, De Vos R, Verstappen F, Puech-Pages V, Becard G, Mulder P, Bouwmeester HJ. 2008. Tomato strigolactones are derived from carotenoids and their biosynthesis is promoted by phosphate starvation. *New Phytol.* 178:863–874.

López-Ráez JA, Charnikhova T, Fernández I, Bouwmeester H, Pozo MJ. 2011a. Arbuscular mycorrhizal symbiosis decreases strigolactone production in tomato. *J Plant Physiol.* 168:294–297.

López-Ráez JA, Pozo MJ, García-Garrido JM. 2011b. Strigolactones: A cry for help in the rhizosphere. *Botany.* 89(8):513–522.

López-Ráez JA, Fernández I, García JM, Berrio E, Bonfante P, Walter MH, Pozo MJ. 2015. Differential spatio-temporal expression of carotenoid cleavage dioxygenases regulates apocarotenoid fluxes during AM symbiosis. *Plant Science.* 230:59–69.

López-Ráez JA, Shirasu K, Foo E. 2017. Strigolactones in plant interactions with beneficial and detrimental organisms: The Yin and Yang. *Trends Plant Sci.* 22:527–537.

Lumba S, Bunsick M, McCourt P. 2017. Chemical genetics and strigolactone perception. *F1000Research.* 6:975.

Lumbroso A, Villedieu-Percheron E, Zurwerra D, Screpanti C, Lachia M, Dakas PY, Castelli L, Paul V, Wolf HC, Sayer D, Beck A, Rendine S, Fonné-Pfister R, De Mesmaeker A. 2016. Simplified strigolactams as potent analogues of strigolactones for the seed germination induction of *Orobanche cumana* Wallr. *Pest Manag Sci.* 72(11):2054–2068.

Lv S, Zhang Y, Li C, Liu Z, Yang N, Pan L, et al. 2018. Strigolactone-triggered stomatal closure requires hydrogen peroxide synthesis and nitric oxide production in an abscisic acid-independent manner. *New Phytol.* 217(1):290–304.

Machin DC, Hamon-Josse M, Bennet T. 2020. Fellowship of the rings: A saga of strigo-lactones and other small signals. *New Phytol.* 225:621–636, doi:10.1111/nph.16135.

Martin FM, Uroz S, Barker DG. 2017. Ancestral alliances: Plant mutualistic symbioses with fungi and bacteria. *Science.* 356(6340):eaad4501.

Marzec M, Muszynska A, Gruszka D. 2013. The role of strigolactones in nutrient-stress responses in plants. *Int J Mol Sci.* 14(5):9286–9304.

Matthys C, Walton A, Struk S, Stes E, Boyer FD, Gevaert K, et al. 2016. The whats, the wheres and the hows of strigolactone action in the roots. *Planta.* 243(6):1327–1337.

Matusova R, Rani K, Verstappen FWA, Franssen MCR, Beale MH, Bouwmeester HJ. 2005. The strigolactone germination stimulants of the plant-parasitic *Striga* and *Orobanche* spp. are derived from the carotenoid pathway. *Plant Physiol.* 139:920–934.

Mayzlish-Gati E, De-Cuyper C, Goormachtig S, Beeckman T, Vuylsteke M, Brewer PB, Beveridge C, Yermiyahu U, Kaplan Y, Enzer S, Wininger N, Resnick N, Cohen M, Kapulnik Y, Koltai H. 2012. Strigolactones are involved in root response to low phosphate conditions in Arabidopsis. *Plant Physiol.* 160(3):1329–1341.

Mouchel CF, Leyser O. 2007. Novel phytohormones involved in long-range signaling. *Curr Opin Plant Biol.* 10(5):473–476.

Othira JO, Omolo JO, Wachira FN, Onek LA. 2012. Effectiveness of arbuscular mycorrhizal fungi in protection of maize (*Zea mays* L.) against witchweed (*Striga hermonthica* DelBenth) infestation. *J Agric Biotech Sustain Dev.* 4:37–44.

Pandya-Kumar N, Shema R, Kumar M, Mayzlish-Gati E, Levy D, Zemach H, Belausov E, Wininger S, Abu-Abied M, Kapulnik Y, Koltai H. 2014. Strigolactone analog GR24 triggers changes in PIN2 polarity, vesicle trafficking and actin filament architecture. *New Phytol.* 202:1184–1196.

Parniske M. 2008. Arbuscular mycorrhiza: The mother of plant root endosymbioses. *Nat Rev Microbiol.* 6:763–775.

Pavan S, Schiavulli A, Marcotrigiano AR, Bardaro N, Bracuto V, Ricciardi F, Charnikhova T, Lotti C, Bouwmeester H, Ricciardi L. 2016. Characterization of low-strigolactone germplasm in pea (*Pisum sativum* L.) resistant to crenate broomrape (*Orobanche crenata* Forsk.). *Mol Plant-Microbe Interact.* 29(10):743–749.

Peláez-Vico MA, Bernabéu-Roda L, Kohlen W, Soto MJ, López-Ráez JA. 2016. Strigolactones in the Rhizobium-legume symbiosis: Stimulatory effect on bacterial surface motility and down-regulation of their levels in nodulated plants. *Plant Science.* 245:119–127.

Piisilä M, Keceli MA, Brader G, Jakobson L, Jõesaar I, Sipari N, et al. 2015. The F-box protein MAX2 contributes to resistance to bacterial phytopathogens in *Arabidopsis thaliana. BMC Plant Biol.* 15(1):53.

Prandi C, Occhiato EG, Tabasso S, Bonfante P, Novero M, Scarpi D, et al. 2011. New potent fluorescent analogues of strigolactones: Synthesis and biological activity in parasitic weed germination and fungal branching. *Eur J Org Chem.* 2011(20–21):3781–3793.

Rameau C, Bertheloot J, Leduc N, Andrieu B, Foucher F, Sakr S. 2015. Multiple pathways regulate shoot branching. *Front Plant Sci.* 5:741.

Rasmussen A, Mason MG, De Cuyper C, Brewer PB, Herold S, Agusti J, Geelen D, Greb T, Goormachtig S, Beeckman T, Beveridge CA. 2012. Strigolactones suppress adventitious rooting in *Arabidopsis* and pea. *Plant Physiol.* 158:1976–1987.

Ruyter-Spira C, Kohlen W, Charnikhova T, van Zeijl A, van Bezouwen L, de Ruijter N, Cardoso C, Lopez-Raez JA, Matusova R, Bours R, Verstappen F, Bouwmeester H. 2011. Physiological effects of the synthetic strigolactone analog GR24 on root system architecture in Arabidopsis: Another belowground role for strigolactones? *Plant Physiol.* 155:721–734.

Salvioli A, Ghignone S, Novero M, Navazio L, Venice F, Bagnaresi P, Bonfante P. 2016. Symbiosis with an endobacterium increases the fitness of a mycorrhizal fungus, raising its bioenergetic potential. *ISME J.* 10:130–144.

Samejima H, Sugimoto Y. 2018. Recent research progress in combatting root parasitic weeds. *Biotechnol Biotechnol Equipm.* 32(2):221–240

Sasse J, Simon S, Gübeli C, Liu GW, Cheng X, Friml J, et al. 2015. Asymmetric localizations of the ABC transporter PaPDR1 trace paths of directional strigolactone transport. *Curr Biol.* 25(5):647–655.

Scaffidi A, Waters MT, Sun YK, Skelton BW, Dixon KW, Ghisalberti EL, Flematti GR, Smith SM. 2014. Strigolactone hormones and their stereoisomers signal through two related receptor proteins to induce different physiological responses in Arabidopsis. *Plant Physiol.* 165:1221–1232.

Schwartz SH, Qin X, Loewen MC. 2004. The biochemical characterization of two carotenoid cleavage enzymes from Arabidopsis indicates that a carotenoid-derived compound inhibits lateral branching. *J Biol Chem.* 279(45):46940–46945.

Seto Y, Sado A, Asami K, Hanada A, Umehara M, Akiyama K, Yamaguchi S. 2014. Carlactone is an endogenous biosynthetic precursor for strigolactones. *Proc Natl Acad Sci.* 111(4):1640–1645.

Shabek N, Ticchiarelli F, Mao H, Hinds TR, Leyser O, Zheng N. 2018. Structural plasticity of D3–D14 ubiquitin ligase in strigolactone signalling. *Nature.* 563(7733):652.

Simons JL, Napoli CA, Janssen BJ, Plummer KM, Snowden KC. 2007. Analysis of the decreased apical dominance genes of petunia in the control of axillary branching. *Plant Physiol.* 143:697–706.

Sorefan K, Booker J, Haurogne K, Goussot M, Bainbridge K, Foo E, Chatweld S, Ward S, Beveridge C, Rameau C, Leyser O. 2003. *MAX4* and *RMS1* are orthologous dioxygenase-like genes that regulate shoot branching in Arabidopsis and pea. *Gene Dev.* 17:1469–1474.

Sosa-Hernández MA, Leifheit EF, Ingraffia R, Rillig MC. 2019. Subsoil arbuscular mycorrhizal fungi for sustainability and climate-smart agriculture: A solution right under our feet? *Front Microbiol.* 10:744.

Soto MJ, Fernández-Aparicio M, Castellanos-Morales V, García-Garrido JM, Ocampo JA, Delgado MJ, Vierheilig H. 2010. First indications for the involvement of strigolactones on nodule formation in alfalfa (*Medicago sativa*). *Soil Biol Biochem.* 42(2):383–385.

Stauder R, Welsch R, Camagna M, Kohlen W, Balcke GU, Tissier A, Walter MH. 2018. Strigolactone levels in dicot roots are determined by an ancestral symbiosis-regulated clade of the *PHYTOENE SYNTHASE* gene family. *Front Plant Sci.* 9:255.

Stirnberg P, van de Sande K, Leyser HMO. 2002. MAX1 and MAX2 control shoot lateral branching in Arabidopsis. *Development.* 129:1131–1141.

Sun H, Tao J, Liu S, Huang S, Chen S, Xie X, et al. 2014. Strigolactones are involved in phosphate-and nitrate-deficiency–induced root development and auxin transport in rice. *J Exp Bot.* 65(22):6735–6746.

Tan M, Li G, Chen X, Xing L, Ma J, Zhang D, Ge H, Han M, Sha G, An N. 2019. Role of cytokinin, strigolactone and auxin export on outgrowth of axillary buds in apple. *Front Plant Sci.* 10:616.

Toh S, Holbrook-Smith D, Stokes ME, Tsuchiya Y, McCourt P. 2014. Detection of parasitic plant suicide germination compounds using a high-throughput Arabidopsis HTL/KAI2 strigolactone perception system. *Chem Biol.* 21:988–998.

Toh S, Holbrook-Smith D, Stogios PJ, Onopriyenko O, Lumba S, Tsuchiya Y, et al. 2015. Structure–function analysis identifies highly sensitive strigolactone receptors in Striga. *Science.* 350:203–207.

Tsuchiya Y, McCourt P. 2009. Strigolactones: A new hormone with a past. *Curr Opin Plant Biol.* 12:556–561.

Tsuchiya Y, Yoshimura M, Sato Y, Kuwata K, Toh S, Holbrook-Smith D, et al. 2015. Probing strigolactone receptors in *Striga hermonthica* with fluorescence. *Science.* 349:864–868.

Tsuchiya Y, Yoshimura M, Hagihara S. 2018. The dynamics of strigolactone perception in *Striga hermonthica*: A working hypothesis. *J Exp Bot.* 69(9):2281–2290.

Tsuchiya Y. 2018. Small molecule toolbox for strigolactone biology. *Plant Cell Physiol.* 59(8):1511–1519.

Turnbull C, Booker J, Leyser O. 2002. Micrografting techniques for testing long-distance signalling in *Arabidopsis. Plant J.* 32:255–262.

Umehara M, Hanada A, Yoshida S, Akiyama K, Arite T, Takeda-Kamiya N, Magome H, Kamiya Y, Shirasu K, Yoneyama K, Kyozuka J, Yamaguchi S. 2008. Inhibition of shoot branching by new terpenoid plant hormones. *Nature.* 455:195–200.

Umehara M, Hanada A, Magome H, Takeda-Kamiya N, Yamaguchi S. 2010. Contribution of strigolactones to the inhibition of tiller bud outgrowth under phosphate deficiency in rice. *Plant Cell Physiol.* 51:1118–1126.

Umehara M. 2011. Strigolactone, a key regulator of nutrient allocation in plants. *Plant Biotechnol.* 28(5):429–437.

van Rongen M, Bennett T, Ticchiarelli F, Leyser O. 2019. Connective auxin transport contributes to strigolactone-mediated shoot branching control independent of the transcription factor BRC1. *PLoS Genet.* 15(3):e1008023.

van Zeijl A, Liu W, Xiao TT, Kohlen W, Yang WC, Bisseling T, et al. 2015. The strigo-lactone biosynthesis gene DWARF27 is co-opted in rhizobium symbiosis. *BMC Plant Biol.* 15:260.

Vogel JT, Walter MH, Giavalisco P, Lytovchenko A, Kohlen W, Charnikhova T, Simkin AJ, Goulet C, Strack D, Bouwmeester HJ, Fernie AR, Klee HJ. 2010. SlCCD7 controls strigolactone biosynthesis, shoot branching and mycorrhiza-induced apocarotenoid formation in tomato. *Plant J.* 61:300–311.

Vurro M, Prandi C, Baroccio F. 2016. Strigolactones: How far is their commercial use for agricultural purposes? *Pest Manag Sci.* 72:2026–2034.

Walter MH. 2013. Role of carotenoid metabolism in the arbuscular mycorrhizal symbiosis. In: *Molecular Microbial Ecology of the Rhizosphere*, F. de Bruijn, ed. (Chichester: Wiley-Blackwell), pp. 513–524.

Walter MH, Hans J, Strack D. 2002. Two distantly related genes encoding 1-deoxy-D-xylulose 5-phosphate synthases: Differential regulation in shoots and apocarotenoid-accumulating mycorrhizal roots. *Plant J.* 31:243–254.

Walter MH, Floss DS, Strack D. 2010. Apocarotenoids: Hormones, mycorrhizal metabolites and aroma volatiles. *Planta.* 232:1–17.

Walter MH, Strack D. 2011. Carotenoids and their cleavage products: Biosynthesis and functions. *Nat Prod Rep.* 28:663–692.

Walter MH, Stauder R, Tissier A. 2015. Evolution of root-specific carotenoid precursor pathways for apocarotenoid signal biogenesis. *Plant Sci.* 233:1–10.

Wang JY, Haider I, Jamil M, Fiorilli V, Saito Y, Mi J, Baz L, Kountche BA, Jia KP, Guo X, Balakrishna A, Ntui VO, Reinke B, Volpe V, Gojobori T, Blilou I, Lanfranco L, Bonfante P, Al-Babili S. 2019. The apocarotenoid metabolite zaxinone regulates growth and strigolactone biosynthesis in rice. *Nature Commun.* 10(1):810.

Wang Y, Bouwmeester HJ. 2018. Structural diversity in the strigolactones. *J Exp Bot.* 69(9):2219–2230.

Waters MT, Gutjahr C, Bennett T, Nelson DC. 2017. Strigolactone signaling and evolution. *Annu Rev Plant Biol.* 68:291–322.

Xie X, Yoneyama K, Kisugi T, Nomura T, Akiyama K, Asami T, Yoneyama K. 2015. Strigolactones are transported from roots to shoots, although not through the xylem. *J Pestic Sci.* 40(4):214–216.

Yao R, Ming Z, Yan L, Li S, Wang F, Ma S, et al. 2016. DWARF14 is a non-canonical hormone receptor for strigolactone. *Nature.* 536(7617):469–473.

Yoneyama K, Yoneyama K, Takeuchi Y, Sekimoto H. 2007a. Phosphorus deficiency in red clover promotes exudation of orobanchol, the signal for mycorrhizal symbionts and germination stimulant for root parasites. *Planta.* 225:1031–1038.

Yoneyama K, Xie X, Kusumoto D, Sekimoto H, Sugimoto Y, Takeuchi Y, Yoneyama K. 2007b. Nitrogen deficiency as well as phosphorus deficiency in sorghum promotes the production and exudation of

5-deoxystrigol, the host recognition signal for arbuscular mycorrhizal fungi and root parasites. *Planta.* 227:125–132.

Yoneyama K, Xie X, Kim HI, Kisugi T, Nomura T, Sekimoto H, Yokota T, Yoneyama K. 2012. How do nitrogen and phosphorus deficiencies affect strigolactone production and exudation? *Planta.* 235:1197–1207.

Yoneyama K, Mori N, Sato T, Yoda A, Xie X, Okamoto M, et al. 2018a. Conversion of carlactone to carlactonoic acid is a conserved function of MAX1 homologs in strigolactone biosynthesis. *New Phytol.* 218(4):1522–1533.

Yoneyama K, Xie XN, Yoneyama K, Kisugi T, Nomura T, Nakatani Y, Akiyama K, McErlean CSP. 2018b. Which are the major players, canonical or non-canonical strigolactones? *J Exp Bot.* 69:2231–2239.

Yoneyama K, Xie X, Yoneyama K, Nomura T, Takahashi I, Asami T, Mori N, Akiyama K, Kusajima M, Nakashita H. 2019. Regulation of biosynthesis, perception, and functions of strigolactones for promoting arbuscular mycorrhizal symbiosis and managing root parasitic weeds. *Pest Manag Sci.* 75:2353–2359.

Yoshida S, Kameoka H, Tempo M, Akiyama K, Umehara M, Yamaguchi S, Hayashi H, Kyozuka J, Shirasu K. 2012. The D3 F-box protein is a key component in host strigolactone responses essential for arbuscular mycorrhizal symbiosis. *New Phytol.* 196(4):1208–1216.

Zhang Y, van Dijk ADJ, Scaffidi A, Flematti GR, Hofmann M, Charnikhova T, Verstappen F, Hepworth J, van der Krol S, Leyser O, Smith SM, Zwanenburg B, Al-Babili S, Ruyter-Spira, Bouwmeester HJ. 2014. Rice cytochrome P450 MAX1 homologs catalyze distinct steps in strigolactone biosynthesis. *Nat Chem Biol.* 10:1028–1033.

Zwanenburg B, Blanco-Ania D. 2018. Strigolactones: New plant hormones in the spotlight. *J Exp Bot.* 69(9):2205–2218.

Zwanenburg B, Pospíšil T. 2013. Structure and activity of strigolactones: New plant hormones with a rich future. *Mol Plant.* 6(1):38–62.

Index

Page numbers in *italic* indicate a figure and page numbers in **bold** indicate a table on the corresponding page.

Index

Page numbers in *italic* indicate a figure and page numbers in **bold** indicate a table on the corresponding page.